Essential Concepts of Chemistry

Essential Concepts of Chemistry

Editor: Johana Meyer

NY RESEARCH PRESS

New York

Published by NY Research Press
118-35 Queens Blvd., Suite 400,
Forest Hills, NY 11375, USA
www.nyresearchpress.com

Essential Concepts of Chemistry
Edited by Johana Meyer

International Standard Book Number: 978-1-63238-658-8 (Hardback)

Cataloging-in-Publication Data

Essential concepts of chemistry / edited by Johana Meyer.
     p. cm.
Includes bibliographical references and index.
ISBN 978-1-63238-658-8
1. Chemistry. I. Meyer, Johana.
QD39.2 .E87 2019
540--dc23

Contents

Preface

Chemistry is the scientific study of compounds, their composition, structure, properties and reactions. Compounds are chemical substances formed by chemical bonds between two or more identical molecules. It also studies the bonding responsible for the formation of these compounds. The factors of energy and entropy are important in the study of chemical processes. Chemical substances are classified on the basis of their structure, phase and composition. They can be analyzed using the tools of spectroscopy and chromatography. Chemistry is divided into a number of significant sub-disciplines such as analytical chemistry, biochemistry, inorganic chemistry, organic chemistry, etc. There has been rapid progress in this field and its applications are finding their way across multiple industries such as extraction of metals from ores, fermentation of beer and wine, manufacture of glass and alloys, etc. This book is a compilation of chapters that discuss the most vital concepts and emerging trends in the field of chemistry. The data presented herein also explores some of the most innovative concepts and elucidates the unexplored aspects of this discipline. This book will serve as a reference to chemical engineers, researchers, experts and students.

This book is the end result of constructive efforts and intensive research done by experts in this field. The aim of this book is to enlighten the readers with recent information in this area of research. The information provided in this profound book would serve as a valuable reference to students and researchers in this field.

At the end, I would like to thank all the authors for devoting their precious time and providing their valuable contribution to this book. I would also like to express my gratitude to my fellow colleagues who encouraged me throughout the process.

Editor

Progress towards a process for the recycling of nickel metal hydride electric cells using a deep eutectic solvent

Mark R.StJ. Foreman[1]*

*Corresponding author: Mark R.StJ. Foreman, Department of Chemistry and Chemical Engineering, Chalmers University of Technology, Gothenburg, Sweden
Email: Foreman@chalmers.se
Reviewing editor: Alexandra Martha Zoya Slawin, University of St. Andrews, UK

Abstract: Solvent extraction experiments relating to the recycling of the transition metals and lanthanides in nickel metal hydride cells are presented. The metal extraction is occurring from a deep eutectic solvent which is formed from chemicals suitable for use in food and related products. While it has been shown that the water content of the DES has a large effect on the extraction of transition metals by a mixture of chloride ionic liquid (Aliquat 336) and an aromatic solvent, the water content has a smaller effect on the solvent extraction of lanthanides with a solution of di(2-ethylhexyl) hydrogen phosphate (DEHPA) in a saturated aliphatic hydrocarbon. This study suggests that an industrial scale solvent extraction process for the recycling of metals from nickel hydride electrical cells will be feasible.

Subjects: Inorganic chemistry; Solvent Extraction; Waste & Recycling

Keywords: nickel, cobalt; iron; lanthanides; deep eutectic solvents; ionic liquids; solvent extraction; Aliquat 336; Di(2-ethylhexyl) hydrogen phosphate (DEHPA)

1. Introduction

In the past, a large fraction of municipal solid waste was placed in landfills and thus buried, this waste would have contained the vast majority of the items discarded by the general public. In order

ABOUT THE AUTHOR

Mark R StJ Foreman is an associate professor in the department of Chemistry and Chemical Engineering at Chalmers University of Technology in Sweden. His research interests include both nuclear chemistry and the recycling of materials. For ca. 16 years he has had an interest in the solvent extraction of metals.

PUBLIC INTEREST STATEMENT

When a product or an object is no longer useful it becomes waste. Landfilling of waste is often undesirable as it places many valuable materials out of reach of society, it also can result in contamination of the environment. Additionally, the landfilling and incineration of waste industrial and automotive batteries are illegal in some states (The Waste Batteries & Accumulators Regulations 2009, 2009, Section 56). Recycling is part of the movement away from landfilling waste, the goal of a recycling process should be to obtain as much valuable material as possible from waste using the least amount of energy, producing as little secondary waste as possible and in such a way that leaves the least amount of unrecycled residual waste as possible. In this paper some early steps toward an improved industrial process based on an ionic liquid mimic for the recycling of disused nickel metal hydride electrical cells are presented. This process is designed to produce pure streams of metals from the waste while using as little energy as possible.

to prevent further proliferation of landfill sites, to reduce society's dependence on primary resources and to minimize the overall cost of waste disposal, both industry and the general public are encouraged to sort their waste into different classes to facilitate recycling and reuse. The public are encouraged to submit used electrical cells and batteries for recycling separately from their other wastes as these electrochemical devices often contain valuable and harmful materials. For example, lithium ion batteries often contain cobalt (Xu et al., 2008) which is an attractive target for recycling. Cobalt is an attractive target for recycling as it is an important material whose supply is limited. Cobalt, the lanthanides, and some other elements were identified as being critical elements by the European Commission (European Commission, 2010). The toxic metals such as mercury and cadmium in disused electrical cells need to be managed correctly to prevent the environmental contamination and human disease, one of the purposes of the recycling sector is to concentrate these toxic metals into a form suitable for reuse or a small volume of waste which is suitable for long-term storage.

While it is possible to recover nickel and cobalt from nickel metal hydride cells by heating in a furnace (pyrometalurgy) (Müller & Friedrich, 2006), the valuable lanthanides will be lost into the slag. While it might be possible to recover the lanthanides from the slag such a process will require a large input of energy and the slag would require leaching in a mineral acid to liberate the lanthanides (Kim et al., 2014). An alternative to recycling using pyrometallury is to use hydrometallurgy in which metals are liberated from ores or wastes using aqueous reagents, the wanted metals are then harvested from the aqueous solution using methods which include liquid–liquid extraction (Rydberg, Musikas, & Choppin, 1992), ion exchange on solid materials (Alexandratos, 2009), electrodeposition (Moskalyk & Alfantazi, 2002) and precipitation (Kyle, Breuer, Bunney, & Pleysier, 2012; Chmielewski, Urbanski, & Migdal, 1997). Already it has been shown that it is possible to recover cobalt and other metals from nickel metal hydride cells by dissolution in hydrochloric acid (8 M) followed by liquid–liquid extraction with a concentrated (70% v/v) solution of trialkyl phosphine oxides (Cyanex 923) together with tributyl phosphate (10% v/v) was able to separate cobalt and the lanthanides from the nickel containing mixture (Larsson, Ekberg, & Ødegaard-Jensen, 2012). While the Cyanex 923 process might be able to separate cobalt, lanthanides, and nickel, it does suffer from several disadvantages, firstly Cyanex 923 is able to extract hydrochloric acid from aqueous solutions into the organic phase (Sarangi, Padhan, Sarma, Park, & Das, 2006). This loss of hydrochloric acid will result in the contamination of stripping (back extraction) solutions with hydrochloric acid and will compel the operator of a plant to continually add more hydrochloric acid to the leaching process. A second disadvantage is the fact that the air above 8 M solution of hydrochloric acid (26% w/w) will contain a large amount of hydrogen chloride. This can lead to a range of corrosion issues inside a recycling plant.

In recent times, ionic liquids and the related deep eutectic solvents (Abbott, Boothby, Capper, Davies, & Rasheed, 2004) have been the subject of considerable interest (Abbott, Frisch, & Ryder, 2011). One of the perceived advantages of these solvents is that often their vapor pressures are very low, if our consideration is limited for a moment to corrosive inorganic compounds then it is clear that because of the noninfinite dissociation constants of hydrochloric acid that it is possible for an aqueous solution to contain undissociated hydrogen chloride which is available to leave the solution at the surface and thus enter the air. In the case of an ionic liquid formed from an organic cation and chloride anions (such as Aliquat 336), it is not possible for a volatile chloride to be formed at room temperature. Additionally, in the case of a deep eutectic formed from choline chloride even if the chloride anion was to combine with the choline cation, the resulting chlorospecies would be unlikely to be volatile because the alcohol group anchors it into the liquid through hydrogen bonding. While it would be foolish to blindly equate involatility with a chemical being harmless, a great reduction in the ability of a reagent to form a corrosive atmosphere is desirable in a solvent extraction plant. As it is known that deep eutectics formed from choline chloride are able to dissolve a range of metal oxides (Abbott, Capper, Davies, McKenzie, & Obi, 2006), it was decided that solvent extraction from such a deep eutectic with conventional solvent extraction reagents would be investigated for its potential to create a new industrial process.

2. Results and discussion

It was found that metal salts such as cobalt(II) chloride were soluble in the deep eutectic solvent formed by the combination of choline chloride and glycolic acid. The solution of cobalt(II) in this medium was deep blue in color indicating that the cobalt is likely to have a tetrahedral coordination environment similar to that of cobalt in an aqueous medium containing a high concentration of chloride anions. It was found that when such a solution was shaken with a solution of di(2-ethylhexyl) phosphoric acid (DEHPA) in an aliphatic kerosene (Solvent 70) that the organic layer remained colorless suggesting that no cobalt extraction occurred. The reagent DEHPA was selected as it is well known to be able to extract a range of metals by means of an ion-exchange mechanism as shown in Scheme 1.

The extraction of metals by DEHPA can be reversed by shaking the metal-loaded solution of DEHPA with an aqueous solution of an acid. As an organic phase based upon DEHPA failed to extract cobalt.

However, when the solution of cobalt was shaken with a solution of Aliquat 336 (30% v/v) in toluene, the organic phase became blue in color suggesting that cobalt extraction was occurring. It was soon understood that the rate of extraction from the DES layer into the organic layer was very slow compared with the rate at which an analogous system with an aqueous phase reached equilibrium. While papers in which metal extraction using pure ionic liquids have been reported (Wellens, Thijs, Moller, & Binnemans, 2013). It was decided that the high viscosities of pure ionic liquids such as Aliquat 336 would render them unsuitable as organic phases for solvent extraction. Additionally, the use of a pure ionic liquid in experiments would stymie any attempt to use slope analysis to determine how many formula units of ionic liquid are required per metal atom. As many laboratory and industrial solvent extraction machines, such as mixer settlers, are designed to operate using organic liquids similar to kerosene it was decided that the properties of the ionic liquid should be altered to make them compatible with the solvent extraction equipment. This was done so by the addition of carefully selected organic molecular solvents to create free flowing ionic liquid-based organic phases. As such mixtures of ionic liquids and organic solvents have been used for the extraction of metals for at least four decades (Stronski & Nahlik, 1972), rather than pursuing an ideal of a process based upon pure ionic liquids it was decided that adapting the nature of the liquid phases to suit existing solvent extraction equipment was a better goal.

It was found that by the addition of a small amount of water to lower deep eutectic phase, the viscosity of that phase was decreased greatly and solvent extraction experiments could be rapidly brought to equilibrium. Using a series of mixtures of sodium chloride solution and the deep eutectic it was shown that the distribution ratio when extracting with 30% Aliquat 336 (v/v) in toluene at 40°C of many transition metals is a function of the water content of the deep eutectic solvent layer. While the relationship between the zinc and cadmium distribution ratios (D_{Zn} and D_{Cd}) and the water content could be modeled with a simple exponential expression (Figure 1) the relationship between the cobalt, copper, iron, and manganese distribution ratios (D_{Co}, D_{Cu}, D_{Fe}, and D_{Mn}) (Figures 2 and 3) and water content was complex.

$$M^{n+}_{(aq)} + nDEHPA\text{-}H_{(org)} \rightleftharpoons [M(DEHPA)_n]_{(org)} + nH^+_{(aq)}$$

Scheme 1. Extraction equilibrium for a metal being extracted by DEHPA.

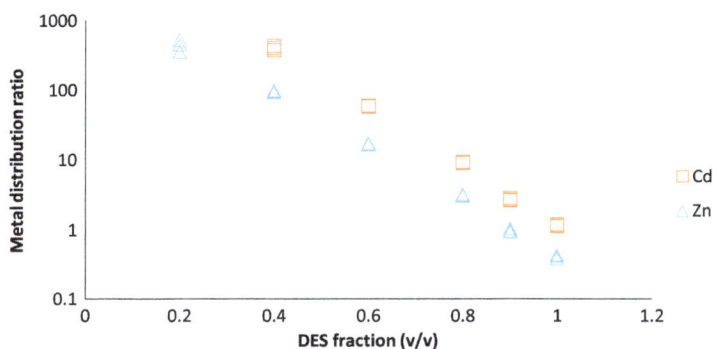

Figure 1. The relationship between the water content of the deep eutectic phase and the distribution ratios for cadmium and zinc.

When the water content of the system was very low, the extraction of cobalt and copper metals was suppressed; when the water content was increased, the copper distribution ratios increased again but the cobalt distribution ratios decreased as the water content approached 80%; with water contents above 80% the cobalt distribution ratio increased again.

While the iron and manganese distribution ratios had a different relationship with the water content of the lower phase as can be seen in Figure 3.

As it was shown that industrially useful distribution ratios for key metals could be obtained, a search was made for safer diluents. While toluene might not be considered a "green solvent" it is important to note that it is a simple reagent which is available throughout the world at a reasonable cost. While some grades of diluents might be safer, a proprietary mixture is less likely to be available decades into the future so many of the proof of principle tests were made with toluene or other similar aromatic diluents. A range of diluents were screened using a lower layer containing DES 80% (v/v) and 20% water at 40°C using 30% (v/v) solutions of Aliquat 336 as the organic phase. In this experiment, toluene, Solvesso 150ND (S150), Solvesso 200ND (S200), solvent 70 (modified with 15% decanol [v/v]) (S70D), and a fatty acid methyl ester (FAME) mixture from sunflower cooking oil were compared. It was found that with both Solvesso 150ND (S150) and Solvesso 200ND (S200), the distribution ratios of the wanted transition metals were higher than those obtained with toluene, the modified solvent 70 and the FAME biodiesel (Table 1). When the experiment was attempted using Escaid 110 the organic phase separated into two layers when it was loaded with a moderate amount of metals (assumed to be Cd, Co, Cu, Fe, Mn, and Zn). Isopar L and Escaid 120 were not able to dissolve the Aliquat 336.

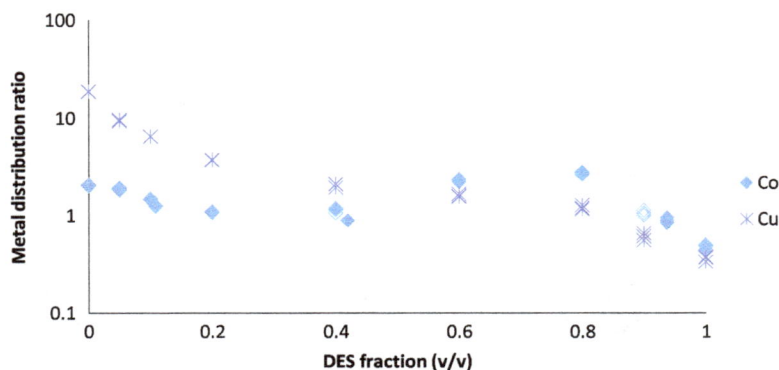

Figure 2. The relationship between the water content of the deep eutectic phase and the distribution ratios for copper and cobalt.

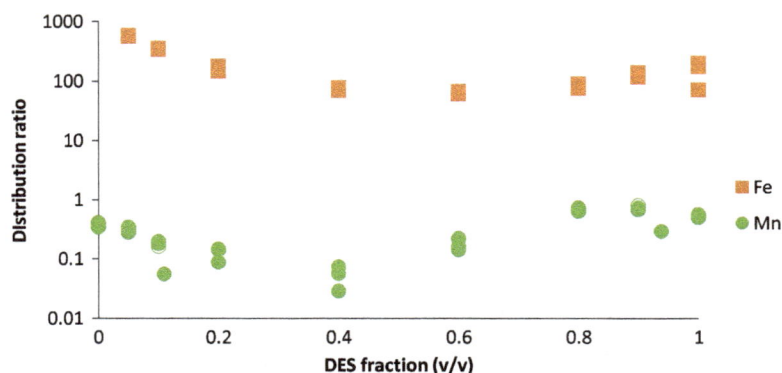

Figure 3. The relationship between the water content of the deep eutectic phase and the distribution ratios for iron and manganese.

Metal	Diluent				
	Toluene	Solvesso150 ND	Solvesso200 ND	Solvent 70 and decanol	Sunflower oil FAME biodiesel
Al	0	0	0	0	0
Cd	7,8	8,6	12,4	2,5	5.0
Ce	0,001	0,03	0,03	0,05	0.1
Co	1,55	1,66	2,14	0,62	0.88
Cu	0,69	0,90	0,82	0,37	0.27
Fe	16,5	13,8	20,1	3,7	1.1
La	0	0	0	0	0
Mn	0,28	0,37	0,50	0,18	0.17
Nd	0	0,0216	0,031	0,040	0.1
Pb	0,24	0,18	0,23	0,003	0.18
Pr	0	0	0	0	0.1
Zn	2,67	2,83	3,65	0,94	1.8

Table 1. Distribution ratios obtained at 40°C using 30% (v/v) solutions of Aliquat 336 in different dilutents

Of these solvents, Solvesso200ND was selected as the best solvent for a process as the cobalt distribution ratio was the highest and no third phase formation was observed. The water content test was repeated using Solvesso200ND as the diluent using deionized water in place of the sodium chloride solution (Figure 4) it was still found that the extraction of cadmium and zinc was favored by a high water content, while the iron extraction was favored by a high chloride content of the lower phase. The cobalt and copper distribution ratios peaked at *ca.* 80%.

This suggests that the dominant driving force for the extraction of iron is different to that of zinc/cadmium. For iron, the dominant factor is the chloride concentration suggesting the position of the equilibrium as shown in Scheme 2.

$$[Fe(H_2O)_6]^{3+}{}_{(aq)} + 3Cl^-{}_{(aq)} + R_4NCl_{(org)} \rightleftharpoons R_4N^+{}_{(org)} + [FeCl_4]^-{}_{(org)} + 6H_2O_{(aq)}$$

Scheme 2. Extraction equilibrium for iron(III) being extracted by a quaternary ammonium salt such as Aliquat 336.

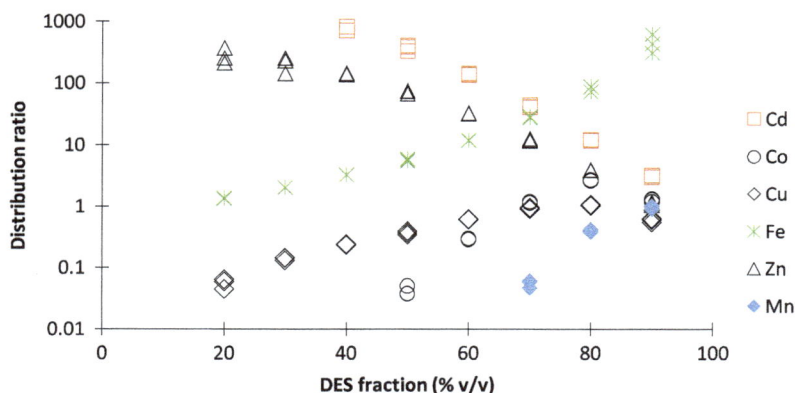

Figure 4. The relationship between the water content of the deep eutectic phase and the distribution ratios for cadmium, cobalt, copper, iron, manganese and zinc when no attempt is made to maintain a constant chloride concentration.

Is driven to the right by a high chloride concentration. While in the case of the zinc and cadmium some other effect dominates. One possibility was that the increase in entropy when the number of free molecules and ions which occurs in the following extraction reaction was the important factor. With the data which has been obtained so far, it is not possible to state with certainty the reason for the difference in the behavior of the iron and zinc/cadmium. It is possible that when the water activity (and concentration) is low that highly charged anion cadmium and zinc chloro complexes such as $[CdCl_5]^{3-}$ and $[ZnCl_6]^{4-}$ could be present in the lower phase. In the solid state, $[CdCl_5]^{3-}$ and $[CdCl_6]^{4-}$ have been observed (Bouchene, Bouacida, Berrah, & Roisnel, 2014) suggesting that such complexes could form in a concentrated chloride medium where the water content is low. While the formation of an extractant/metal complex is normally associated with metal extraction, it is important to note that an excessive concentration of a charged ligand can result in a suppression of extraction because of the formation of charged complexes. For example, the extraction of promethium(III) by acetylacetone into benzene is surpressed by the formation of $[Pm(acac)_4]^-$ complexes when the concentration of acetylacetonate anions is high (Rydberg & Albinsson, 1989).

To test the hypothesis, the iron and cadmium/zinc extractions are mainly determined by different factors, a variable temperature experiment was performed. In this experiment, mixtures of the two liquids were shaken at different temperatures before phase separation and sampling. It was found that extraction of manganese and iron was favored by an increase in temperature while the zinc, lead, and cadmium extractions were favored by a reduction in temperature (Figure 5). The copper and cobalt extractions were less sensitive to changes in temperature which is advantageous for process development.

It is important to stress that the distribution ratio of a metal depends on a series of equilibria which in turn depend on the enthalpy and entropy of these reactions. With the results obtained so far, it would be premature to state the reason why the different elements have different relationships between their distribution ratios and the temperature and water content of the DES phase. It might be possible using microcalorimetry and variable temperature EXAFS to obtain an insight into the extraction chemistry.

The hypothesis that Aliquat 336 is required for the metal extraction was tested by shaking a series of different concentrations of the extractant in toluene with a mixture of metals in the wet (20% water v/v) deep eutectic solvent. It was found that as the Aliquat 336 concentration in the organic phase was increased, the metal distribution ratios increased (Figure 6). During the experiments, no extraction of lanthanides or nickel by Aliquat 336 in chloride media was observed. This can be rationalized by the facts that lanthanides have a low affinity for chloride anions and that the crystal field stablization energy of nickel(II) octahedral and square planar complexes inhibits the formation of tetrahedral $[NiCl_4]^{2-}$ complexes in aqueous media.

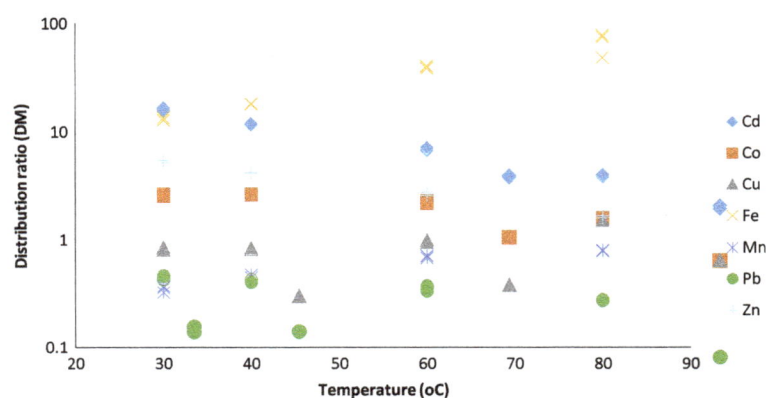

Figure 5. The relationship between the temperature and the distribution ratios for cadmium, cobalt, copper, iron, manganese, lead and zinc for an extraction from the deep eutectic solvent into a solution of Aliquat 336.

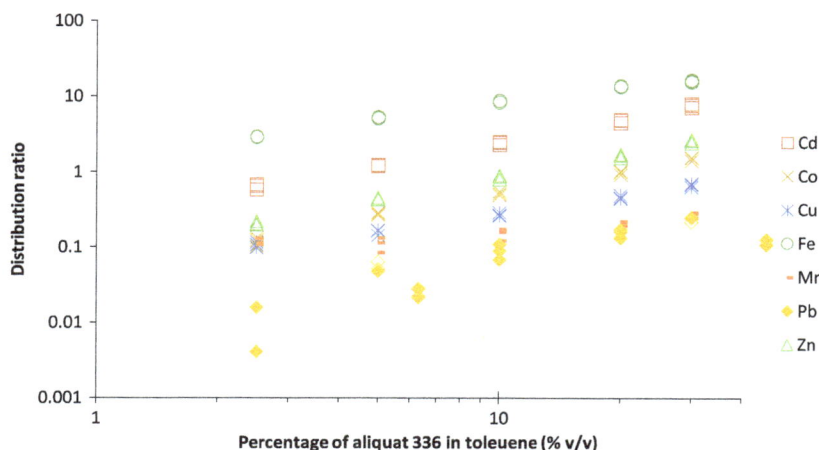

Figure 6. The relationship between the concentration of Aliquat 336 in toluene and the distribution ratios for cadmium, cobalt, copper, iron, manganese, lead, and zinc for an extraction from the deep eutectic solvent.

It is well known that transition metals such as cobalt, copper, and iron can be stripped from an organic phase formed from Aliquat 336 by contacting the organic phase with dilute aqueous chloride solutions. As this is so well established and occurs in a part of the process separate from the deep eutectic solvent, the stripping will not be discussed in detail here.

While the extraction of cobalt and other transition metals from the deep eutectic solvent is not fully understood with the information in this paper, it is possible to create the core of a process for cobalt recovery from a cobalt rich leach liquor obtained by the digestion of nickel metal hydride electrical cells. As the disused electrical cells contain large amounts of lanthanides, the extraction of lanthanides from the deep eutectic solvent was considered.

A series of mixtures of lanthanides and other metals in a range of mixtures of the deep eutectic solvent and aqueous sodium chloride solution were shaken with DEHPA (30% v/v) in solvent 70. It was found that the lanthanide and manganese distribution ratios changed only slightly as the composition of the lower phase was changed (Figure 7). While the aluminum extraction was favored by a low water content of the lower phase, the extraction of the iron was favored when the water content of the lower phase was high.

By using a series of different concentrations of DEHPA in solvent 70, the effect of changing the extractant concentration on the distribution ratios of iron, manganese, and the lanthanides was tested. It was found that while the slopes were smaller than three, the results were broadly in

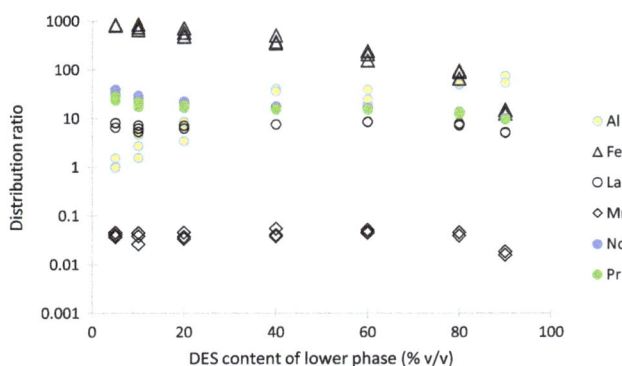

Figure 7. The relationship between the water content of the lower deep eutectic solvent phase and the distribution ratios for aluminum, iron, manganese, and the lanthanides when the organic phase was a solution of DEHPA in solvent 70.

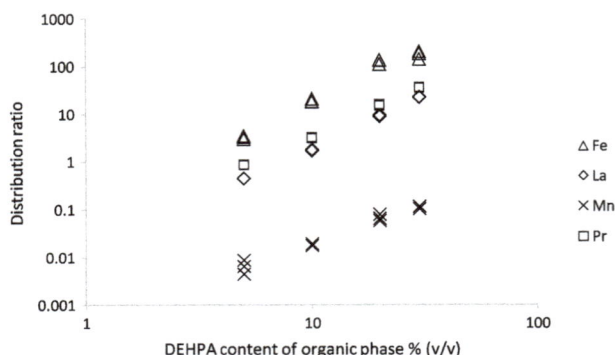

Figure 8. The relationship between the DEHPA concentration in the upper phase and the distribution ratios for iron, lanthanum, manganese and praseodymium when the organic phase was a solution of DEHPA in solvent 70.

agreement with those obtained by others (Peppard, Mason, Maier, & Driscoll, 1957). In general, an increase in the DEHPA concentration causes an increase in the metal distribution ratios (Figure 8).

One of the long-term goals is to create a process which uses hydrotreated vegetable oil as a diluent for the DEHPA extraction, as part of the effort to enable a smooth transition to such a diluent a series of alternative grades of kerosene were tested using a 8:2 mixture of the deep eutectic solvent and water. It was found that with four different grades of aliphatic kerosene (Solvent 70, Escaid 110, Isopar L and Escaid 120) similar results were obtained (Table 2). However, when the aliphatic diluent was replaced with an alkylarene (*tert*-butylbenzene TBB), the distribution ratios of all the metals were reduced.

It is likely that a mixture of n-alkanes formed by the hydrogenation and hydrodeoxygenation of vegetable oil (Huber, O'Connor, & Corma, 2007) will be suitable for use as a diluent for the lanthanide extraction. While the stripping of lanthanides from organic phases containing DEHPA is well established, it was shown by shaking a series of solutions of metals in dilute sulfuric acid that such a

Table 2. Distribution ratios obtained by shaking DEHPA (30% v/v) in different diluents with a solution of metal salts in the deep eutectic solvent

Metal	Diluent				
	TBB	Solvent 70	Escaid 110	Escaid 120	Isopar L
Iron	39	206	204	248	232
Lanthanum	4.7	22.9	23.5	27.5	27.5
Neodymium	7.4	37.5	37.4	45.1	43.9
Praseodymium	7.5	36.3	37.2	43.2	42.6
Manganese	0.034	0.13	0.14	0.15	0.15

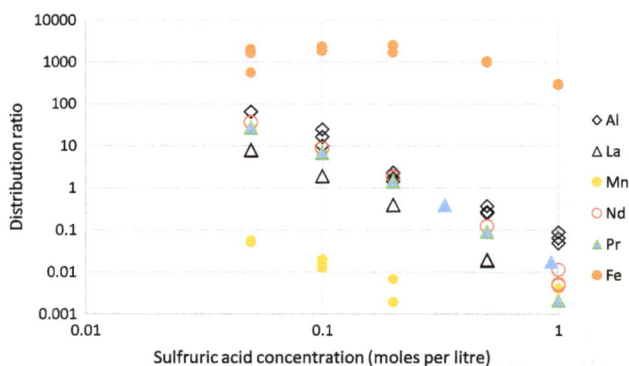

Figure 9. A graph of the distribution ratios of aluminum, iron, manganese, and the lanthanides against the initial sulfuric acid concentration in the aqueous phase.

stripping is thermodynamically possible using the organic phase used in the extraction experiments. However, the iron(III) has such a high affinity for the DEHPA that it is not possible to prevent its extraction into the organic phase with sulfuric acid (1 M) (Figure 9).

3. Conclusions

It was shown that cobalt and some other transition metals can be extracted from the deep eutectic solvent by an organic phase containing Aliquat 336, an organic phase which is nonflammable could be formulated. In a series of experiments, it was shown that an industrial process for the extraction of cobalt and the lanthanides from the deep eutectic of lactic acid and choline chloride is plausible. The selectivity of the extraction reagents observed in the experiments was similar to that which can be expected from experiments using more conventional aqueous media. Further experiments will be required to create and demonstrate the process.

4. Experimental

All reagents were purchased from Aldrich and were used as received unless stated otherwise. All pipetting was done using the forward pipetting method using piston driven air displacement pipettes made by Gilson. Solvent extraction experiments were performed by pipetting equal volumes of an organic phase and either an aqueous phase or a deep eutectic solvent phase into a glass vial (3.5 ml). It was normal to add the aqueous or deep eutectic solvent phase before the addition of the organic phase to the vial. With the exception of the part of the study in which different diluents are used for the extractions and the experiment in which the DEHPA concentration was altered the phase ratio was checked by measuring the mass of the empty shaking vial, then again after the addition of the first liquid and then finally after the addition of the last liquid layer. Using the densities of the liquids, the phase ratio was then calculated and used to correct the concentration of the metals within the organic layer. When deep eutectic phases, ionic liquids or other viscous liquids were pipetted; additional care was taken as these fluids require a greater time to flow out of the pipette tip. Additionally, filter tips were used to reduce the likelihood of the pipette becoming contaminated with a splash of the liquid. It was found that the moment when the push button of the pipette is released after the liquid had been ejected from the tip was the time when splashing of liquid upward was most likely to occur. After the liquids had been dispensed into the vial, it was sealed with a push on polyethylene cap before being shaken using a IKA Vibrax VXR basic machine equipped with a thermostated sample holder feed with warm water from a Grant TC120 circulating water bath. Unless otherwise stated all samples were shaken at 40°C. After shaking the samples were centrifuged (Heraeus Labofuge 200) at 4,000 rotations per minute for at least five minutes to ensure good phase disengagement. Samples (200–500 µl) of the lower phase were taken using a Gilson pipette in the following manner. A pipette bearing a tip was set to the required volume, the push button depressed to the first stop before the tip was inserted into the lower phase. Through additional pressure on the button one or two bubbles of air were expelled from the nozzle of the tip. The pressure on the push button was slowly relaxed allowing liquid to enter the tip. The pipette and tip were then removed from the shaking vial, using tissue paper the outside of the tip was wiped to remove any trace of the upper phase before the liquid inside the tip was dispensed into a preweighed polyethylene vial (17 ml). To the polyethylene vials was added dilute (0.1 M) nitric acid containing ruthenium (1–2 ppm). In early tests, a ×20 dilution was used but this was associated with the formation of carbon on torch components in the ICP machine (axial view), to reduce this problem the dilution was increased to ×50. If the Kistiakowsky–Wilson rules (Akhavan, 1998) or similar rules apply to the conditions in the ICP machine's plasma then a reduction in the amount of organic matter entering the torch will result in a reduction in the formation of carbon. The ruthenium was present as an internal standard in case the operation of the ICPOES machine (Thermo Scientific iCAP6500) was subject to some variation. Normally the internal standard was not needed, but was present as a contingency against a pump or nebulizer malfunction.

The metal content of the lower phase before the extraction experiments was measured under the same conditions using samples feed into the ICP machine which have similar dilution factors to ensure that the measurements of the metals were made with the same matrix. The concentration of metals within the organic phase was calculated by measurement of the difference between the lower phase before and after the extraction.

A metal stock was prepared by dissolving the following metal salts, aluminum chloride hexahydrate (18.2 g), cadmium nitrate tetrahydrate (5.2 g), cerium chloride heptahydrate (6.4 g), cobalt chloride hexahydrate (8.1 g), iron(III) chloride (8.6 g), lanthanum chloride heptahydrate (5.8 g), neodymium chloride (1.7 g), praseodymium chloride (1.3 g), and zinc chloride (4.0 g) in water, to this mixture was added acetic acid (2.5 ml) and the mixture diluted to a total volume of 250 ml. This solution was normally diluted by a factor of 20 with water and the deep eutectic solvent obtained by combining choline chloride with two equivalents of lactic acid. The resulting mixture was used for the majority of solvent extraction experiments. When copper, nickel, or manganese was required in solvent extraction experiments then an aqueous solution containing these metals (16 grams per liter) as the sulfate, chloride, and chloride were diluted by a factor of 40 by its addition to a combination of water and the deep eutectic solvent. For those experiments in which lead(II) was required a deep eutectic solvent bearing the other metals which had been stored for many days in contact with solid lead(II) chloride was used.

Sunflower oil FAME biodiesel was formed by the reaction of sunflower oil with methanol using sodium hydroxide as the catalyst. After the reaction the glycerol rich layer was separated and the product washed with water before use.

Funding
The research leading to these results has received funding from the European Community's Seventh Framework Programe ([FP7/2007–2013] under grant agreement number 603482 (Project named CoLaBats)). Mark Foreman is grateful to Exxon for their gift of Solvesso 150ND, Solvesso 200ND, Escaid 110, Isopar L and Escaid 120.

Author details
Mark R.StJ. Foreman[1]
E-mail: Foreman@chalmers.se
ORCID ID: http://orcid.org/0000-0002-1491-313X
[1] Department of Chemistry and Chemical Engineering,
  Chalmers University of Technology, Gothenburg, Sweden.

References
Abbott, A. P., Boothby, D., Capper, G., Davies, D. L., & Rasheed, R. K. (2004). Deep eutectic solvents formed between choline chloride and carboxylic acids: Versatile alternatives to ionic liquids. *Journal of the American Chemical Society, 126*, 9142–9147. http://dx.doi.org/10.1021/ja048266j

Abbott, A. P., Capper, G., Davies, D. L., McKenzie, K. J., & Obi, S. U. (2006). Solubility of metal oxides in deep eutectic solvents based on choline chloride. *Journal of Chemical and Engineering Data, 51*, 1280–1282. http://dx.doi.org/10.1021/je060038c

Abbott, G., Frisch, J. Hartley, & Ryder, K. S. (2011). Processing of metals and metal oxides using ionic liquids. *Green Chemistry, 13*, 471–481. http://dx.doi.org/10.1039/c0gc00716a

Akhavan, J. (1998). *The chemistry of explosives*. Cambridge: Royal Soceity of Chemistry.

Alexandratos, S. D. (2009). Ion-exchange resins: A retrospective from industrial and engineering chemistry research. *Industrial & Engineering Chemistry Research, 48*, 388–398. http://dx.doi.org/10.1021/ie801242v

Bouchene, R., Bouacida, S., Berrah, F., & Roisnel, T. (2014). A simple hydrated salt with a complex hydrogen-bonding network: Tetrakis(5-amino-3-carboxy-4H-1,2,4-triazol-1-ium) hexachloridocadmate(II) tetrahydrate. *Acta Crystallographica Section C: Structural Chemistry, 70*, 672–676.

Chmielewski, A. G., Urbanski, T. S., & Migdal, W. (1997). Separation technologies for metals recovery from industrial wastes. *Hydrometallurgy, 45*, 333–344. http://dx.doi.org/10.1016/S0304-386X(96)00090-4

European Commission. (2010). *Critical raw materials for the EU: Report of the ad-hoc working group on defining critical raw materials*. Brussels.

Huber, G. W., O'Connor, P., & Corma, A. (2007). Processing biomass in conventional oil refineries: Production of high quality diesel by hydrotreating vegetable oils in heavy vacuum oil mixtures. *Applied Catalysis A-General, 329*, 120–129. http://dx.doi.org/10.1016/j.apcata.2007.07.002

Kim, C.-J., Yoon, H.-S., Chung, K. W., Lee, J.-Y., Kim, S.-D., Shin, S. M., ... Kim, S.-H. (2014). Leaching kinetics of lanthanum in sulfuric acid from rare earth element (REE) slag. *Hydrometallurgy, 146*, 133–137. http://dx.doi.org/10.1016/j.hydromet.2014.04.003

Kyle, J. H., Breuer, P. L., Bunney, K.G., & Pleysier, R. (2012). Review of trace toxic elements (Pb, Cd, Hg, As, Sb, Bi, Se, Te) and their deportment in gold processing Part II: Deportment in gold ore processing by cyanidation. *Hydrometallurgy, 111*, 10–21. http://dx.doi.org/10.1016/j.hydromet.2011.09.005

Larsson, K., Ekberg, C., & Ødegaard-Jensen, A. (2012). Using Cyanex 923 for selective extraction in a high centration chloride medium on nickel metal hydride battery waste. *Hydrometallurgy, 129*, 35–42. http://dx.doi.org/10.1016/j.hydromet.2012.08.011

Moskalyk, M. M., & Alfantazi, A. M. (2002). Nickel laterite processing and electrowinning practice. *Minerals Engineering, 15*, 593–605. http://dx.doi.org/10.1016/S0892-6875(02)00083-3

Müller, T., & Friedrich, B. (2006). Development of a recycling process for nickel-metal hydride batteries. *Journal of Power Sources, 158*, 1498–1509.

Peppard, D. F., Mason, G. W., Maier, J. L., & Driscoll, W. J. (1957). Fractional extraction of the lanthanides as their di-alkyl orthophosphates. *Journal of Inorganic and Nuclear Chemistry, 4*, 334–343. http://dx.doi.org/10.1016/0022-1902(57)80016-5

Rydberg, J., & Albinsson, Y. (1989). Solvent-extraction studies on lanthanide acetylacetonates. 1. Promethium complexes investigated by the Akufve-Lisol technique. *Solvent Extraction and Ion Exchange, 7*, 577–612. http://dx.doi.org/10.1080/07360298908962326

Rydberg, J., Musikas, C., & Choppin, G. R. (Eds.). (1992). *Principles and practices of solvent extraction.* New York, NY: Dekker.

Sarangi, K., Padhan, E., Sarma, P. V. R. B., Park, K. H., & Das, R. P. (2006). Removal/recovery of hydrochloric acid using Alamine 336, Aliquat 336, TBP and Cyanex 923. *Hydrometallurgy, 84*, 125–129. http://dx.doi.org/10.1016/j.hydromet.2006.03.063

Stronski, I., & Nahlik, E. (1972). Influence of hammetts coefficients on distribution ratio of lanthanide and transplutonide ion-Pairs with Aliquat 336. *Radiochimica Acta, 18*, 47–50.

The Waste Batteries and Accumulators Regulations 2009. (2009). *Statutory instrument number 890.* London: Stationery Office.

Wellens, S., Thijs, B., Moller, C., & Binnemans, K. (2013). Separation of cobalt and nickel by solvent extraction with two mutually immiscible ionic liquids. *Physical Chemistry Chemical Physics, 15*, 9663–9669. http://dx.doi.org/10.1039/c3cp50819f

Xu, J., Thomas, H. R., Francis, R. W., Lum, K. R., Wang, J., & Liang, B. (2008). A review of processes and technologies for the recycling of lithium-ion secondary batteries. *Journal of Power Sources, 177*, 512–527.

2

Lactic acid-mediated tandem one-pot synthesis of 2-aminothiazole derivatives: A rapid, scalable, and sustainable process

Mohan Reddy Bodireddy[1], P.Md. Khaja Mohinuddin[1], Trivikram Reddy Gundala[1] and N.C. Gangi Reddy[1]*

*Corresponding author: N.C. Gangi Reddy, Department of Chemistry, School of Physical Sciences, Yogi Vemana University, Kadapa, Andhra Pradesh 516 003, India

E-mail: ncgreddy@ yogivemanauniversity.ac.in

Reviewing editor: George Weaver, University of Loughborough, UK

Abstract: Environmentally benign and biodegradable lactic acid is identified as alternative solvent and catalyst for the tandem one-pot synthesis of Hantzsch 2-aminothiazole derivatives (**4**) from readily available aralkyl ketones (**1**) through *in situ* regioselective α-bromination using *N*-bromosuccinimide (**2**) followed by heterocyclization using thiourea (**3**) at 90–100°C. The major advantages of the present method include short reaction times (10–15 min), practical, simple to perform, easy work-up, good yield of products (up to 96%), productive for large-scale applications, free from apply of α-bromoketones (lachrymator) as substrates, avoids column purification. Hence, the present method meets with the concepts of both Wender's "ideal synthesis" and sustainable chemical process.

Subjects: Environmental Chemistry; Medicinal & Pharmaceutical Chemistry; Organic Chemistry

Keywords: 2-aminothiazole; lactic acid; NBS; thiourea, α-bromination; scale-up process

ABOUT THE AUTHORS

N.C. Gangi Reddy was born in 1981 in Naravakati Palle village, Kadapa, Andhra Pradesh, INDIA. He has been awarded the PhD degree from Sri Venkateswara University, Tirupati, INDIA in April 2007. Later, he joined as an assistant professor in the Department of Chemistry, Yogi Vemana University, Kadapa in June 2007. His research interests are design and synthesis of medicinally valuable organic compounds and development of catalyst-based synthetic methodologies. He published more than 25 research papers in journals of international repute. He completed two major research projects as principal investigator.

PUBLIC INTEREST STATEMENT

The reported method provides wide scope and quick access to Hantzsch 2-aminothiazole derivatives in good to excellent isolated yields within a short period of time from readily available aralkyl ketones through *in situ* regioselective α-bromination using NBS and subsequent heterocyclization using thiourea in presence of environmentally benign and biodegradable lactic acid at 90–100°C in a single step operation. Major advantages of the present method include scale-up process, non-explosive, easy to perform, simple work-up, easy isolation of products, improved worker safety and use of environmentally benign, non-volatile and biodegradable lactic acid as an alternative solvent and catalyst. Further, the present protocol is free from (i) column purification, (ii) the use of hazardous solvents and (iii) the isolation of lachrymatory α-bromoketones. Hence, the present method meets the concept of Wender's "ideal synthesis". Finally, it is concluded that the present method is an attractive addition to the sustainable chemical processes.

1. Introduction

The growing interest in developing simple, more convenient methods for the synthesis of medicinally important thiazole moiety has great demand both in academic, chemical, and pharmaceutical domains (Dondoni, 1985). Consequently, extensive use of conventional solvents and hazardous catalysts became mandatory for their preparation which is leading to environmental pollution. Besides, these are mostly prepared from α-bromoketones that are difficult to store and not easily accessible. Further, these α-bromoketones are lachrymatory and cause other severe health hazards to the operating chemists. As a result, the growing interest in developing simple, safe, more convenient, and sustainable scale-up methods for the synthesis of these thiazole synthetic precursors from readily available ketones has great demand in academic, chemical, and pharmaceutical domains. The Hantzsch thiazole synthesis (Hantzsch & Weber, 1887) is a powerful synthetic tool for the construction of the five-membered 2-aminothiazole derivatives. Even though it was introduced more than one century ago, still the development of new Hantzsch-based methods is warranted. As a result, numerous methods have been reported with the improvement of many aspects of the original synthetic protocol for their syntheses from either ketones or α-brominated ketones or other key starting materials (Jacques & Vernin, 2008). 2-aminothiazole derivatives possess broad range of pharmacological activities (Biagetti et al., 2010; González Cabrera et al., 2011; Jaen et al., 1990; Parekh, Juddhawala, & Rawal, 2013; Patt et al., 1992; Shah, Shah, Patel, & Patel, 2012; Shao et al., 2013; Singh et al., 2014; Spector, Liang, Giordano, Sivaraja, & Peterson, 1998; Yang et al., 2010) and plant growth activities (An Tran, Anil Kumar, Jung-Ae, Lee, & Park, 2015). Recently, 2-aminothiazole derivatives are identified as a prodrug for the treatment of type 2 diabetes (**I** and **II**) (Erion et al., 2005), anti-tuberculosis agent (**III**) (Makam & Kannan, 2014), and anti-Parkinsonian agent-pramipexole (**IV**) (Chau, Cooper, & Schapira, 2013) as shown in Figure 1. Consequently, thiazole moiety is considered as "Important Structural motif" and the desirable properties of 2-amino thiazole derivatives render them attractive targets for new drug discovery. There are many reports on the synthesis of thiazole derivatives from ketones or their derivatives (Kumar, Minh An, Lee, Park, & Lee, 2015; Heravi, Poormohammad, Beheshtiha, & Baghernejad, 2011; Huang, Zhu, & Zhang, 2002; Janardhan, Krishnaiah, Rajitha, & Crooks, 2014; Kidwai, Chauhan, & Bhatnagar, 2011; Meshram, Thakur, Madhu Babu, & Bangade, 2012; Nitta et al., 2012; Potewar, Ingale, & Srinivasan, 2008; Zhuravel, Kovalenko, Vlasov, & Chernykh, 2005; Zhu et al., 2012) and thiourea or substituted thiourea. However, most of the reported methods suffer from one or more disadvantages including use of hazardous and toxic reagents, long reaction times, low selectivity of the products, use of toxic and volatile solvents as well as catalysts, environmentally hazardous processes, low yields and difficulties in work-up and isolation of products and oppressive operational procedures.

On the other hand, now-a-days, recycle and reduce of solvent usage or change to other solvents with better environmental profiles (Jessop, 2011; Kerton, 2009) become prominent. Accordingly, we found that environmentally benign lactic acid (Yang, Tan, & Gu, 2012) can act as catalyst cum solvent for *in situ* regioselective α-bromination and subsequent heterocyclization in the synthesis of pharmacologically active 2-aminothiazole derivatives. The present method is scale-up process and also meets the concept of "ideal synthesis" as described by Wender, Handy, and Wright (1997).

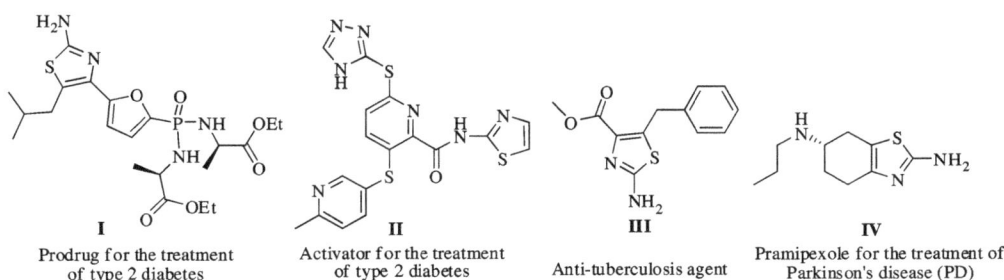

I
Prodrug for the treatment of type 2 diabetes

II
Activator for the treatment of type 2 diabetes

III
Anti-tuberculosis agent

IV
Pramipexole for the treatment of Parkinson's disease (PD)

Figure 1. Representative examples of biologically active 2-aminothiazole derivatives.

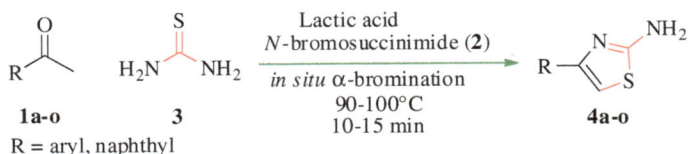

Scheme 1. Lactic acid as catalyst and solvent for the one-pot synthesis of Hantzsch 2-aminothiazole derivatives.

Herein, we report a simple, more convenient, practical method in which lactic acid acts as green catalyst and solvent for the one-pot synthesis of Hantzsch 2-aminothiazole derivatives (**4**) within 10–15 min from readily available aralkyl ketones (**1**) through *in situ* regioselective α-bromination using N-bromosuccinimide (**2**) followed by heterocyclization using thiourea (**3**) at 90–100°C (Scheme 1).

2. Results and discussion

It is planned to develop a simple, more convinient method for the preparation of Hantzsch 2-amino thiazole derivatives (**4**). For this purpose, initially we have choosen to synthesize the 4-(4-bromophe-nyl)thiazol-2-amine (**4a**) using 4′-bromoacetophenone (**1a**). It is used as model substrate for the optimization of reaction conditions as discusssed below.

2.1. Selection of suitable catalyst and solvent

For the success of the present hypothesis, an acidic organic liquid which can function as a catalyst and solvent is needed to identify for the accomplishment of both *in situ* regioselective α-bromination and heterocyclization in a single step operation, as the solubility of key reactants and their α-bromination and subsequent heterocyclization become easy in the presence of acidic organic liq-uid compared to solid catalysts alone. At the same time, it is evident that both the α-bromination and hetrocyclization processes proceed in the presence of acidic catalysts.

Toward this direction, a reaction is carried out using 4′-bromoacetophenone (**1a**) and N-bromosuccinimide (**2**) followed by addition of thiourea (**3**) in the presence of acetic acid (entry 1, Table 1) and the isolated yield of desired product (**4a**) is 30% at RT (entry 1, Table 1) and 58% at 90–100°C. To increase the yield of **4a**, the same reaction is carried out in lactic acid and the obtained yield of product (**4a**) is increased to 45% at RT (entry 3). The same reaction at 90–100°C provided 96% yield (entry 4, Table 1). It may be due to more acidic nature of lactic acid compared to acetic acid. The evaluation of suitable acidic organic liquid which acts both as catalyst and solvent dis-closed that lactic acid is the best option to obtain maximum yield (96%) of the desired product (**4a**) (entry 4, Table 1) compared to acetic acid (entry 2, Table 1).

Table 1. Screening of suitable acidic organic liquid (act as catalyst and solvent) for the synthesis of 4-(4-bromophenyl)thiazol-2-amine (4a)[a]

Entry	Catalyst[b]	Temperature	Time (min)	Product	Yield[c] (%)
1	Acetic acid	RT	7 h	**4a**	30
2		90–100°C	90 min		58
3	Lactic acid	RT	1.2 h	**4a**	45
4		90–100°C	13 min		96

Notes: [a]Reaction conditions: 4′-bromoacetophenone (**1a**) (25.0 mmol), N-bromosuccinimide (**2**) [(30.0 mmol) is added in 4 portions], thiourea (**3**) (30.0 mmol) and dual functional weak organic acid (20 mL) at 90–100°C.

[b]Act as catalyst and solvent.

[c]Isolated yield.

Later, other reaction conditions such as effect of brominating agent and mode of addition of brominating agent on the course of *in situ* regioselective α-bromination and also the effect of temperature both on *in situ* regioselective α-bromination and subsequent heterocyclization is investigated and the results obtained are discussed below.

In the present method, the *in situ* α-bromination reaction plays a vital role as the yield of target 2-aminothiazoles (**4**) is directly proportional to the yield of *in situ* generated α-brominated ketone (**5**). For this reason, various brominating agents and their mode of addition are studied. In particular, greater attention has been paid to improve the efficiency of α-bromination of ketones in terms of regioselectivity to generate α-monobrominated ketones exclusively compared to α-dibrominated ketones and also to achieve maximum conversion. The results obtained are presented in Tables 2 and 3.

2.2. Screening of suitable brominating agent

The effect of brominating agent is studied on the course of *in situ* regioselective α-bromination and the results obtained are presented in Table 2. Accordingly, a reaction is carried out using molecular bromine and obtained lower yield (55%) of product **5a** (entry 1, Table 2). The toxicity, difficulties in handling and low selectivity of molecular bromine encouraged us to use other user-friendly and readily available brominating agents such as HBr-H$_2$O$_2$, dioxane dibromide, *N*-bromosuccinimide (NBS) and CuBr$_2$ which provided 60, 76, 97, and 81% yields of product **5a**, respectively (entries 2–5, Table 2).

The study revealed that NBS is the best brominating agent as it gave maximum yield (97%) of product **5a** when added in 4 portions (entry 3, Table 3).

2.3. Effect of temperature

In general, the solubility of reactants and products, rate of reaction, efficiency and selectivity of catalyst are highly dependent on operating temperature of the reaction. Hence, we desired to study the effect of temperature on the course of both *in situ* α-bromination (**A**) and subsequent heterocyclization (**B**) process. Accordingly, reactions are conducted at various ranges of temperature and the results obtained are presented in Table 4.

Table 2. Effect of brominating agent on the course of in situ α-bromination reaction[a]

Entry	Brominating agent	Time (min)	Temperature (°C)	Product (5a)	Yield[b] (%)	Selectivity (%) (5a:6a)[c]
1	Br$_2$	25	10–15	**5a**	55	55:16
2	HBr-H$_2$O$_2$	31	RT	**5a**	60	60:13
3	Dioxane dibromide	20	90–100	**5a**	76	76:12
4	*N*-bromosuccinimide	12	90–100	**5a**	97	97:02
5	CuBr$_2$	25	90–100	**5a**	81	81:10

Notes: [a]Reaction conditions: 4′-bromoacetophenone (**1a**) (25.0 mmol), brominating agent (**2**) [(30.0 mmol) is added in 4 portions] and lactic acid (20 mL) at 90–100°C.

[b]Isolated yield.

[c]Unreacted substrate.

Table 3. Effect of mode of addition of N-bromosuccinimide (NBS) on course of in situ α-bromination reaction[a]

Entry	Mode of NBS addition (Portions)	Time (min)	Product	Yield[b] (%)	Selectivity (%) (5a:6a)[c]
1	Once	7	**5a**	67	67:16
2	Two	9	**5a**	76	76:14
3	Four	12	**5a**	97	97:02

Notes:[a]Reaction conditions: 4'-bromoacetophenone (**1a**) (25.0 mmol), N-bromosuccinimide (**2**) (30.0 mmol) is added in 4 portions and lactic acid (20 mL) at 90–100°C.

[b]Isolated yield.

[c]Unreacted substrate.

For instance, when the reaction is carried out at room temperature, the rate of α-bromination (**A**) process is slow (3 h) and provided lower yield (53%) of product **5a** (entry 1, Table 4) and subsequent heterocyclization (**B**) using thiourea (**3**) provided only 40% yield of final product **4a** (entry 1). This may be due to partial solubility of the reactants. To improve the solubility of reactants and yield of α-bromoketone (**5a**) further, the reaction is conducted at different temperatures, for example at

Table 4. Effect of temperature on in situ regioselective α-bromination and subsequent heterocyclization[a]

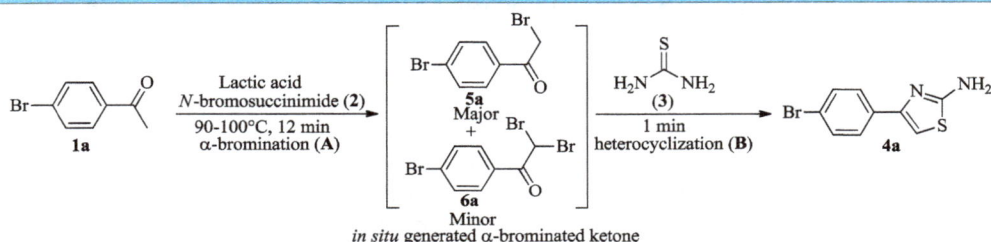

in situ generated α-brominated ketone

Entry	Type of reaction α-bromination (A)[b] Heterocyclization (B)[c]	Temperature (°C)	Time	Product	Yield[d] (%)	Selectivity (%) (5a:6a)[e]
1	A	RT	4 h	**5a**	53	53:18
	B		30 min	**4a**	40	–
2	A	40–50	3 h	**5a**	65	65:15
	B		15 min	**4a**	58	–
3	A	50–60	70 min	**5a**	69	69:13
	B		6 min	**4a**	62	–
4	A	70–80	50 min	**5a**	75	75:10
	B		4 min	**4a**	71	–
5	A	80–90	45 min	**5a**	84	84:6
	B		5 min	**4a**	82	–
6	A	90–100	12 min	**5a**	97	97:2
	B		1 min	**4a**	96	–

Notes: [a]Reaction conditions: 4'-bromoacetophenone (**1a**) (25.0 mmol), N-bromosuccinimide (**2**) (30.0 mmol) is added in 4 portions), thiourea (**3**) (30.0 mmol) and lactic acid (20 mL) at specified temperature.

[b]The yields of α-brominated product (**5a**) increases with increase of No. of portions of NBS and temperature.

[c]Isolated yield of final product (**4a**) depends on yield of *in situ* generated regioselective α-brominated product (**5a**).

[d]Isolated yield.

[e]Regioselectivity of *in situ* α-bromination [Unreacted substrate (**1a**)].

Table 5. Tandem one-pot synthesis of 2-amino thiazole derivatives (4a-o) from readily available aralkyl ketones (1a-o)[a]

Entry	Substrate (1)	Product (4)	Time (min)	Yield[b] (%)
1	1a	4a	13	96
2	1b	4b	14	85
3	1c	4c	13	95
4	1d	4d	14	94
5	1e	4e	15	74
6	1f	4f	15	78
7[c]	1g	4g	14	53
8[c]	1h	4h	14	46
9	1i	4i	15	92
10	1j	4j	15	84
11	1k	4k	15	77

(Continued)

Table 5. *(Continued)*

Entry	Substrate (1)	Product (4)	Time (min)	Yield[b] (%)
12	**1l**	**4l**	14	79
13	**1m**	**4m**	13	88
14	**1n**	**4n**	15	81
15	**1o**	**4o**	14	90

Notes: [a]Reaction conditions: Aralkyl ketone (**1a-o**) (25.0 mmol), *N*-bromosuccinimide (**2**) [(30.0 mmol) is added in 4 portions], thiourea (**3**) (30.0 mmol) and lactic acid (20 mL) at 90–100°C.

[b]Isolated yield.

[c]45.0 mmol of *N*-bromosuccinimide is used.

40–50, 50–60, 70–80, 80–90, and 90–100°C provided 65, 69, 75, 84, and 97% isolated yield of α-bromoketone (**5a**) (entries 2–6).

While we consider heterocyclization (**B**) step alone, this stage is also slightly influenced by the operating temperature. For example, at room temperature, the yield of product **4a** (40%) is decreased (entry 1, Table 4) significantly due to the partial solubility of thiourea (**3**) in lactic acid. When the same reaction is conducted at 40–50°C, improved yield (58%) of product **4a** is obtained compared to RT (entry 2). But, when the reaction is operated at 50–60, 70–80, 80–90, and 90–100°C provided yields of 62, 71, 82, and 96% of product **4a**, respectively (entries 3–6, Table 4).

These results indicate that the isolated yield of product **4a** is directly proportional to the percentage of formation of *in situ* regioselective α-bromoketone **5a** (entries 2–6, Table 4).

When we consider the overall reaction i.e. both *in situ* regioselective α-bromination (**A**) and heterocyclization (**B**), the optimum operating temperature is 90–100°C for maximum conversion and isolated yield (96%) of target product **4a** (entry 7, Table 4). Based on our present study, we found that in case of laboratory scale, the optimum temperature is 90–100°C for the synthesis of 2-aminothiazoles (**4**).

2.4. Scope of the method

The substrate scope of the present method is studied with the help of above optimized reaction conditions using different types of aralkyl ketones (**1a-o**) as key starting substrates and the results obtained are presented in Table 5. The study disclosed that the substrate with fluoro, chloro, and bromo groups at *para*-position provided excellent yields (90–96%) of the products **4a, 4c, 4d** (entries 1, 3, and 4, Table 5), but *meta* monohalogenated substrate, for example 1-(3-bromophenyl) ethanone (**1b**) provided good yield (85%)of the product **4b** (entry 2, Table 5). Whereas the dihalogenated substrates, 1-(2,4-dichlorophenyl)ethanone (**1e**) and 1-(3,4-dichlorophenyl) ethanone (**1f**) provided moderate yields of products **4e** and **4f** in 74% and 78%, respectively (entries 5 and 6, Table 5). It is found that the isolated yield of the respective products depends on the position of halogen(s) on aromatic ring of aralkyl ketone (entries 1–6, Table 5).

Interestingly, substrates, 1-(4-nitrophenyl)ethanone (**1g**) and 1-(3-nitrophenyl)ethanone (**1h**) with high deactivating groups ($-NO_2$) provided lower yields of product **4g** (53%) and **4h** (46%) (entries 7 and 8, Table 5) compared to all other substrates used in the present study. This may be due to the presence of strong electron withdrawing groups on the aromatic ring cause to reduce the percentage of formation of *in situ* α-brominated ketones which play vital role on yield of final product. Simple aralkyl ketones, for instance acetophenone (**1i**) and propiophenone (**1l**) provided good yields of the products **4i** (92%) and **4l** (84%) (entries 9 and 12, Table 5) and the substrates with moderate activating groups, for example 1-(p-tolyl)ethanone (**1j**), 1-(4-ethylphenyl)ethanone (**1k**) provided acceptable yields (79–84%) of respective products (**4j and 4k**) (entries 10–11, Table 5). But, the substrate with high activating group, for example 1-(4-methoxyphenyl)ethanone (**1m**) provided good yield (88%) of the product **4m** (entry 13, Table 5). Interestingly, in case of acenaphthanones (**1n** and **1o**), the 1-acetyl naphthalene (**1n**) gave lower yield (81%) of product **4n** when compared to 2-acetyl naphthalene (**1o**) which afforded higher yield (90%) of product **4o** (entries 14 and 15, Table 5). It may be due to steric effects.

3. Conclusions

In summary, the developed method provides wide scope and quick access to Hantzsch 2-aminothiazole derivatives (**4**) in good yields within 10–15 min from readily available aralkyl ketones (**1**) through *in situ* regioselective α-bromination using NBS (**2**) and subsequent heterocyclization using thiourea (**3**) in the presence of lactic acid at 90–100°C in a single step operation when compared to an alternative two-step procedures or other reported one-step methods. Major advantages of the present method include scale-up process, non-explosive, easy to perform, simple work-up, easy isolation of products, improved worker safety and use of environmentally benign, non-volatile and biodegradable lactic acid as alternative solvent and catalyst. Further, the current protocol is free from (i) column purification, (ii) the use of hazardous solvents, and (iii) the isolation of lachrymatory α-bromoketones. Finally, it is concluded that the present method is an attractive addition to the sustainable chemical processes.

4. Materials and method

4.1. Materials

Melting points of various obtained products are determined and uncorrected. ^{1}H NMR spectra are recorded on a Varian 400 MHz and ^{13}C NMR spectra on a Jeol/AL-100 MHz. Chemical shifts were expressed in parts per million (ppm), coupling constants are expressed in Hertz (Hz). Splitting patterns describe apparent multiplicities and were designated as s (singlet), d (doublet), t (triplet), q (quartet), m (multiplet). High-resolution mass spectra (HRMS) and compound purity data are acquired on a Exactive Orbitrap Mass Spectrometer (ThermoScientific, Waltham, MA, USA) equipped with electro spray ionization (ESI) source. Thin-layer chromatography is performed on 0.25 mm Merck silica gel plates and visualized with UV light. Column chromatography is performed on silica gel. Chemicals and solvents are purchased from Sigma Aldrich and Merck. Isolated compounds are identified on the basis of spectroscopic data (^{1}H & ^{13}C NMR, Mass and HRMS).

4.2. Method for the synthesis of 2-aminothiazole derivatives (4a-o)

In a 100-mL 3 necked round bottom flask, aralkyl ketone (**1**) (25.0 mmole) and lactic acid (20 mL) are taken. The substrate (**1**) is partially soluble in lactic acid at room temperature. The temperature of the reaction mass is raised to 90–100°C. At this temperature, the reaction mass became homogeneous and N-bromosuccinimide (**2**) (30.0 mmol) is added in 4 portions (7.5 × 4 = 30 mmol). On addition of each portion of NBS (3.0 mmol), the color of the reaction mass is changed from colorless to orange red. This indicates that the bromonium ion is released from NBS (**2**). Later, within few minutes, the disappearance of the orange red color is observed. The same situation is observed when another 3 portions of NBS (**2**) is added. After the completion of α-bromination as per TLC, thiourea (**3**) (30.0 mmol) is added and stirred at the same temperature for 1 min. Then, the reaction mass is slowly cooled down to RT. As the temperature of the reaction mass is below 80°C, the final product (**4**) is slowly thrown out from the lactic acid. At room temperature, 20 mL of water is added for the precipitation of product (**4**). Then, it is filtered-off and washed with 10 mL of water. The crude solid product (**4**) is collected and to this 100 mL of cold water is added and quenched with NaHCO$_3$.[1] Again, it is filtered-off and washed twice with water (2 × 25 mL) for the removal of by-product and inorganic salts.[2] The pure solid product (**4**) is collected and dried in vacuum oven at ambient temperature and the isolated yields of respective products are presented in Table 5. All the prepared products (**4a-o**) are characterized by physical and spectroscopic data (^{1}H &^{13}C NMR, Mass, and HRMS).

4.3. Characterization data of the corresponding compounds are as follows

4.3.1. 4-(4-bromophenyl) thiazol-2-amine (4a)
Off-white solid, yield: 96%; m.p. 180–181°C; ^{1}H NMR (400 MHz, DMSO-d$_6$, δ/ppm): 7.73 (2H, dd, J = 6.8 Hz, J = 2.0 Hz, arom H), 7.69 (2H, br s, -NH$_2$), 7.61 (2H, dd, J = 6.8 Hz, J = 2.0 Hz, arom H), 7.16 (1H, s, thiazole H); ^{13}C NMR (100 MHz, DMSO-d$_6$, δ/ppm): 169.12, 144.39, 131.59 (3C), 127.61 (2C), 121.09, 102.91; HRMS (ESI): calcd for C$_9$H$_8$BrN$_2$S [M + H]$^+$ 254.9586, found 254.9596;[M + H + 2]$^+$256.9572.

4.3.2. 4-(3-bromophenyl) thiazol-2-amine (4b)
Yellow solid, yield: 85%; m.p. 162–163°C; ^{1}H NMR (400 MHz, DMSO-d$_6$, δ/ppm): 7.98 (1H, s, arom H), 7.79 (1H, d, J = 7.6 Hz, arom H), 7.44 (1H, d, J = 6.8 Hz, arom H), 7.32 (1H, t, J = 8.0 Hz, arom H), 7.16 (1H, s, thiazole H), 7.15 (2H, br s, -NH$_2$); ^{13}C NMR (100 MHz, DMSO-d$_6$, δ/ppm): 168.35, 148.06, 137.11, 130.62, 129.73, 128.15, 124.32, 121.99, 103.10; HRMS (ESI): calcd for C$_9$H$_8$BrN$_2$S [M + H]$^+$ 254.9507, found 254.9596; [M + H + 2]$^+$ 256.9550.

4.3.3. 4-(4-chlorophenyl) thiazol-2-amine (4c)
Off-white solid, yield: 95%; m.p. 166–167°C; ^{1}H NMR (400 MHz, DMSO-d$_6$, δ/ppm): 7.80 (2H, d, J = 8.4 Hz, arom H), 7.41 (2H, d, J = 8.8 Hz, arom H), 7.09 (2H, s, -NH$_2$), 7.07(1H, s, thiazole H); ^{13}C NMR (100 MHz, DMSO-d$_6$, δ/ppm): 168.39, 148.56, 133.71, 131.57, 128.45 (2C), 127.20(2C), 102.28; HRMS (ESI): calcd for C$_9$H$_8$N$_2$ClS [M + H]$^+$ 211.0091, found 211.0082.

4.3.4. 4-(4-fluorophenyl) thiazol-2-amine (4d)
Off-white solid, yield: 94%; m.p. 101–102°C; ^{1}H NMR (400 MHz, DMSO-d$_6$, δ/ppm): 8.8 (2H, br s, -NH$_2$), 7.82 (2H, m, arom H), 7.34 (2H, t, J = 6.8 Hz, arom H), 7.21 (1H, s, thiazole H); ^{13}C NMR (100 MHz, DMSO-d$_6$, δ/ppm): 170.33, 162.51(1C, d, ^{1}J$_{C-F=}$246.2 Hz), 137.66, 128.17, (2C, d, ^{3}J$_{C-F}$ = 8.3 Hz), 125.20 (1C, d, ^{4}J$_{C-F}$ = 3.3 Hz), 116.08 (2C, d,^{2}J$_{C-F}$ = 21.4 Hz), 102.82; HRMS (ESI): calcd for C$_9$H$_8$N$_2$FS [M + H]$^+$ 195.0387, found 195.0377.

4.3.5. 4-(2, 4-dichlorophenyl) thiazol-2-amine (4e)
Off-white solid, yield: 74%; m.p. 189–191°C; ^{1}H NMR (400 MHz, DMSO-d$_6$, δ/ppm): 8.6 (2H, Br s, -NH$_2$), 7.79 (1H, d, J = 2.4 Hz, arom H), 7.71 (1H, d, J = 8.4 Hz, arom H), 7.57 (1H, dd, J = 8.4 Hz, J = 2.0 Hz, arom H), 7.14 (1H, s, thiazole H); ^{13}C NMR (100 MHz, DMSO-d$_6$, δ/ppm): 169.21, 135.40, 134.92, 132.87 (2C), 129.65, 127.81, 127.58, 108.13; MS (ESI) m/z: [M + H]$^+$ 245.00, [M + H + 2]$^+$ 247.00.

4.3.6. 4-(3, 4-dichlorophenyl) thiazol-2-amine (4f)
Off-white solid, yield: 78%; m.p. 193–194°C; ^{1}H NMR (400 MHz, DMSO-d$_6$, δ/ppm): 8.016 (1H, d, J = 2.0 Hz, arom H), 7.78 (1H, dd, J = 8.8 Hz, J = 1.6 Hz, arom H), 7.61 (1H, d, J = 8.4 Hz, arom H), 7.23 (1H, s, arom H), 7.16 (2H, br s, -NH$_2$); ^{13}C NMR (100 MHz, DMSO-d$_6$, δ/ppm): 168.58, 146.95, 135.18, 131.38, 130.59, 129.44, 127.19, 125.44, 103.76; HRMS (ESI): calcd for C$_9$H$_7$N$_2$Cl$_2$S [M + H]$^+$ 244.9702, found 244.9691;[M + H + 2]$^+$246.9658.

4.3.7. 4-(4-nitrophenyl) thiazol-2-amine (4g)

Orange solid, yield: 53%; m.p. 285–286°C; ^{1}H NMR (400 MHz, DMSO-d$_6$, δ/ppm): 8.23 (2H, d, J = 8.8 Hz, arom H), 8.04 (2H, d, J = 8.8 Hz, arom H), 7.42 (1H, s, thiazole H), 7.25 (2H, br s, -NH$_2$); ^{13}C NMR (100 MHz, DMSO-d$_6$, δ/ppm): 168.79, 146.85, 146.02, 140.28, 126.32 (2C), 124.00 (2C), 106.67; HRMS (ESI): calcd for C$_9$H$_8$O$_2$N$_3$S [M + H] $^+$ 222.0332, found 222.0341.

4.3.8. 4-(3-nitrophenyl) thiazol-2-amine (4h)

Yellow solid, yield: 46%; m.p. 282–283°C; ^{1}H NMR (400 MHz, DMSO-d$_6$, δ/ppm): 8.60 (1H, s, arom H), 8.23 (1H, d, J = 7.6 Hz, arom H), 8.14 (1H, d, J = 8.0 Hz, arom H), 7.69 (1H, t, J = 8.0 Hz, arom H), 7.38 (1H, s, thiazole H);^{13}C NMR (100 MHz, DMSO-d$_6$, δ/ppm): 169.03, 148.19, 144.93, 134.93, 131.69, 130.19, 122.19, 120.05, 104.65; MS (ESI) m/z:[M + H]$^+$ 222.00.

4.3.9. 4-phenylthiazol-2-amine (4i)

Off-white solid, yield: 92%, m.p. 149–151°C; ^{1}H NMR (400 MHz, DMSO-d$_6$, δ/ppm): 8.6 (2H, br s, -NH$_2$), 7.73 (2H, d, J = 7.6 Hz, arom H), 7.48 (2H, t, J = 8.0 Hz, arom H), 7.41 (1H, d, J = 7.6 Hz, arom H), 7.21 (1H, s, thiazole H); ^{13}C NMR (100 MHz, DMSO-d$_6$, δ/ppm): 170.30, 138.53, 129.54, 129.10 (2C), 128.46, 125.80 (2C), 103.06; HRMS (ESI): calcd for C$_9$H$_9$N$_2$S [M + H]$^+$ 177.0481, found 177.0486.

4.3.10. 2-Amino-4-(4-methylphenyl) thiazole (4j)

Off-white solid, yield: 84%; m.p. 136–137°C; ^{1}H NMR (400 MHz, DMSO-d$_6$, δ/ppm): 8.7 (2H, br s, -NH$_2$), 7.62 (2H, d, J = 8.4 Hz, arom H), 7.29 (2H, d, J = 8.0 Hz, arom H), 7.15 (1H, s, thiazole H), 2.34 (3H, s, -CH$_3$); ^{13}C NMR (100 MHz, DMSO-d$_6$, δ/ppm): 170.27, 139.28, 138.46, 129.61 (2C), 125.66 (3C), 102.05, 20.88; HRMS (ESI): calcd for C$_{10}$H$_{11}$N$_2$S [M + H]$^+$ 191.0637, found 191.0644.

4.3.11. 4-(4-ethylphenyl) thiazol-2-amine (4k)

Off-white solid, yield: 77%; m.p. 140–141°C; ^{1}H NMR (400 MHz, DMSO-d$_6$, δ/ppm): 8.75 (2H, br s, -NH$_2$), 7.67 (2H, d, J = 8.4 Hz, arom H), 7.32 (2H, d, J = 8.6 Hz, arom H), 7.14 (1H, s, thiazole H), 2.64 (2H, q, J = 7.6 Hz, -CH$_2$-), 1.19 (3H, t, J = 7.6 Hz, -CH$_3$); ^{13}C NMR (100 MHz, DMSO-d$_6$, δ/ppm): 168.14, 149.97, 142.71, 132.57, 127.82 (2C), 125.56 (2C), 100.58, 27.92, 15.50; HRMS (ESI): calcd for C$_{11}$H$_{13}$N$_2$S [M + H]$^+$ 205.07940, found 205.07976.

4.3.12. 5-methyl-4-phenylthiazol-2-amine (4l)

Off-white solid, yield: 79%, m.p. 118–120°C; ^{1}H NMR (400 MHz, DMSO-d$_6$, δ/ppm): 8.96 (2H, br s, -NH$_2$), 7.57–7.48 (5H, m, arom H), 2.28 (3H, s, -CH$_3$); ^{13}C NMR (100 MHz, DMSO-d$_6$, δ/ppm): 179.46, 167.66, 133.12, 129.43, 128.95 (2C), 128.45 (2C), 114.83, 11.70; HRMS (ESI): calcd for C$_{10}$H$_{11}$N$_2$S [M + H]$^+$191.0637, found 191.0642.

4.3.13. 4-(4-methoxyphenyl) thiazol-2-amine (4m)

Off-white solid, yield: 88%; m.p. 205–206°C; ^{1}H NMR (400 MHz, DMSO-d$_6$, δ/ppm): 8.9 (2H, br s, -NH$_2$) 7.71 (2H, d, J = 8.8 Hz, arom H), 7.085 (1H, s, thiazole H), 7.05 (H, d, J = 8.8 Hz, arom H), 3.81 (3H, s, -OCH$_3$); ^{13}C NMR (100 MHz, DMSO-d$_6$, δ/ppm): 170.38, 160.12, 138.14, 127.26 (2C), 120.91, 114.49 (2C), 100.49, 55.42; HRMS (ESI): calcd for C$_{10}$H$_{11}$ON$_2$S [M + H]$^+$ 207.0587, found 207.0593.

4.3.14. 4-(naphthalen-1-yl) thiazol-2-amine (4n)

Brown solid, yield: 81%; m.p. 164–165°C; ^{1}H NMR (400 MHz, DMSO-d$_6$, δ/ppm): 9.06 (2H, br s, -NH$_2$), 8.10–8.02 (4H, m, arom H), 7.69 -7.60 (3H, m, arom H), 7.07 (1H, s, arom H); ^{13}C NMR (100 MHz, DMSO-d$_6$, δ/ppm): 168.05, 149.86, 133.48, 133.39, 130.74, 128.13, 127.96, 126.61, 126.23, 125.89, 125.79, 125.41, 104.99; HRMS (ESI): calcd for C$_{13}$H$_{11}$N$_2$S [M + H]$^+$ 227.06375, found 227.06416.

4.3.15. 4-(naphthalen-2-yl) thiazol-2-amine (4o)

Brown solid, yield: 90%; m.p. 151–152°C; ^{1}H NMR (400 MHz, DMSO-d$_6$, δ/ppm): 8.32 (1H, s, arom H), 7.96–7.87 (4H, m, arom H), 7.52–7.47 (2H, m, arom H), 7.17 (1H, s, arom H), 7.13 (2H, br s, -NH$_2$); ^{13}C NMR (100 MHz, DMSO-d$_6$, δ/ppm): 169.44, 153.37, 133.01, 132.74, 132.36, 128.36, 127.80, 127.49, 126.58, 126.42, 125.66, 123.53, 103.63; HRMS (ESI): calcd for C$_{13}$H$_{11}$N$_2$S [M + H]$^+$ 227.0637, found 227.0645.

4.3.16. 2-bromo-1-(4-bromophenyl)ethanone (5a)

Off-white solid, yield: 97%; m.p. 108–110°C; ^{1}H NMR (400 MHz, CDCl$_3$): δ = 7.53 (d, J = 8.8 Hz, 2H, arom H), 7.64 (d, J = 8.4 Hz, 2H, arom H), 4.39 (s, 2 H, -CH$_2$-).2011/37C/52/BRNS/2264

Funding
The authors acknowledge the financial support provided by the Department of Atomic Energy-Board of Research in Nuclear Sciences (DAE-BRNS) (Bhabha Atomic Research Centre), Mumbai, India through a major research project (grant number 2011/37C/52/BRNS/2264).

Author details
Mohan Reddy Bodireddy[1]
E-mail: bmmreddys@gmail.com
P.Md. Khaja Mohinuddin[1]
E-mail: khajamohinuddin786@gmail.com
Trivikram Reddy Gundala[1]
E-mail: gtvreddy@gmail.com
N.C. Gangi Reddy[1]
E-mail: ncgreddy@yogivemanauniversity.ac.in

[1] Department of Chemistry, School of Physical Sciences, Yogi Vemana University, Kadapa, Andhra Pradesh 516 003, India.

References

An Tran, N. M., Anil Kumar, M., Jung-Ae, K., Lee, K. D., & Park, S. (2015). Synthesis, anticancer and antioxidant activity of novel carbazole-based thiazole derivatives. *Phosphorus, Sulfur, and Silicon and the Related Elements, 190,* 1–9.

Biagetti, M., Leslie, C. P., Mazzali, A., Seri, C., Pizzi, D. A., Bentley, J., ... Caberlotto, L. (2010). Synthesis and structure–activity relationship of N-(3-azabicyclo [3.1.0] hex-6-ylmethyl)-5-(2-pyridinyl)-1, 3-thiazol-2-amines derivatives as NPY Y5 antagonists. *Bioorganic & Medicinal Chemistry Letters, 20,* 4741–4744.

Chau, K. Y., Cooper, J. M., & Schapira, A. H. (2013). Pramipexole reduces phosphorylation of α-synuclein at serine-129. *Journal of Molecular Neuroscience, 51,* 573–580. http://dx.doi.org/10.1007/s12031-013-0030-8

Dondoni, A. (1985). New perspectives in thiazole chemistry. *Phosphorus and Sulfur and the Related Elements, 24,* 1–38. http://dx.doi.org/10.1080/03086648508073395

Erion, M. D., van Poelje, P. D., Dang, Q., Kasibhatla, S. R., Potter, S. C., Reddy, M. R., ... Lipscomb, W. N. (2005). MB06322 (CS-917): A potent and selective inhibitor of fructose 1, 6-bisphosphatase for controlling gluconeogenesis in type 2 diabetes. *Proceedings of the National Academy of Sciences, 102,* 7970–7975. http://dx.doi.org/10.1073/pnas.0502983102

González Cabrera, D. G., Douelle, F., Feng, T.-S., Nchinda, A. T., Younis, Y., White, K. L., ... Chibale, K. (2011). Novel orally active antimalarial thiazoles. *Journal of Medicinal Chemistry, 54,* 7713–7719. http://dx.doi.org/10.1021/jm201108k

Hantzsch, A., & Weber, J. H. (1887). Ueber Verbindungen des Thiazols [Pyridins der Thiophenreihe]. *Berichte der deutschen chemischen Gesellschaft, 20,* 3118–3132. http://dx.doi.org/10.1002/(ISSN)1099-0682

Heravi, M. M., Poormohammad, N., Beheshtiha, Y. S., & Baghernejad, B. (2011). Efficient synthesis of 2,4-disubstituted thiazoles under grinding. *Synthetic Communications, 41,* 579–582. http://dx.doi.org/10.1080/00397911003629440

Huang, X., Zhu, Q., & Zhang, J.-Z. (2002). Synthesis of a new polymer-supported reagent-poly [4-hydroxy (tosyloxy) iodo] styrene and its application to the synthesis of 2-amino-4-arylthiazoles. *Chinese Journal of Chemistry, 20,* 1411–1414.

Jacques, V. M., & Vernin, G. (2008). *Chemistry of heterocyclic compounds: Thiazole and its derivatives* (Part One, Vol. 34). Wiley-VCH. ISBN: 978-0-471-03993-8.

Jaen, J. C., Wise, L. D., Caprathe, B. W., Tecle, H., Bergmeier, S., Humblet, C. C., ... Pugsley, T. A. (1990). 4-(1,2,5,6-Tetrahydro-1-alkyl-3-pyridinyl)-2-thiazolamines: A novel class of compounds with central dopamine agonist properties. *Journal of Medicinal Chemistry, 33,* 311–317. http://dx.doi.org/10.1021/jm00163a051

Janardhan, B., Krishnaiah, V., Rajitha, B., & Crooks, P. A. (2014). Sodium fluoride as an efficient catalyst for the synthesis of 2,4-disubstituted-1,3-thiazoles and selenazoles at ambient temperature. *Chinese Chemical Letters, 25,* 172–175.

Jessop, P. G. (2011). Searching for green solvents. *Green Chemistry, 13,* 1391–1398. http://dx.doi.org/10.1039/c0gc00797h

Kerton, F. M. (2009). Alternative solvents for green chemistry. In J. Clark (Ed.), *RSC green chemistry series* (pp. 1–218). Cambridge. ISBN 978-0-85404-163-3

Kidwai, M., Chauhan, R., & Bhatnagar, D. (2011). Eco-friendly synthesis of 2-aminothiazoles using Nafion-H as a recyclable catalyst in PEG-water solvent system. *Journal of Sulfur Chemistry, 32,* 37–44. http://dx.doi.org/10.1080/17415993.2010.533773

Kumar, M., Minh An, T. N., Lee, I. J., Park, S., & Lee, K. D. (2015). Synthesis and bioactivity of novel phenothiazine-based thiazole derivatives. *Phosphorus, Sulfur, and Silicon and the Related Elements, 190,* 1160–1168. http://dx.doi.org/10.1080/10426507.2014.978324

Makam, P., & Kannan, T. (2014). 2-Aminothiazole derivatives as antimycobacterial agents: Synthesis, characterization, *in vitro* and in silico studies. *European Journal of Medicinal Chemistry, 87,* 643–656. http://dx.doi.org/10.1016/j.ejmech.2014.09.086

Meshram, H. M., Thakur, P. B., Madhu Babu, B., & Bangade, V. M. (2012). Convenient and simple synthesis of 2-aminothiazoles by the reaction of α-halo ketone carbonyls with ammonium thiocyanate in the presence of N-methylimidazole. *Tetrahedron Letters, 53,* 5265–5269. http://dx.doi.org/10.1016/j.tetlet.2012.07.080

Nitta, A., Fujii, H., Sakami, S., Satoh, M., Nakaki, J., Satoh, S., ... Kawai, H. (2012). Novel series of 3-amino-N-(4-aryl-1, 1-dioxothian-4-yl) butanamides as potent and selective dipeptidyl peptidase IV inhibitors. *Bioorganic & Medicinal Chemistry Letters, 22,* 7036–7040.

Parekh, N. M., Juddhawala, K. V., & Rawal, B. M. (2013). Antimicrobial activity of thiazolyl benzenesulfonamide-condensed 2, 4-thiazolidinediones derivatives. *Medicinal Chemistry Research, 22,* 2737–2745. http://dx.doi.org/10.1007/s00044-012-0273-x

Patt, W. C., Hamilton, H. W., Taylor, M. D., Ryan, M. J., Taylor, Jr. D. G., Connolly, C. J. C., ... Olson, S. C. J. (1992). Structure-activity relationships of a series of 2-amino-4-thiazole-containing renin inhibitors. *Journal of Medicinal Chemistry, 35,* 2562–2572. http://dx.doi.org/10.1021/jm00092a006

Potewar, T. M., Ingale, S. A., & Srinivasan, K. V. (2008). Catalyst-free efficient synthesis of 2-aminothiazoles in water at ambient temperature. *Tetrahedron, 64,* 5019–5022. http://dx.doi.org/10.1016/j.tet.2008.03.082

Shah, N. K., Shah, N. M., Patel, M. P., & Patel, R. G. (2012). Synthesis, characterization and antimicrobial activity of some new biquinoline derivatives containing a thiazole moiety. *Chinese Chemical Letters, 23,* 454–457.

Shao, H., Shi, S., Huang, S., Hole, A. J., Abbas, A. Y., Baumli, S., ... Wang, S. (2013). Substituted 4-(Thiazol-5-yl)-2-(phenylamino)pyrimidines are highly active CDK9 inhibitors: Synthesis, X-ray crystal structures, structure–activity relationship, and anticancer activities. *Journal of Medicinal Chemistry, 56*, 640–659. http://dx.doi.org/10.1021/jm301475f

Singh, S., Prasad, N. R., Chufan, E. E., Patel, B. A., Wang, Y.-J., Chen, Z.-S., ... Talele, T. T. (2014). Design and synthesis of human ABCB1 (P-Glycoprotein) inhibitors by peptide coupling of diverse chemical scaffolds on carboxyl and amino termini of (S)-valine-derived thiazole amino acid. *Journal of Medicinal Chemistry, 57*, 4058–4072. http://dx.doi.org/10.1021/jm401966m

Spector, F. C., Liang, L., Giordano, H., Sivaraja, M., & Peterson, M. G. (1998). Inhibition of herpes simplex virus replication by a 2-amino thiazole via interactions with the helicase component of the UL5-UL8-UL52 complex. *Journal of Virology, 72*, 6979–6987.

Wender, P. A., Handy, S. L., & Wright, D. L. (1997). Towards the ideal synthesis. *Chemistry and Industry, 1997*, 765–769.

Yang, B. V., Weinstein, D. S., Doweyko, L. M., Gong, H., Vaccaro, W., Huynh, T., ... Barrish, J. C. (2010). Dimethyl-diphenyl-propanamide derivatives as nonsteroidal dissociated glucocorticoid receptor agonists. *Journal of Medicinal Chemistry, 53*, 8241–8251. http://dx.doi.org/10.1021/jm100957a

Yang, J., Tan, J.-N., & Gu, Y. (2012). Lactic acid as an invaluable bio-based solvent for organic reactions. *Green Chemistry, 14*, 3304–3317. http://dx.doi.org/10.1039/c2gc36083g

Zhu, Y.-P., Yuan, J.-J., Zhao, Q., Lian, M., Gao, Q.-H., Liu, M.-C., ... Wu, A.-X. (2012). I2/CuO-catalyzed tandem cyclization strategy for one-pot synthesis of substituted 2-aminothiozole from easily available aromatic ketones/α, β-unsaturated ketones and thiourea. *Tetrahedron, 68*, 173–178. http://dx.doi.org/10.1016/j.tet.2011.10.074

Zhuravel, I. O., Kovalenko, S. M., Vlasov, S. V., & Chernykh, V. P. (2005). Solution-phase synthesis of a combinatorial library of 3-[4- (coumarin-3-yl)-1,3-thiazol-2-ylcarbamoyl] propanoic acid Amides. *Molecules, 10*, 444–456. http://dx.doi.org/10.3390/10020444

3

One-pot, solvent-free, and efficient synthesis of 2,4,6-triarylpyridines using CoCl$_2$.6H$_2$O as a recyclable catalyst

Mahmood Kamali[1*]

*Corresponding author: Mahmood Kamali, Faculty of Chemistry, Kharazmi University, 49-Mofetteh Ave., Tehran, Iran
E-mails: kamali.mahmood@ymail.com; Mkamali@khu.ac.ir
Reviewing editor: Chris Smith, University of Reading, UK

Abstract: A one-pot, three components coupling of aryl aldehyde, acetophenone, and ammonium acetate was performed to afford the corresponding 2,4,6-triarylpyridines (**TAP$_{1-17}$**). The **TAP$_{1-17}$** were synthesized in the presence of cobalt(II) chloride hexahydrate (**CoCl$_2$.6H$_2$O**) via an improved Chichibabin pyridine synthesis protocol. This study has shown that CoCl$_2$.6H$_2$O promotes this reaction in comparison to other transition metal salt such as with FeCl$_3$, NiCl$_2$.6H$_2$O, CuCl$_2$.2H$_2$O, CdCl$_2$.H$_2$O, SbCl$_3$, SnCl$_2$.2H$_2$O, and ZnCl$_2$. This method has several advantages, for example, excellent yields, short reaction times, easy work up, and solvent-free condition. Also, this catalyst was recyclable for four consecutive runs.

Subjects: Chemistry; Environmental Chemistry; Organic Chemistry; Physical Sciences

Keywords: 2,4,6-triarylpyridine; chichibabin pyridine synthesis; kröhnke pyridines; cobalt(ii) chloride hexahydrate

1. Introduction

Pyridine ring systems are of interest because of their wide range of pharmalogical activities such as antimalarial, vasodilator, anesthetic, anticonvulsant, antiepileptic, and agrochemicals such as fun gicidal, pesticidal, and herbicidal (Enyedy, Sakamuri, Zaman, Johnson, & Wang, 2003; Kim et al., 2004; Klimesová, Svoboda, Waisser, Pour, & Kaustová, 1999; Pillai et al., 2003). Recent studies have highlighted the biological activity of triarylpyridines as a pyridine derivative, providing impetus for further studies in utilizing this scaffold in new therapeutic drug classes (Bonse, Richards, Ross, Lowe, & Kraut-Siegel, 2000; Lowe et al., 1999; Zhao et al., 2001, 2004). Due to their π-stacking ability, triar ylpyridines are commonly used as building blocks in supramolecular chemistry (Cave, Hardie, Roberts, & Raston, 2001; Constable et al., 2000; Jetti, Nagia, Xue, & Mak, 2001; Watson, Bampos, & Sanders, 1998). Therefore, there has been increasing interest to develop new methods for the syn thesis of 2,4,6-triarylpyridines, Kröhnke pyridines. Previously, 2,4,6-triarylpyridines have been

ABOUT THE AUTHOR

Mahmood Kamali obtained his PhD in organic chemistry in 2011 from Kharazmi University, Tehran, Iran. Now he is an assistant professor in the faculty of chemistry, Kharazmi University. His main research focus is design of synthesis of new organic compounds and new synthetic methodologies (in macrocyclic, polymeric, and other fields for targeted applications).

PUBLIC INTEREST STATEMENT

Pyridine ring systems, such as triarylpyridines, Kröhnke pyridines, are of interest because of their wide range of pharmaceutical activities. Also due to their π-stacking ability, triarylpyridines are used building blockers in supramolecular chemistry. Therefore, their synthesis has attracted continuous interest to develop methods for the synthesis of 2,4,6-triarylpyridines. The current work reports an efficient new catalyst (**CoCl$_2$.6H$_2$O**) and its comparison with some other metal chlorides as catalysts in the preparation of such products.

prepared by the condensation of 1,5-diketones with formamide-formic acid (Chubb, Hay, & Sandin, 1953) and by other synthetic procedures including the Chichibabin method (Frank & Seven, 1949; Zecher & Kröhnke, 1961), and reaction of N-phenacylpyridinium salts with α,β-unsaturated ketones in the presence of ammonium acetate (Kröhnke, 1976; Kröhnke & Zecher, 1962). Recently, several new improved methods and procedures for preparation of 2,4,6-triarylpyridines have been reported, for example, the reaction of α-ketoketene dithioacetals with methyl ketones in the presence of NH_4OAc (Potts, Cipullo, Ralli, & Theodoridis, 1981), the reaction of N-phosphinylethanimines with aldehydes (Kobayashi, Kakiuchi, & Kato, 1991), solvent-free reaction of chalcones with ammonium acetate (Adib, Tahermansouri, Koloogani, Mohammadi, & Bijanzadeh, 2006). Also, there are a number of methods reported for synthesis of these compounds using various catalysts, for example, Preyssler-type heteropolyacid ($H_{14}[NaP_5W_{30}O_{110}]$) (Heravi, Bakhtiari, Daroogheha, & Bamoharram, 2007), $HClO_4$-SiO_2 (Nagarapu, Peddiraju, & Apuri, 2007), $AlPO_4$ (Rajput, Subhashini, & Shivaraj, 2010), $Bi(OTf)_3$ (Shinde, Labade, Gujar, Shingate, & Shingare, 2012), I_2 (Ren & Cai, 2009), ionic liquid ($[HO_3S(CH_2)_4MIM][HSO_4]$) (Davoodnia, Bakavoli, Moloudi, Tavakoli-Hoseini, & Khashi, 2010), nanoparticles (Safari, Zarnegar, & Borujeni, 2013; Shafiee & Moloudi, 2011), and without catalyst (Tu et al., 2005; Wang, Yang, Song, & Wang, 2015).

Herein, we would like to report an efficient procedure for the preparation of 2,4,6-triarylpyridines through a one-pot condensation reaction including aldehydes, acetophenones, and NH_4OAc in the presence of cobalt(II) chloride hexahydrate (**CoCl$_2$.6H$_2$O**) under solvent-free conditions.

Table 1. Synthesis of TAP$_1$ by different catalysts, under condition reaction[a]

Catalyst	Isolated yield (%)
–	10
FeCl$_3$	–
CoCl$_2$.6H$_2$O	**90**
NiCl$_2$.6H$_2$O	38
CuCl$_2$.2H$_2$O	53
ZnCl$_2$	68
CdCl$_2$.H$_2$O	65
SnCl$_2$.2H$_2$O	67
SbCl$_3$	56

[a]Benzaldehyde (1 mmol), acetophenone (2 mmol), NH$_4$OAc (1.5 mmol), catalyst, 20% mol, Solvent Free, 120°C, 5 h.

Table 2. Synthesis of TAP$_1$ under different conditions for optimization of reactions by CoCl$_2$.6H$_2$O as catalyst

Temperature (°C) of React.	Catalyst (mol%)	CoCl$_2$.6H$_2$O as catalyst	
		Time (h)	Isolated yield (%)
90	20	4	35
100	20	4	83
110	20	4	90
120	20	4	90
110	20	5	90
110	20	3	78
110	0.5	4	55
110	1	4	75
110	**2.5**	**4**	**89**
110	5	4	90
110	10	4	90

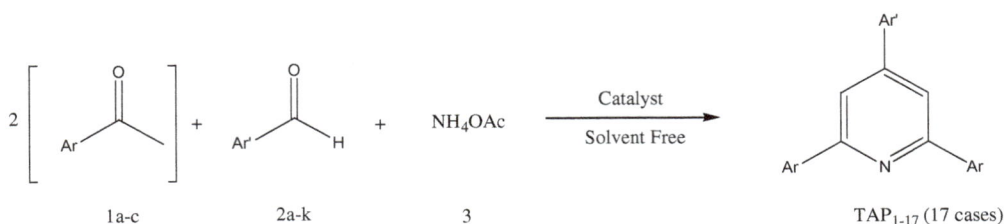

Scheme 1. Synthesis of 2,4,6-triarylpyridine (TAP[1–17]).

2. Results and discussion

In order to study the efficiency of new methods, acetophenone (1), benzaldehyde (2), ammonium acetate (3), and a range of different metal salt were investigated and were heated to give 2,4,6-triphenylpyridine (**TAP$_1$**) (Scheme 1), under solvent-free conditions. Initially, the reactions were carried out using different catalysts (CoCl$_2$.6H$_2$O, FeCl$_3$, NiCl$_2$.6H$_2$O, CuCl$_2$.2H$_2$O, CdCl$_2$.H$_2$O, SbCl$_3$, SnCl$_2$.2H$_2$O). **CoCl$_2$.6H$_2$O** was selected as the best catalyst of those investigated with an initial yield of 90% (Table 1). The reaction was performed at different temperatures, times, and differing amounts of **CoCl$_2$.6H$_2$O**. The results from this study are presented in Table 2, whereby the best yields were obtained when the temperature was at 110°C with 4 h reaction time and 2.5 mol% of **CoCl$_2$.6H$_2$O**.

Several activated and deactivated aromatic aldehydes, and acetophenones derivatives underwent the reaction to give the corresponding **TAP**s in high yields. The results are shown in Table 3. The experimental procedure was very simple, convenient, and had the ability to tolerate a variety of other functional groups such as methoxy, nitro, hydroxyl, and halides under the reaction conditions (Table 3).

Interestingly, the catalyst can be recycled for four consecutive runs without significant loss of activity (Table 4). For this purpose, after completion of the reaction, the reaction mixture was cooled to room temperature, and then, water was added. The precipitated solid was isolated by filtration;

Table 3. Details 2,4,6-triarylpyridine synthesis

Entry	Ar	Ar'	Product	Isolated yield (%)	mp°C	
					Found	Lit.
1	Ph	Ph	TAP$_1$	89	135–137	134–135[a]
2	Ph	4-Cl-Ph	TAP$_2$	91	124–127	124–126[b]
3	Ph	4-NO$_2$-Ph	TAP$_3$	92	196–198	195–197[b]
4	Ph	2-Me-Ph	TAP$_4$	86	122–124	120–122[a]
5	Ph	4-Me-Ph	TAP$_5$	87	121–123	123–124[a]
6	Ph	4-HO-Ph	TAP$_6$	89	194–196	197[c]
7	Ph	4-MeO-Ph	TAP$_7$	90	99–101	98[c]
8	Ph	4-Br-Ph	TAP$_8$	92	103–105	102–104[d]
9	Ph	2-Thienyl	TAP$_9$	84	162–164	165–166[e]
10	Ph	2-Furyl	TAP$_{10}$	83	169–170	170–171[a]
11	4-Cl-Ph	Ph	TAP$_{11}$	84	177–189	188–190[f]
12	4-Cl-Ph	2-Cl-Ph	TAP$_{12}$	76	165–169	168–170[g]
13	4-Me-Ph	Ph	TAP$_{13}$	90	159–160	159–160[h]
14	4-Me-Ph	4-MeO-Ph	TAP$_{14}$	86	154–156	156–157[h]
15	4-Me-Ph	4-Me-Ph	TAP$_{15}$	89	178–179	178–180[h]
16	4-Me-Ph	4-Cl-Ph	TAP$_{16}$	91	199–201	200–202[i]
17	4-MeO-Ph	4-NO$_2$-Ph	TAP$_{17}$	92	142–144	143–144[j]

[a]Adib et al. (2006); [b]Ren and Cai (2009); [c]Heravi et al. (2007); [d]Shinde et al. (2012); [e]Kobayashi et al. (1991); [f]Chiu, Tang, and Ellingboe (1998); [g]Safari et al. (2013); [h]Maleki et al. (2010); [i]Kröhnke and Zecher (1962); [j]Shafiee and Moloudi (2011).

Table 4. Recycled of $CoCl_2.6H_2O$ in the synthesis of TAP_1 reactions

Catalyst type	Runs					
	1	2	3	4	5	6
Product yield (%)	89	88	86	82	80	75

the catalyst was recovered from the filtrate by evaporation of the water at room temperature, and reused for the similar reaction.

3. Experimental

All reactions were carried out in an efficient hood. The starting materials were purchased from Merck and Fluka chemical companies. Melting points were determined with a Branstead Electrothermal model 9200 apparatus and are uncorrected. IR spectra were recorded on a Perkin Elmer RX1 Fourier transform infrared spectrometer. The ^1H and ^{13}C NMR spectra were recorded in DMSO-d_6 on Bruker Avance 300-MHz spectrometers. Elemental analyses were carried out by a Perkin Elmer 2400 series II CHN/O analyzer.

3.1. Synthesis of TAP_1 as general procedure

A mixture of benzaldehyde (0.21 mL, 2 mmol), acetophenone (0.47 mL, 4 mmol), NH_4OAc (0.23 gr, 3 mmol), and **$CoCl_2.6H_2O$** (0.12 gr, 2.5 mol%) was heated on oil bath with stirring at 110°C for 4 h (Tables 1 and 2). After cooling, the reaction mixture was poured in ice water (10 mL) and the precipitated solid was collected by filtration, washed with distilled water (40 mL), and dried. The crude product was recrystallized from 95% ethanol (10 mL) to give the corresponding pure product (**TAP_1**). Colorless crystals in 89% yield, mp 135–137°C, IR (KBr) ν: 3,071, 1,585, 1,583, 1,496, 1,476, 1,384, 1,054, 1,011, 742, 665 cm^{-1}. ^1H NMR (300 MHz, DMSO-d_6): δ: 7.40–7.60 (9H, m), 8.03 (d, $J = 7.6$ Hz, 2H), 8.17 (s, 2H), 8.28 (d, $J = 7.6$ Hz, 2H), 8.35 (d, $J = 7.3$ Hz, 2H) ppm. ^{13}C NMR (75 MHz, DMSO-d_6): δ: 117.2, 127.4, 127.7, 128.8, 129.0, 129.4, 129.5, 139.0, 139.5, 150.2 and 157.3 ppm. Anal. Calcd for $C_{23}H_{17}$ N: C, 89.87; H, 5.57; N, 4.56. Found: C, 89.53; H, 5.49; N, 4.89.

4. Conclusion

In conclusion, we have successfully developed a quick, convenient, and efficient method for the synthesis of **TAP**s under solvent-free conditions. The environmental advantages include omitting organic solvent, generality and simplicity of procedure, shorter reaction time, simple workup, reusable catalyst condition, and pure products in excellent yields.

Funding

Mahmood Kamali appreciates the Research Council of the Kharazmi University for financial support.

Author details

Mahmood Kamali[1]

E-mails: kamali.mahmood@ymail.com, Mkamali@khu.ac.ir

[1] Faculty of Chemistry, Kharazmi University, 49-Mofetteh Ave., Tehran, Iran.

References

Adib, M., Tahermansouri, H., Koloogani, S. A., Mohammadi, B., & Bijanzadeh, H. R. (2006). Kröhnke pyridines: An efficient solvent-free synthesis of 2,4,6-triarylpyridines. *Tetrahedron Letters, 47*, 5957–5960. doi:10.1016/j.tetlet.2006.01.162

Bonse, S., Richards, J. M., Ross, S. A., Lowe, G., & Kraut-Siegel, R. L. (2000). (2,2′:6′,2″-Terpyridine)platinum (II) complexes are irreversible inhibitors of Trypanosoma cruzi trypanothione reductase but not of human glutathione reductase. *Journal of Medicinal Chemistry, 43*, 4812–4821. doi:10.1021/jm000219o

Cave, G. W. V., Hardie, M. J., Roberts, B. A., & Raston, C. L. (2001). A versatile six-component molecular capsule based on benign synthons selective confinement

of a heterogeneous molecular aggregate. *European Journal of Organic Chemistry, 2001*, 3227–3231. doi:10.1002/1099-0690(200109)

Chiu, C. F., Tang, Z. L., & Ellingboe, J. W. (1998). Solid-phase synthesis of 2,4,6-trisubstituted pyridines. *Journal of Combinatorial Chemistry, 1*, 73–77. doi:10.1021/cc980005g

Chubb, F., Hay, A. S., & Sandin, R. B. (1953). The Leuckart reaction of some 1,5-diketones. *Journal of the American Chemical Society, 75*, 6042–6044. doi:10.1021/ja01119a508

Constable, E. C., Housecroft, C. E., Neuburger, M., Phillips, D., Raithby, P. R., Schofield, E., ... Zimmermann, Y. (2000). Development of supramolecular structure through alkylation of pendant pyridyl functionality. *Journal of the Chemical Society, Dalton Transactions, 2000*, 2219–2228. doi:10.1039/B000940G

Davoodnia, A., Bakavoli, M., Moloudi, R., Tavakoli-Hoseini, N., & Khashi, M. (2010). Highly efficient, one-pot, solvent-free synthesis of 2,4,6-triarylpyridines using a Brønsted-acidic ionic liquid as reusable catalyst. *Monatshefte für Chemie-Chemical Monthly, 141*, 867–870. doi:10.1007/s00706-010-0329-x

Enyedy, I. J., Sakamuri, S., Zaman, W. A., Johnson, K. M., & Wang, S. I. (2003). Pharmacophore-based discovery of substituted pyridines as novel dopamine transporter inhibitors. *Bioorganic & Medicinal Chemistry Letters, 13,* 513–517. doi:10.1016/S0960-894X(02)00943-5

Frank, R. L., & Seven, R. P. (1949). Pyridines. IV. A study of the Chichibabin synthesis. *Journal of the American Chemical Society, 71,* 2629–2635. doi:10.1021/ja01176a008

Heravi, M. M., Bakhtiari, K., Daroogheha, Z., & Bamoharram, F. F. (2007). An efficient synthesis of 2,4,6-triarylpyridines catalyzed by heteropolyacid under solvent-free conditions. *Catalysis Communications, 8,* 1991–1994. doi:10.1016/j.catcom.2007.03.028

Jetti, K. R. R., Nagia, A., Xue, F., & Mak, T. C. W. (2001). Polar host–guest assembly mediated by halogen···π interaction: Inclusion complexes of 2,4,6-tris(4-halophenoxy)-1,3,5-triazine (halo = chloro, bromo) with trihalobenzene (halo = bromo, iodo). *Chemical Communications,* 919–920. doi: 10.1039/B102150H

Kim, B. Y., Ahn, J. B., Lee, H. W., Kang, S. K., Lee, J. H., Shin, J. S., ... Yoon, S. S. (2004). Synthesis and biological activity of novel substituted pyridines and purines containing 2,4-thiazolidinedione. *European Journal of Medicinal Chemistry, 39,* 433–447. doi:10.1016/j.ejmech.2004.03.001

Klimesová, V., Svoboda, M., Waisser, K., Pour, M., & Kaustová, J. (1999). New pyridine derivatives as potential antimicrobial agents. *Farmaco, 54,* 666–672. doi:10.1016/S0014-827X(99)00078-6

Kobayashi, T., Kakiuchi, H., & Kato, H. (1991). On the reaction of N-(diphenylphosphinyl)-1-phenylethanimine with aromatic aldehydes giving 4-aryl-2,6-diphenylpyridine derivatives. *Bulletin of the Chemical Society of Japan, 64,* 392–395. doi:10.1246/bcsj.64.392

Kröhnke, F. (1976). The specific synthesis of pyridines and oligopyridines. *Synthesis, 1976*(1), 1–24. doi:10.1055/s-1976-23941

Kröhnke, F., & Zecher, W. (1962). Syntheses using the Michael adddition of phridinium salts. *Angewandte Chemie International Edition in English, 1,* 626–632. doi:10.1002/anie.196206261

Lowe, G., Droz, A. S., Vilaivan, T., Weaver, G. W., Tweedale, L., Pratt, J. M., ... Croft, S. L. (1999). Cytotoxicity of (2,2':6',2"-terpyridine)platinum(II) complexes to Leishmania donovani, Trypanosoma cruzi, and Trypanosoma brucei. *Journal of Medicinal Chemistry, 42,* 999–1006. doi:10.1021/jm981074c

Maleki, B., Azarifar, D., Veisi, H., Hojati, S. F., Salehabadi, H., & Neja Yami, R. (2010). Wet 2,4,6-trichloro-1,3,5-triazine (TCT) as an efficient catalyst for the synthesis of 2,4,6-triarylpyridines under solvent-free conditions. *Chinese Chemical Letters, 21,* 1346–1349. doi:10.1016/j.cclet.2010.06.028

Nagarapu, L., Peddiraju, A. R., & Apuri, S. (2007). HClO4–SiO2 as a novel and recyclable catalyst for the synthesis of 2,4,6-triarylpyridines under solvent-free conditions. *Catalysis Communications, 8,* 1973–1976. doi:10.1016/j.catcom.2007.08.003

Pillai, A. D., Rathod, P. D., Franklin, P. X., Patel, M., Nivarsarkar, M., Vasu, V., ... Sudarsanam, V. (2003). Novel drug designing approach for dual inhibitors as anti-inflammatory agents: implication of pyridine template. *Biochemical and Biophysical Research Communications, 301,* 183–186. doi:10.1016/S0006-291X(02)02996-0

Potts, K. T., Cipullo, M. J., Ralli, P., & Theodoridis, G. (1981). Ketene dithio acetals as synthetic intermediates. Synthesis of unsaturated 1,5-diketones. *Journal of the*

Rajput, P., Subhashini, N. J. P., & Shivaraj, J. (2010). Synthesis of 2,4,6-triarylpyridines using AlPO4 under solvent-free conditions. *Journal of Scientific Research, 2,* 337–342. doi:10.3329/jsr.v2i2.3859

Ren, Y. M., & Cai, C. (2009). Three-components condensation catalyzed by molecular iodine for the synthesis of 2,4,6-triarylpyridines and 5-unsubstituted-3,4-dihydropyrimidin-2(1H)-ones under solvent-free conditions. *Monatshefte für Chemie-Chemical Monthly, 140,* 49–52. doi:10.1007/s00706-008-0011-8

Safari, J., Zarnegar, Z., & Borujeni, M. B. (2013). Mesoporous nanocrystalline MgAl$_2$O$_4$: A new heterogeneous catalyst for the synthesis of 2,4,6-triarylpyridines under solvent-free conditions. *Chemical Papers, 67,* 688–695. doi:10.2478/s11696-013-0361-5

Shafiee, M. R. M., & Moloudi, R. (2011). Solvent-free preparation of 2,4,6-triaryl pyridines using silver(I) nitrate adsorbed on silica gel nanoparticles (AgNO3-Nano SiO2) as an efficient catalyst. *Letters in Organic Chemistry, 8,* 717–721. doi:10.2174/157017811799304214

Shinde, P. V., Labade, V. B., Gujar, J. B., Shingate, B. B., & Shingare, M. S. (2012). Bismuth triflate catalyzed solvent-free synthesis of 2,4,6-triaryl pyridines and an unexpected selective acetalization of tetrazolo[1,5-a]-quinoline-4-carbaldehydes. *Tetrahedron Letters, 53,* 1523–1527. doi:10.1016/j.tetlet.2012.01.059

Tu, S., Li, V., Shi, F., Fang, F., Zhu, S., Wei, X., & Zong, Z. (2005). An efficient improve for the kröhnke reaction: one-pot synthesis of 2,4,6-triarylpyridines using raw materials under microwave irradiation. *Chemistry Letters, 34,* 732–733. doi:10.1246/cl.2005.732

Wang, M., Yang, Z., Song, Z., & Wang, Q. (2015). Three-component one-pot synthesis of 2,4,6-triarylpyridines without catalyst and solvent. *Journal of Heterocyclic Chemistry, 52,* 907–910. doi:10.1002/jhet.2132

Watson, Z. C., Bampos, N. J., & Sanders, K. M. (1998). Mixed cyclic trimers of porphyrins and dioxoporphyrins: geometry vs. electronics in ligand recognition. *New Journal of Chemistry, 22,* 1135–1138. doi:10.1039/A805504A

Zecher, W., & Kröhnke, F. (1961). Eine neue synthese substituierter pyridine, I. Grundzüge der synthese [A new synthesis of substituted pyridines, I. Broad range synthesis]. *Chemische Berichte, 94,* 690–697. doi:10.1002/cber.19610940317

Zhao, L. X., Kim, T. S., Ahn, S. H., Kim, T. H., Kim, E. K., Cho, W. J., ... Lee, E. S. (2001). Synthesis, topoisomerase I inhibition and antitumor cytotoxicity of 2,2':6',2"-, 2,2':6',3"- and 2,2':6',4"-Terpyridine derivatives. *Bioorganic & Medicinal Chemistry Letters, 11,* 2659–2662. doi:10.1016/S0960-894X(01)00531-5

Zhao, L. X., Moon, Y. S., Basnet, A., Kim, E. K., Jahng, Y., Park, J. G., ... Lee, E. S. (2004). Synthesis, topoisomerase I inhibition and structure–activity relationship study of 2,4,6-trisubstituted pyridine derivatives. *Bioorganic & Medicinal Chemistry Letters, 14,* 1333–1337. doi:10.1016/j.bmcl.2003.11.084

4

Regioselective synthesis of imines (2-*N*-amine-3-*N*-(phenylmethylene)-5-pyridine) in water under microwave irradiation

Nohana C. Ramos[1], Aurea Echevarria[1], Arthur Valbon[1], Adailton J. Bortoluzzi[2], Guilherme P. Guedes[1] and Cláudio E. Rodrigues-Santos[1]*

*Corresponding author: Cláudio E. Rodrigues-Santos, Departamento de Química, Instituto de Ciências Exatas, Universidade Federal Rural do Rio de Janeiro, 23890-900, Seropédica, RJ, Brazil
E-mail: claudioers@ufrrj.br
Reviewing editor: Chris Smith, University of Reading, UK

Abstract: The eco-friendly synthesis of imines was reported in this work using water as solvent under microwave irradiation. This protocol achieved excellent yields in short time. Water plays an important role in this reaction, in contrast with some previous studies.

Subjects: Chemical Spectroscopy; Chemistry; Organic Chemistry

Keywords: schiff base; water; microwave

1. Introduction

The imine group is present in several compounds that have demonstrated great relevance due to their biological properties, such as antibacterial, antifungal properties, and their inhibition of the enzyme DNA-topoisomerase I, an important target against cancer (Asadi, Asadi, Torabi, & Lotfi, 2012; Ganguly, Chakraborty, Banerjee, & Choudhuri, 2014; Jhaumeer-Laulloo, Gupta Bhowon, Mungur, Fawzi Mahomoodally, & Hussein Subratty, 2012; Lee et al., 2014; Qin, Long, Panunzio, & Biondi, 2013). In the field of new materials, imines, and their metal clusters, are present in the formation of thermostable polymers, as synthetic intermediates, corrosion inhibitors, and high-performance catalysts (Chen, Xiang, & Feng, 2012; Coulthard, Unsworth, & Taylor, 2015; Dogan & Kaya, 2013; Doi, Kawai, Murayama, Kashida, & Asauna, 2016; Kielland, Escudero-Adán, Martínez Belmonte, & Kleij, 2013; Seifzdeh, Basharnavaz, & Bezaatpour, 2014; Zarei, 2014). The synthesis of imines is performed with aldehydes or ketones and a primary amine and an apparatus to remove water, such as a Dean Stark trap, or a dehydration agent to increase the yield. Recently, the literature registered the use of catalysts, such as montmorillonite, $ZnCl_2$, and $TiCl_4$ (Qin et al., 2013). Unfortunately, the synthesis of Schiff bases is not facile and in some cases extremely slow, such as when there are withdrawing groups linked to benzylamines. Previously, 2-*N*-amine-3-*N*-(phenylmethylene)-pyridine was prepared from benzaldehyde and 2,3-diaminepyridine in THF reflux for 4 h, more 18 h of stirring at room temperature, using molecular sieves (Khanna, Weier, Lentz, Swenton, & Lankin, 1995). Recently, the same compound was obtained by stirring in ethanol at room temperature for 48 h (Jhaumeer-Laulloo et al., 2012). The long reaction time and the use of THF as a solvent are not part of 12 green rules. Our research group believes in the words written by Deligeorgiev et al. (2010), "Green Chemistry is not a new branch of science. It is a new philosophical approach that, through the introduction and expansion of its principles could lead to a substantial development in chemistry, the chemical

ABOUT THE AUTHORS

Our group has developed projects in organic synthesis focusing on two points, Green chemistry: utilizing new catalytic, solvent-free, microwave and ultrasound, and in the synthesis of bioactive compounds. This paper is related to the first point.

PUBLIC INTEREST STATEMENT

Nowadays, cheap, nontoxic, and safe source is a fundamental point in the production of materials, this results in the reduction of environmental impacts. In the context, our research has been developing new compounds from these points, and good results have already been achieved.

industry and environmental protection." In this context, water is an eco-friendly solvent because it has the following properties: (a) cheap; (b) nontoxic; (c) non-flammable; (d) safe; and (e) the products are usually isolated by filtration (Esteves-Sousa et al., 2012; Panda & Jain, 2012; Rodrigues-Santos & Echevarria, 2007, 2011). The involvement of water in the formation of imine has caused controversy among authors. Saggiomo and Lüning (2009) stated important considerations about this matter. The research group indicated that water did not act as a catalyst or solvent, and the imines formed during or after workup. However, Cordes and Jencks (1962) and Kommi, Kumar, Chakraborti, Chebolu, and Chadraborti (2013) published an article suggesting that the water molecules participated in the formation of the imine. Our work describes an efficient synthesis of imines, employing water as solvent under microwave irradiation, corroborating with the Codes and Kommi et al.

2. Results and discussion

To find the best reaction conditions, the reaction of compound **1a** and **2a** was performed under several conditions, entry 3–8 (Table 1).

As it can been observed in Table 1, it was not possible to obtain good yields from toluene when using alternative methods, such as ultrasound or microwave irradiation. However, the use of water was successful for this reaction. After 1-h reflux, a yellow solid was obtained in 98% yield. Ultrasound irradiation in water also achieved good yield (96%), but the best method was obtained by microwave irradiation in water (98%), it decreased the reaction time by sixfold compared to traditional reflux conditions. It is important to report that no oil pump vacuum was used to help in the formation of the product, as indicated by Saggiomo and Lüning (2009). The solid was washed with water and dried. Thus, microwave irradiation in water was selected to obtain the **3a–n** derivatives. Good results (54–95%) were obtained, as observed in Table 2, even when an electron withdrawing group was present. In addition, the reaction time was significantly reduced to 7–10 min, with the exception of compound **3n** (30 min).

The attack of the amine group (5-bromo-2,3-diaminepyridine) could produce the isomer **A** (3-amine, *E*-iosmer) and/or isomer **B** (2-amine, *E*-isomer) (Figure 1).

After analysis of the ^{1}H NMR chemical shifts, NOESY spectroscopy and crystal structures obtained by X-ray diffraction, only the *E*-isomer **A** was observed; therefore, this reaction was regioselective, with exception of compound **3m**, which presented both isomers (**A** and **B**) due to the electron

Table 1. Reaction times, yields, temperature and method for the synthesis of 2-N-amine-3-N-(2-hydroxy-phenylmethylene)-5-bromo-pyridine

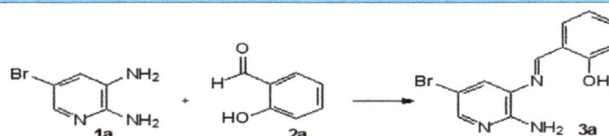

Entry	Solvent	Time (min)	Method	Temp. (°C)	Yield (%)[a]
1	Toluene[b]	60	Reflux	100	52
2	Toluene[b]	60	Microwave	100	62
3	Toluene[b]	60	Ultrasound	40	58
4	Water	60	Reflux	100	98
5	Water	60	Microwave	100	90
6	Water	60	Ultrasound	40	96
7	Water	20	Microwave	100	98
8	Water	10	Microwave	100	95

[a]Yield isolated imine.

[b]Dried solvent.

Table 2. Reaction times and yields for compounds 3a–n obtained using water and MW irradiation

Compound	Substituents						Yield (%)
	X_1	X_2	R_1	R_2	R_3	R_4	
3a[a]	N	C	Br	OH	H	H	95
3b[a]	N	C	Br	H	OCH_3	OH	85
3c[a]	N	C	Br	H	H	OH	86
3d[a]	N	C	H	H	OH	OH	87
3e[a]	N	N	Br	OCH_3	–	OCH_3	64
3f[a]	N	C	Br	H	$-O-CH_2-O-$		78
3g[a]	N	C	Br	H	H	CH_3	84
3h[a]	N	C	Br	2-methxycinnamal			85
3i[a]	N	C	Br	H	H	CF_3	78
3j[a]	N	C	H	H	$-O-CH_2-O-$		64
3k[a]	N	C	Br	H	OH	OH	86
3l[b]	N	C	Br	H	H	H	54
3m[a]	N	C	Br	H	H	NO_2	74
3n[c]	CH	C	CO_2H	–	H	H	78

[a]Reaction time: 10 min.
[b]Reaction time: 7 min.
[c]Reaction time: 30 min.

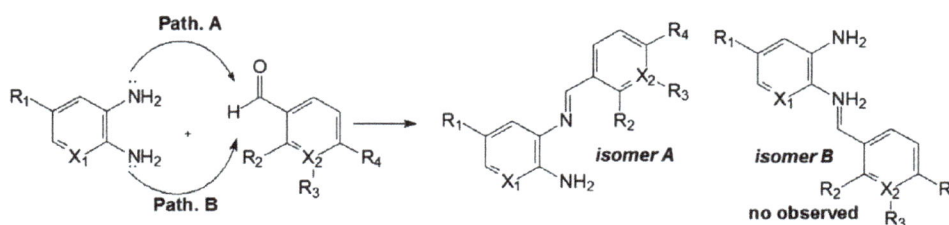

Figure 1. Isomers that can be formed from this class of compounds.

strength-withdrawing effect of the nitro group ($\sigma_p = 0.78$) *para* to the aldehyde. The isomers were observed by TLC and were separated by fractioned recrystallization. Due to the electronic effects, ^{1}H NMR spectra showed the differences of the NH_2 and H-imine groups for each isomer. Isomer **A** had an NH_2 (H-7) group downfield ($\delta = 6.41$) and an H-imine (H-9) group upfield ($\delta = 8.91$). The opposite was observed in the NH_2 and H-imine groups of isomer **B** (NH_2, $\delta = 6.03$ and H-imine, $\delta = 9.33$) (see ESI). The ^{1}H NOESY spectra in DMSO-d_6 revealed the correlation between the H-imine and H-4 for isomer **A**, characterizing conformer **A**. The absence of a correlation between the NH_2 (H-7) and H-imine (H-9) indicated the possibility of conformer **A** for isomer **B** (Figure 2). In the ^{1}H NOESY spectra of the isomers, there were no correlations between the hydrogen aromatic rings, which indicated that there was not formation of the *Z*-isomer. These data were corroborated by the crystal structures.

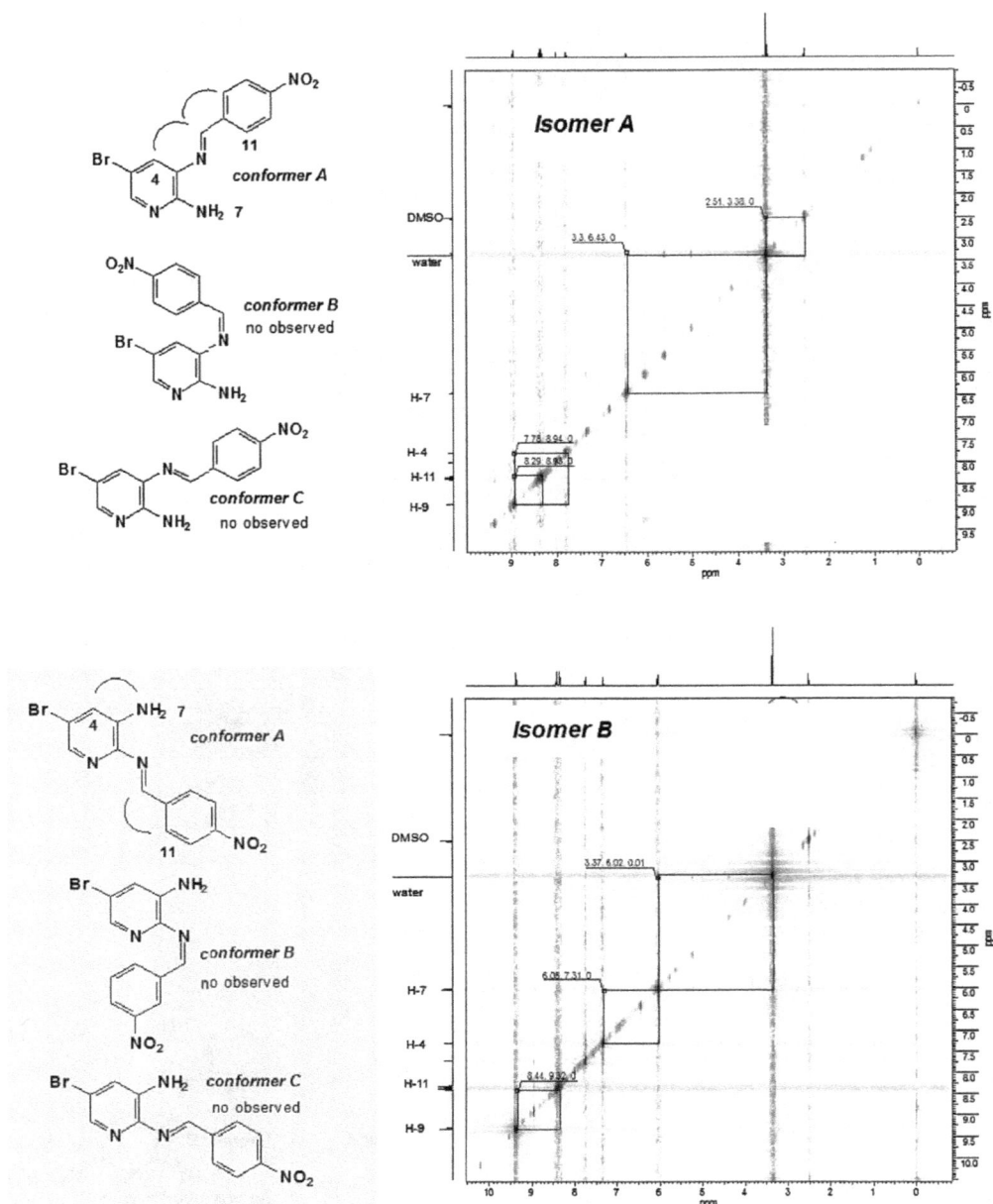

Figure 2. NOESY spectra of 3m isomer A and 3m isomer B and their possible conformations (for the best visualization of the content, see electronic supplementary information, page S20 and S22).

Eight crystal structures were obtained in this study (**3b**, **3b′**, **3c**, **3e**, **3f**, **3g**, **3h**, and **3l**); (see Supplementary data) the structures **3b** and **3b′** represent the same compound but with different conformations (Figure 3). Compound **3b** contained fewer π–π conjugation effects between the **A** and **B** rings; this observation is based on C_6–C_7, C_5–N_3, and the C_6=N_3 bond lengths and the torsion angles among C_5–N_3–C_6–C_7.

3. Conclusions
The easy, quick, and eco-friendly synthesis of imines under microwave irradiation and water was more efficient when compared to other methods described in the literature (Jhaumeer-Laulloo et al., 2012;

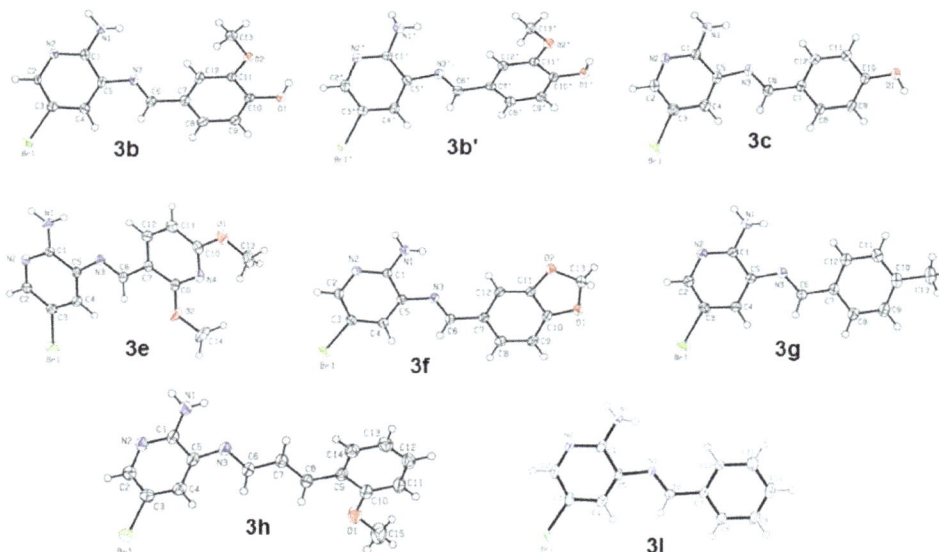

Figure 3. ORTEP plots of compounds 3b, 3b′, 3c, 3e, 3f, 3g, 3h, and 3l.

Khanna et al., 1995). The synthesis was regioselective, with the exception of that of compound 3m, due to the reactivity of the 4-nitrobenzaldehyde.

4. Experimental

4.1. Microwave irradiation
A mixture of 2,3-diaminepyridine, 1a–c (2.66 mmol), and aldehyde, 2a–o, (2.66 mmol), and then water (4 mL) was irradiated (CEM-discover-power 0–300 W, magnetron: 2,450 MHz-open vessel system) at 100°C for indicated in Table 2. The reaction progress was monitored by TLC (EtOAc/hexane, 7:3). After irradiation, the mixture was washed with water (5 × 10 mL) and dried at room temperature. The compounds (3a–n) were recrystallized by ethanol.

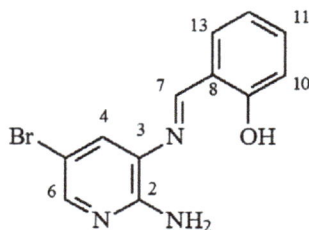

4.2. 2-N-amine-3-N-(2-hydroxy-phenylmethylene)-5-bromo-pyridine (3a)
mp 171–172°C; IR (ν/cm^{-1}); 3,449, 3,266, 3,133, 2,906, 1,617, 1,588, 1,482, 1,439, 1,383, 1,240, 1,383, 1,240, 1,203, 1,098, 1,027, 803, 549; ^{1}H NMR (400 MHz, DMSO-d$_6$) δ 6.18 (s, 2H, NH$_2$), 6.97 (m, 2H, H-9/12), 7.4 (t, 1H, J = 8.0 Hz, H-11), 7.61 (s, 1H, H-4), 7.73 (d, 1H, J = 8.0 Hz, H-13), 7.93 (s, 1H, H-6), 8.87 (s, 1H, H-7), 11.81 (s, 1H, OH); DEPTQ (100 MHz, DMSO-d$_6$) δ 105.5, 116.5, 119.3, 120.1, 127.0, 131.5, 132.0, 133.5, 145.8, 153.4, 159.6, 163.4.

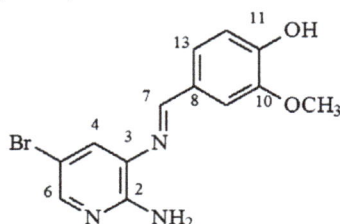

4.3. 2-N-amine-3-N-(4-hydroxy-3-methoxy-phenylmethylene)-5-bromo-pyridine (3b)

mp 207–209°C; IR (ν/cm^{-1}) 3,491, 3,384, 2,924, 2,853, 1,593, 1,490, 1,453, 1,426, 1,241, 1,206, 1,028, 815, 734; ^{1}H NMR (400 MHz, DMSO-d$_6$) δ 3.86 (s, 3H), 6.20 (s, 2H, NH$_2$), 6.87 (d, 1H, J = 8.0 Hz, H-9), 7.35 (d, 1H, J = 8.0 Hz, H-13), 7.54 (s, 1H, H-9), 7.66 (s, 1H, H-4), 7.85 (s, 1H, H-6), 8.55 (s, 1H, H-7), 9.80 (s, 1H, OH); DEPTQ (100 MHz, DMSO-d$_6$) δ 55.6, 105.3, 110.8, 115.1, 124.8, 125.0, 127.8, 132.5, 144.8, 148.0, 150.5, 154.3, 160.4.

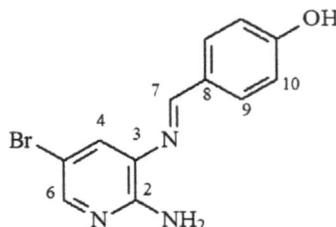

4.4. 2-N-amine-3-N-(4-hydroxy-phenylmethylene)-5-bromo-pyridine (3c)

mp 220–222°C; IR (ν/cm^{-1}) 3,054, 3,009, 1,607, 1,444, 1,404, 1,352, 1,237, 1,171, 837; ^{1}H NMR (500 MHz, DMSO-d$_6$) δ 6.11 (s, 2H, NH$_2$), 6.85 (d, 2H, J = 5.0 Hz, H-10), 7.54 (s, 1H, H-4), 7.85 (m, 3H, H-9/H9'/H6), 8.57 (s, 1H, H-7), 10.17 (s, 1H, OH); DEPTQ (125 MHz, DMSO-d$_6$) δ 105.4, 115.6, 125.0, 127.4, 131.2, 132.6, 144.8, 154.2, 160.3, 160.9.

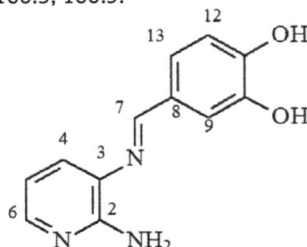

4.5. 2-N-amine-3-N-(3,4-dihydroxy-phenylmethylene)-5-bromo-pyridine (3d)

mp 223–225°C; IR (ν/cm^{-1}) 3,461, 3,400, 3,348, 3,033, 1,607, 1,593, 1,447, 1,393, 1,263, 1,147, 756; ^{1}H NMR (500 MHz, DMSO-d$_6$) δ 6.05 (s, 2H, NH$_2$), 6.54 (dd, J = 5.0 Hz, 1H, H-5), 6.83 (d, 2H, J = 8.0, H-12), 7.24 (d, 1H, J = 8.0 Hz, H-13), 7.28 (d, 1H, J = 8.0 Hz, H-6), 7.43 (s, 1H, H-9), 7.78 (d, J = 4.0 Hz, 1H, H-4), 8.43 (s, 1H, H-7), 9.48 (sl, 2H, 2xOH); DEPTQ (125 MHz, DMSO-d$_6$) δ 112.6, 114.7, 115.4, 112.5, 123.0, 128.0, 131.6, 145.1, 145.6, 149.3, 155.1, 159.2.

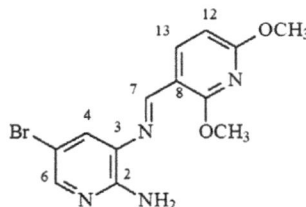

4.6. 2-N-amine-3-N-(2,4-dimethoxy-pyridinylmehylene)-5-bromo-pyridine (3e)

mp 152–153°C; IR (ν/cm^{-1}) 3,452, 3,263, 3,128, 2,944, 1,593, 1,574, 1,466, 1,322, 1,088, 1,009, 806, 564; ^{1}H NMR (500 MHz, DMSO-d$_6$) δ 3.94 (s, 3H, OCH$_3$), 4.00 (s, 3H, OCH$_3$), 6.18 (s, 2H, NH$_2$), 6.51 (d, 1H, J = 8.0 Hz, H-12), 7.51 (s, 1H, H-4), 7.85 (s, 1H, H-6), 8.49 (d, 1H, J = 8.0 Hz, H-13), 8.70 (s, 1H, H-7); DEPTQ (100 MHz, DMSO-d$_6$) δ 53.6, 53.8, 102.8, 105.3, 110.3, 125.0, 132.6, 140.0, 145.1, 154.2, 153.9, 162.1, 164.9.

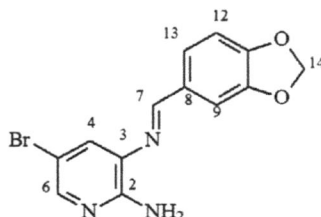

4.7. 2-N-amine-3-N-(3,4-methylenedioxy-phenylmethylene)-5-bromo-pyridine (3f)

mp 175–177°C; IR (ν/cm^{-1}) 3,450, 3,269, 3,141, 2,901, 1,617, 1,590, 1,441, 1,242, 1,099, 908, 805, 551; ^{1}H NMR (400 MHz, DMSO-d$_6$) δ 6.12 (s, 2H, H-14), 6.25 (s, 2H, NH$_2$), 7.04 (d, 1H, J = 8.0 Hz, H-12), 7.41 (d, 1H, J = 8.0 Hz, H-13), 7.59 (s, 1H, H-9), 7.73 (s, 1H, H-4), 7.87 (s, 1H, H-7), 8.62 (s, 1H, H-7); DEPTQ (125 MHz, DMSO-d$_6$) δ 101.7, 105.2, 106.6, 108.2, 125.1, 126.6, 130.9, 131.8, 145.7, 148.0, 150.4, 154.4, 159.7.

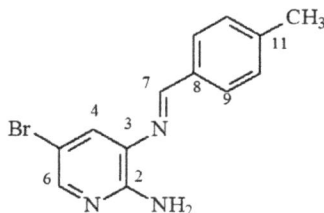

4.8. 2-N-amine-3-N-(4-methy-phenylmethylene)-5-bromo-pyridine (3g)

mp 156–157°C; IR (ν/cm^{-1}) 3,420, 3,115, 3,018, 2,913, 2,854, 1,600, 1,560, 1,462, 1,373, 1,241, 1,165, 808; ^{1}H NMR (500 MHz, DMSO-d$_6$) δ 2.37 (s, 3H), 6.19 (s, 2H, NH$_2$), 7.31 (d, 2H, J = 8.0 Hz, H-10/10′), 7.60 (d, 1H, J = 2.0 Hz, H-4), 7.88 (d, 1H, J = 2.0 Hz, H-6), 7.90 (d, 2H, J = 8.0 Hz, H-9/9′), 8.68 (s, 1H, H-7); DEPTQ (125 MHz, DMSO-d$_6$) δ 21.2, 105.3, 125.3, 129.1, 129.3, 132.0, 133.4, 141.8, 145.5, 154.3, 160.6.

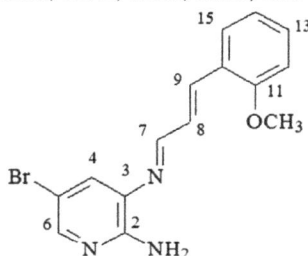

4.9. 2-N-amine-3-N-(4-methoxy-cinnamylmethylene)-5-bromo-pyridine (3h)

mp 173–174°C; IR (ν/cm^{-1}) 3,457, 3,200, 3,127, 2,919, 2,835, 1,610, 1,592, 1,467, 1,240, 1,020, 760, 571; ^{1}H NMR (500 MHz, DMSO-d$_6$) δ 3.87 (s, 3H, OCH$_3$), 6.05 (s, 2H, NH$_2$), 7.09 (d, 1H, J = 8.0 Hz, H-12), 7.15 (dd, 1H, J = 9.00 and 16.0 Hz, H-8), 7.38 (t, 1H, J = 7.00 Hz, H-14), 7.53 (d, 1H, J = 2.0 Hz, H-4), 7.58 (d, 1H, J = 16.0 Hz, H-9), 7.67 (dd, J = 2.0 and 7.0 Hz, H-15), 7.86 (d, 1H, J = 2.0 Hz, H-6), 8.51 (d, 1H, J = 9.0 Hz, H-7); DEPTQ (125 MHz, DMSO-d$_6$) δ 55.6, 105.5, 111.7, 120.8, 123.6, 125.2, 127.8, 128.8, 131.3, 132.5, 139.5, 145.3, 154.1, 157.3, 163.2.

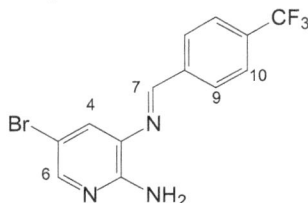

4.10. 2-N-amine-3-N-(4-trifluoromethyl-phenylmethylene)-5-bromo-pyridine (3i)

mp 108–112°C; IR (ν/cm^{-1}) 3,466, 3,268, 3,134, 1,616, 1,559, 1,468, 1,382, 1,318, 1,160, 1,108, 1,061, 837; ^{1}H NMR (500 MHz, DMSO-d$_6$) δ 6.35 (s, 2H, NH$_2$), 7.70 (d, 1H, H-4), 7.86 (d, 2H, J = 8.0 Hz, H-10), 7.94 (d, 1H, H-6), 8.23 (d, 2H, J = 8.0 Hz, H-10), 8.85 (s, 1H, H-7), DEPTQ (125 MHz, DMSO-d$_6$) δ 105.3, 125.7, 125.9, 129.8, 129.9, 131.2, 139.7, 146.7, 154.6, 159.4.

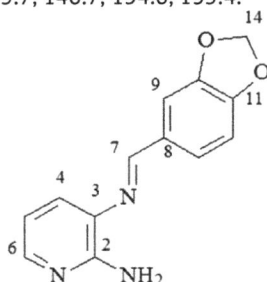

4.11. 2-N-amine-3-N-(3,4-methylenedioxy-phenylmethylene)-pyridine (3j)

mp 140–141°C; IR (ν/cm⁻¹) 3,471, 3,122, 1,624, 1,603, 1,274, 1,171, 1,150, 745, 728; ¹H NMR (500 MHz, DMSO-d₆) δ 6.00 (s, 2H, H-14), 6.10 (s, 2H, NH₂), 6.53 (dd, 1H, J = 5.0 and 7.5 Hz, H-5), 7.02 (d,1H, J = 7.5 Hz, H-13), 7.34 (dd, 1H, J = 8.0, H-12), 7.40 (dd, 1H, J = 1.0 and 8.0 Hz, H-4), 7.70 (dd, 1H, J = 1.0 Hz, H-9), 7.80 (dd, 1H, J = 1.0 and 5.0 Hz, H-6), 8.56 (s, 1H, H-7); DEPTQ (125 MHz, DMSO-d₆) δ 101.6, 106.5, 108.2, 112.4, 122.9, 126.0, 130.6, 131.2, 145.7, 148.0, 150.1, 155.4, 158.1.

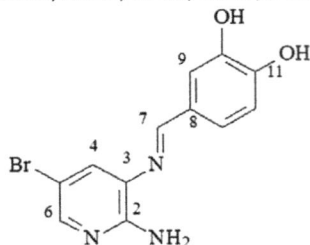

4.12. 2-N-amine-3-N-(3,4-dihydroxy-phenylmethylene)-5-bromo-pyridine (3k)

mp 200–201°C; IR (ν/cm⁻¹) 3,430, 3,347, 3,074, 1,582, 1,533, 1,465, 1,445, 1,359, 1,290, 1,271, 1,171, 1,148, 914, 886, 805, 555; ¹H NMR (400 MHz, DMSO-d₆) δ 6.05 (s, 1H, J = 8.0 Hz, H-13), 6.83 (s, 1H, J = 8.0 Hz, H-12), 7.27 (d, 1H, J = 8.0 Hz, H-13), 7.42 (s, 1H, H-9), 7.51 (s, 1H, H-4), 7.84 (s, 1H, H-6), 8.48 (s, 1H, H-7), DEPTQ (100 MHz, DMSO-d₆) δ 105.5, 115.0, 115.4, 122.8, 125.1, 127.8, 132.8, 144.7, 145.5, 149.6, 154.1, 160.7.

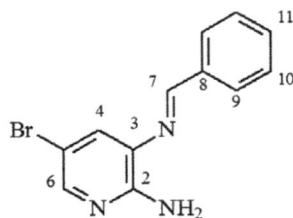

4.13. 2-N-amine-3-N-(phenylmethylene)-5-bromo-pyridine (3l)

mp 120–122°C; IR (ν/cm⁻¹) 3,461, 3,263, 3,140, 2,995, 1,607, 1,570, 1,464, 1,401, 1,373, 1,244, 1,161, 690; ¹H NMR (500 MHz, DMSO-d₆) δ 6.21 (s, 2H, NH₂), 7.48–7.52 (m, 3H, H-10/10′/11), 7.63 (d, 1H, J = 4.0 Hz, H-4), 7.90 (d, 1H, J = 2.0 Hz, H-6), 8.01 (d, 2H, J = 2.0 and 8.0 Hz, H-9/H9′), 8.74 (s, 1H, H-7), DEPQ (100 MHz, DMSO-d₆), 106.4, 119.0, 128.6, 129.3, 130.8, 131.9, 132.8, 134.0, 147.4, 167.3.

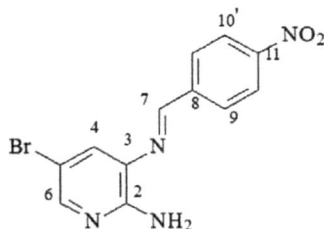

4.14. 2-N-amine-3-N-(4-nitro-phenylmethylene)-5-bromo-pyridine (3m-isomer A)

mp 212–214°C; IR (ν/cm⁻¹) 3,491, 3,385, 1,591, 1,503, 1,456, 1,332, 1,244, 1,101, 913, 845, 740, 683, 567; ¹H NMR (400 MHz, DMSO-d₆) δ 6.41 (s, 2H, NH₂), 7.75 (d, 1H, J = 4.0 Hz, H-4), 7.95 (d, 1H, J = 4.0 Hz, H-6), 8.28 (d, 2H, J = 8.0, H-9/9′), 8.33 (d, 2H, J = 10.0 Hz, H-10/10′), 8.91 (s, 1H, H-7), DEPTQ (100 MHz, DMSO-d₆) δ 114.1, 124.1, 124.6, 128.3, 130.0, 131.1, 135.1, 145.6, 148.8, 152.2.

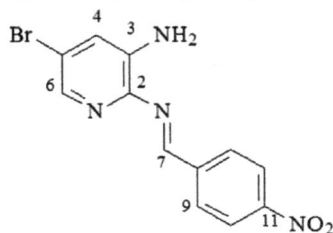

4.15. 2-N-(4-nitro-phenylmethylene)-3-N-amine-5-bromo-pyridine (3m-isomer B)

mp 195–197°C; IR (ν/cm^{-1}) 3,389, 3,053, 2,968, 1,674, 1,611, 15,602, 1,449, 1,279, 1,259, 1,150, 1,150, 1,029, 751, 563, 505; ^{1}H NMR (500 MHz, DMSO-d$_6$) δ 6.01 (s, 2H, NH$_2$), 7.29 (d, 1H, J = 2.0 Hz, H-4), 7.70 (d, 1H, J = 2.0 Hz, H-6), 8.30 (d, 2H, J = 8.0 Hz, H-9/9′), 8.37 (d, 2H, J = 8.0, H-10/10′), 9.33 (s, 1H, H-7); DEPTQ (125 MHz, DMSO-d$_6$) δ 119.9, 123.3, 123.8, 130.3, 134.9, 141.6, 142.4, 142.5, 148.8, 155.9.

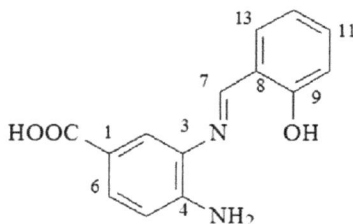

4.16. 4-N-amine-3-N-(2-hydroxy-phenylmethilene)-benzoic acid (3n)

mp 194–195°C; IR (ν/cm^{-1}) 3,498, 3,391, 3,098, 1,590, 1,508, 1,334, 1,241, 1,100, 913, 845, 682; ^{1}H NMR (500 MHz, DMSO-d$_6$) δ 5.88 (s, 2H, NH$_2$), 6.76 (d, 1H, J = 8.0 Hz, H-5), 6.94 (d, 1H, J = 5.0 Hz), 6.96 (d, 1H, J = 5.0 Hz), 7.38 (t, 1H, J = 8.0 Hz, H-11), 7.60 (s, 1H), 7.75 (d, 1H, J = 7.0 Hz), 8.90 (s, 1H); 12.36 (s, 1H, OH); DEPTQ (125 MHz, DMSO-d$_6$) δ 113.9, 116.4, 117.9, 119.8, 120.1, 129.5, 132.0, 133.0, 133.7, 147.3, 159.7, 161.4.

Funding

This study was supported by Universidade Federal de Santa Catarina (UFSC), Universidade Federal Rural do Rio de Janeiro (UFRRJ), and the Brazilian government funding agencies CNPq (Conselho Nacional de Desenvolvimento Científico e Tecnológico), CAPES (Coordenação de Aperfeiçoamento de Pessoal de Nível Superior) and FAPERJ (Fundação de Amparo a Pesquisa do Rio de Janeiro) for financial support and fellowships received.

Author details

Nohana C. Ramos[1]
E-mail: nohanacaruso@gmail.com
ORCID ID: http://orcid.org/0000-0002-4947-1466
Aurea Echevarria[1]
E-mail: echevarr@ufrrj.br
Arthur Valbon[1]
E-mail: arthur-valbon@hotmail.com
ORCID ID: http://orcid.org/0000-0002-6240-7098
Adailton J. Bortoluzzi[2]
E-mail: adailton.bortoluzzi@ufsc.br
Guilherme P. Guedes[1]
E-mail: guilherme@ufrrj.br
Cláudio E. Rodrigues-Santos[1]
E-mail: claudioers@ufrrj.br

[1] Departamento de Química, Instituto de Ciências Exatas, Universidade Federal Rural do Rio de Janeiro, 23890-900, Seropédica, RJ, Brazil.
[2] Departamento de Química, Universidade Federal de Santa Catarina, 88040-900, Florianópolis, SC, Brazil.

References

Asadi, M., Asadi, Z., Torabi, S., & Lotfi, N. (2012). Synthesis, characterization and thermodynamics of complex formation of some new Schiff base ligands with some transition metal ions and the adduct formation of zinc Schiff base complexes with some organotin chlorides. *Spectrochimica Acta Part A: Molecular and Biomolecular Spectroscopy, 94,* 372–377. http://dx.doi.org/10.1016/j.saa.2012.03.061

Chen, L., Xiang, Y., & Feng, T. (2012). Hybrid compounds of Schiff base Cu, Fe, Co complexes with molybdovanadophoric heteropolyacids: Synthesis, characterization and their catalytic performance to hydroxylation of benzene with H$_2$O$_2$. *Applied Organometallic Chemistry, 26,* 108–113. http://dx.doi.org/10.1002/aoc.v26.3

Cordes, E. H., & Jencks, W. P. J. (1962). On the mechanism of Schiff base formation and hydrolysis. *Journal of the American Chemical Society, 84,* 832–837. http://dx.doi.org/10.1021/ja00864a031

Coulthard, G., Unsworth, W, P., & Taylor, R. J. K. (2015). Propylphosphonic anhydride (T3P) mediated synthesis of β-lactams from imines and aryl-substituted acetic acids. *Tetrahedron Letters, 56,* 3113–3116.

Deligeorgiev, T., Gadjev, N., Vasilev, A., Kaloyanova, St., Vaquero, J. J., Alvarez-Builla, J. (2010). Green chemistry in organic synthesis. *Mini-Reviews in Organic Chemistry, 7,* 44–53.

Dogan, F., & Kaya, I. J. (2013). Thermal decomposition studies of Schiff-base-substitute polyphenol–metal complexes. *Journal of Applied Polymer Science, 128,* 3354–3362.

Doi, T., Kawai, H., Murayama, K., Kashida, H., & Asauna, H. (2016). Synthesis and pharmacologicalporperties of 2-azabicyclo [2.2.2] octane derivatives representing conformational restricted isopethidine analogues. *European Journal of Medicinal Chemistry, 22,* 1–7.

Esteves-Sousa, A., Rodrigues-Santos, C. E., Cistia, C. N. D., Silva, D. R., Sant'Anna, C. M. R., & Echevarria, A. (2012). Solvent-free synthesis, DNA-topoisomerase II activity and molecular docking study of new asymmetrically N,N'-substituted ureas. *Molecules, 17,* 12882–12894.

Ganguly, A., Chakraborty, P., Banerjee, K., & Choudhuri, S. K. (2014). The role of a Schiff base scaffold, N-(2-hydroxy acetophenone) glycinate-in overcoming multidrug resistance in cancer. *European Journal of Pharmaceutical Sciences, 51,* 96–104.

Jhaumeer-Laulloo, S., Gupta Bhowon, M. G., Mungur, S., Fawzi Mahomoodally, M. F., & Hussein Subratty, A. H. (2012). *In vitro* anti-glycation and anti-oxidant properties of synthesized schiff bases. *Medicinal Chemistry, 8,* 409–414. http://dx.doi.org/10.2174/1573406411208030409

Khanna, I. K., Weier, R. M., Lentz, K. T., Swenton, L., & Lankin, D. C. (1995). Facile, regioselective syntheses of N-alkylated 2, 3-diaminopyridines and imidazo[4, 5-b]pyridines. *The Journal of Organic Chemistry, 60,* 960–965. http://dx.doi.org/10.1021/jo00109a029

Kielland, N., Escudero-Adán, E. C., Martínez Belmonte, M. M., & Kleij, A. W. (2013). Unsymmetrical octanuclear Schiff base clusters: Synthesis, characterization and catalysis. *Dalton Transactions, 42,* 1427–1436. http://dx.doi.org/10.1039/C2DT31723K

Kommi, D. N., Kumar, D., Bansal, R., Chebolu, R., Chadraborti, A. K. (2013). "All water chemistry" for a concise total synthesis of the novel class anti-anginal drug (RS), (R), and (S)-ranolazine. *Green Chemistry, 15,* 756–767. http://dx.doi.org/10.1039/c3gc36997h

Lee, S. K., Tan, K. W., Ng, S. W., Ooi, K. K., Ang, K. P., Abdah, M. A. (2014). Zinc (II) complex with a cationic Schiff base ligand: Synthesis, characterization, and biological studies. *Spectrochimica Acta Part A: Molecular and Biomolecular Spectroscopy, 121,* 101–108. http://dx.doi.org/10.1016/j.saa.2013.10.084

Panda, S. S., & Jain, S. C. (2012). "On water" synthesis of spiro-indoles via Schiff bases. *Monatshefte für Chemie— Chemical Monthly, 143,* 1187–1194. http://dx.doi.org/10.1007/s00706-011-0697-x

Qin, W., Long, S., Panunzio, M., & Biondi, S. (2013). Schiff bases: A short survey on an evergreen chemistry tool. *Molecules, 18,* 12264–12289. http://dx.doi.org/10.3390/molecules181012264

Rodrigues-Santos, C. E., & Echevarria, A. (2007). An efficient and fast synthesis of 4-aryl-3,4-dihydrocoumarins by (CF3SO3)3Y catalysis under microwave irradiation. *Tetrahedron Letters, 48,* 4505–4508. http://dx.doi.org/10.1016/j.tetlet.2007.04.144

Rodrigues-Santos, C. E., & Echevarria, A. (2011). Convenient syntheses of pyrazolo[3,4-b]pyridin-6-ones using either microwave or ultrasound irradiation. *Tetrahedron Letters, 52,* 336–340. http://dx.doi.org/10.1016/j.tetlet.2010.11.054

Saggiomo, V., & Lüning, U. (2009). On the formation of imines in water—A comparison. *Tetrahedron Letters, 50,* 4663–4665. http://dx.doi.org/10.1016/j.tetlet.2009.05.117

Seifzdeh, D., Basharnavaz, H., & Bezaatpour, A. (2014). A Schiff base compound as effective corrosion inhibitor for magnesium in acidic media. *Materials Chemistry and Physics, 138,* 794–802.

Zarei, M. (2014). An efficient and green method for the synthesis of 2-azetidinones mediated by propylphosphonic anhydride (T3P®). *Monatshefte für Chemie—Chemical Monthly, 145,* 1495–1499.

5

A green approach to synthesize controllable silver nanostructures from *Limonia acidissima* for inactivation of pathogenic bacteria

E. Chandra Sekhar[1], K.S.V. Krishna Rao[2,3]*, K. Madhusudana Rao[4] and S. Pradeep Kumar[5]

*Corresponding author: K.S.V. Krishna Rao, Polymer Biomaterial Design and Synthesis Laboratory, Department of Chemistry, Yogi Vemana University, Kadapa, Andhra Pradesh, India; Department of Chemical Engineering and Material Science, Wayne State University, Detroit, MI, USA
E-mail: drksvkrishna@yahoo.com
Reviewing editor: Burçak Ebin, Chalmers University of Technology, Sweden

Abstract: Controllable silver nanoparticles were developed by a green approach using extracts of both leaves and bark of *Limonia acidissima* tree. Due to the presence of phytochemical compounds in *L. acidissima* leaves and bark; such as saponins, phenolic compounds, phytosterols, and quinines present in extracts act as reductants, hence the silver nanoparticles were easily produced under mild conditions. The formation and kinetic study of silver nanoparticles were verified by UV–vis spectroscopy. Highly stable and uniform size silver nanoparticles were produced using bark extract reduction than leaf extract and confirmed by dynamic light scattering and transmission electron microscopy analysis. Further we applied antibacterial activity on both *Escherichia coli* and *Bacillus subtilis*. The results suggest that the silver nanoparticles suspension exhibits excellent antibacterial activity. The present study is a simple and eco-friendly approach for production of silver nanoparticles in the large scale up and could be easily commercialized, especially biological applications.

Subjects: Materials Chemistry; Nanoscience & Nanotechnology; Physical Chemistry

Keywords: green synthesis; *Limonia acidissima*; silver nanoparticles; antimicrobial activity and biological applications

ABOUT THE AUTHORS

K.S.V. Krishna Rao, did his BSc, MSc and PhD in Chemistry from Sri Krishnadevaraya University, India. His major fields of research are Drug Delivery, Nanoscience and Nanotechnology for Biomedical Application, Pervaporation, PEM Fuel Cells and separation of toxic metals from IPNs. He did postdoctoral research work at Pusan National University, South Korea. After successful completion of his PDF in Pusan National University, he joined as research professor at Changwon National University, Changwon, South Korea. For his excellent contribution to Chemical Science he got the Young Scientist award in 2009, from DST (Govt. of India), New Delhi, India. Also he received UGC-RAMAN postdoctoral fellowship in 2015 to work at USA, from UGC, India. He is currently an assistant professor in the Department of Chemistry, Yogi Vemana University, Andhra Pradesh, India.

PUBLIC INTEREST STATEMENT

There is a growing necessity of nanotechnology for biomedical application. Green chemistry plays an important role for the production of metal nanoparticles from medicinal plants which have significant including DNA interaction, enzyme inhibition, anti-cancer, and anti-bacterial studies. The present study demonstrates bio-reduction of silver (Ag°) nanoparticles using *L. acidissima* plant materials. *L. acidissima* aqueous leaf and bark extracts appears to be simple, inexpensive, and environmentally eco-friendly. Aqueous extract of leaf and bark could act not only reducing the silver ion (Ag⁺) but also control the size i.e. average sizes ~25 nm and ~12 nm, respectively. Also silver nanoparticle suspension exhibited potential antibacterial activity against *Escherichia coli* and *Bacillus subtilis*.

1. Introduction

Green chemistry is the design of chemical products that reduce the use of hazardous substances, and it applies across the life cycle of a chemical product, including its design, production, use, and decisive disposal (Robert, Morrison Robert, & Boyd, 1992). Green chemistry contributes advantage over physicochemical approaches in terms of its eco-friendly environment and has the ability to re-form biomaterials which are more effective and less toxic and could benefit millions of patients throughout the world. Green chemistry has been engaged effectively in the production of functional-ized products materials that have a strong correlation to the functionalized nanomaterials projected for a range of forthcoming applications; one would suppose effective application of this method for these emerging materials. Green chemistry is progressively seen as a controlling implement that researchers must use to evaluate the environmental impression of nanotechnology (Zhao & Stevens, 1998). And it refers to investigate on materials that are nanometer in size and pertinent to almost every technology and medicine through medicinal plants.

Nanotechnology is a momentous technology which is widely using in miscellaneous areas like medicine, biology, chemistry, catalysis, photonics, electronics, and bio-labeling (Kim et al., 2007). Outstanding physicochemical properties of nanoparticles, including catalytic activity (Rai, Yadav, & Gade, 2009), optical and electronic properties (Sun, 1988; Valiathan, 1998) as well as cytotoxic and antibacterial properties (Talalay, 2001), researchers showed attainment interest towards the expan-sion of novel methods for synthesis of nanoparticles. Their antibacterial effect is applicable to stum-bling block of respiratory enzyme pathways, interacting with the proteins, and modification of DNA (Robinson & Zhang, 2011; Xavier, Auxilia, & lSelvi, 2013). The combination of traditional and modern knowledge can produce better source of the active components for the treatment of diseases with fewer side effects (Ameenah, 2006). Green synthesis is guiltless and eco-friendly method for origi-nated nanoparticles (Adeleh & Hamed, 2012). The use of nanoparticles as potential drug carriers in the treatment of cancer has also been reported (Anish, Templeton, Munshi, & Ramesh, 2013; Lim, Jang, Lee, Haam, & Huh, 2013; Shankar, 2004; Yang, Mu-Yi, Liu, Huang, & Wei, 2012). Nanoparticles can be produced using different types of chemicals and physical methods to bring under control the problem of toxicity in synthesis, medicinal plants have come to have a major role in the synthesis of nanoparticles. Phytochemical compounds such as saponins, phenolic compounds, phytosterols, and quinines present in plant (Manish, Arun, & Ansari, 2009) have both preservative and reductive activ-ity, and are mainly liable for the reduction of silver ions to nanoparticles using chemical and radia-tion methods. At present there has been a progressive need to derive an environmentally exonerable nanoparticle synthetic process that does not involve any toxic chemicals. However, metal nanopar-ticles are synthesized by reducing agents such as hydrazine, 1,2-hexadecanediol, $NaBH_4$, and ascor-bic acid (Kim, Cha, & Gong, 2013; Praveena & Kumar, 2014; Venkateswara Rao & Viswanathan, 2010, 2012) for reduction of silver ions to silver nanoparticles (Ag^+ to Ag^o), which are extremely toxic in nature. Green chemistry offers the replacement of anticipatory approach that calls for safe, efficient, and effective materials from the starting to end of a chemical product. Green synthesis facilitates in reduction or elimination of the substances that are hazardous to human health and the environ-ment (Kumar, Vemula, Ajayan, & John, 2008). Nowadays, there is a considerable interest in using green chemistry for fabrication of metal nanoparticles (Arokiyaraj et al., 2014; Christensen, Vivekanandhan, Misra, & Mohanty, 2011; Gandhi, Sirisha, & Sharma, 2014; Geethalakshmi & Sarada, 2010). Sastry *et al.* (Ahmad, Senapati, Khan, Kumar, & Sastry, 2003; Shankar, Rai, Ahmad, & Sastry, 2004) synthesized stable Ag, Au, and Ag–Au bimetallic nanoparticles using *Azadirachta indica* leaf broth. Also, recent literature suggested that green synthesis of Ag and Au nanoparticles from apiin (Kasthuri, Veerapandian, & Rajendiran, 2009), *Cinnamon zeylanicum* bark extract (Sathishkumar et al., 2009), mushroom extract (Philip, 2009), *Acalypha indica* leaf extracts (Krishnaraj et al., 2010), glucose, and starch (Raveendran, Fu, & Wallen, 2006). Considering the above facts, authors chosen a *Limonia acidissima* plant leaf and bark extract as a reducing agent for green synthesis of silver nanoparticles, which is the easiest, most cost efficient, non-toxic, eco-friendly, efficient method, and rapid method.

L. acidissima appertaining to the family *Rutaceae* is a normal-sized deciduous plant (Geeta Vasudevan, Kedlaya, Deepa, & Ballal, 2001). It is given with many medicinal uses in Ayurveda systems of medicine. The fruits are woody, rough, and used as a substitute for "blood alcohol level" in diarrhea and bacterial diarrhea like shigella dysentery (Sunitha & Krishna Mohan, 2013). The fruits are used for tumors, asthma, wounds, cardiac debility, and hepatitis. Leaves are aromatic and astringent, oil of leaves useful in relieving itching and when mixed with a pinch of black pepper is used as a carminative; fruits of *L. acidissima* are used to prepare tonic, antiscorbutic, alexipharmic, astringent, stomachic, and stimulant (Absar, Eswar, Omer, & Feronia, 2010). Different parts of *L. acidissima* (stem bark, root bark, and unripe fruit-shells) contain antifungal compounds (Adikaram, Yamuna, Leslie, Gunatilaka, & Bandara, 1989; Parrotta, 2011). Hence, phytochemical constituents of leaves and bark of *L. acidissima* have orientin, psoraline, vitexin, xanthotoxin, amides, feronolide, feronone, limonoids, physcion, quinolones, indole, alkaloids (Kithsiri, Jeratne, Bandara, Leslie, & Gunatilaka, 1992; Parthasarathi, Asok Kumar, & Swapnadip, 1989).

To the best of the authors' knowledge there is no report on synthesis of silver nanoparticles utilizing an aqueous leaf and bark extracts of *L. acidissima*. In this work, *L. acidissima* extract in a concentrated aqueous solution of silver nitrate ($AgNO_3$) resulted in reduction of silver ions (Ag^+) and formation of silver nanoparticles.

2. Materials and methods

2.1. Materials
L. Acidissima plant materials (leaves and bark) were collected in the month of 19 October 2014 in Yogi Vemana University Andhra Pradesh; India. Silver nitrate (Analytical grade) was received from MERCK Specialties Pvt. Ltd, INDIA. Throughout the experiment double-distilled water was used and all reagents were used without further purification.

2.2. Preparation of Extracts of L. Acidissima Leaves and bark
The *L. acidissima* plant derivatives (leaves and bark) were collected and immersed in a distilled water bath to remove the surfaces adhered to dust particles; the derivatives were removed from the bath and allowed for drying in a dust-free environment at room temperature for 48 h. These plant derivatives were cut into small pieces for further study. 10 g of leaves/10 g bark were added to 100 mL of distilled water in 500 mL Erlenmeyer flask for 60 min at 70°C. The extract was filtered with Whatmann grade No. 1 filter paper. Collected extract was preserved at 4°C for further experiment.

2.3. The synthesis of silver nanoparticles
Aqueous $AgNO_3$ solution and the aqueous *L. acidissima* plant (leaf and bark) extracts were used for fabrication of Ag nanoparticles through bio-reduction process. Briefly, to synthesize silver nanoparticles, 0.01 M $AgNO_3$ 300 μL for leaf extract (5 mL) and 100 μL for bark extract (5 mL) were added in a bottle. After a few minutes, silver nanoparticles were formed and monitored by brown color of colloidal suspension. Different feed compositions for synthesis of silver nanoparticles are presented in Table 1. The schematic representation of fabrication of silver nanoparticles is shown in Figure 1. Further, formation of silver nanoparticles was verified by UV–vis spectrophotometer at different time intervals, and effect of concentration of extract and concentration of silver ion solution.

2.4. Characterization
UV–vis absorbance spectroscopy: bio-reduction of the nanoparticles by silver nitrate was monitored using a Series 3000 double beam ultraviolet–visible spectrophotometer (UV–vis Spectrophotometer, LAB INDIA, UV-3092). Samples were loaded into a 1 cm path length quartz cuvette for analysis. The UV–vis spectrophotometric readings were recorded at a scanning speed of 0.5 nm intervals and were scanned from 200 to 800 nm. Fourier transforms infrared analysis (PerkinElmer Spectrum Two, UK) spectrophotometer was measured to *L. acidissima* silver nanoparticles. Samples of *L. acidissima* leaf extract silver nanoparticles and *L. acidissima* bark extract silver nanoparticles were dried at 40°C for 2 days, and mixed with KBr in a ratio of 10:200 (w/w) and pressed under vacuum to form pellets.

Table 1. Different feed compositions for synthesis of silver nanoparticles

S. No.	Volume of extract (mL)	Volume of AgNO$_3$ (µL)
Variation of plant leaf extract at constant AgNO$_3$ (Figure 1(c))		
1	2	300
2	3	300
3	4	300
4	5	300
5	6	300
Variation of AgNO$_3$ at constant plant leaf extract (Figure 1(d))		
1	5	200
2	5	250
3	5	300
4	5	350
5	5	400
Variation of plant leaf extract at constant AgNO$_3$ (Figure 2(c))		
1	3	100
2	4	100
3	5	100
4	6	100
Variation of AgNO$_3$ at constant plant bark extract (Figure 2(d))		
1	5	50
2	5	100
3	5	150
4	5	200

Fourier transforms infrared analysis of the samples was recorded in transmittance mode. Dynamic light scattering study: The zeta-sized nanosequence performs size measurements using advancement called dynamic light scattering spectroscopy measures Brownian motion and relates this to the size of the particles. Mean diameter and size circulation of the nanoparticles were determined by dynamic light scattering method using a Brookhaven BI-9000 AT instrument (Brookhaven Instruments Corporation, USA). Transmission electron microscopy: Samples of bio-synthesized silver nanoparticles (5 mL of leaf extract vs. 300 µL of AgNO$_3$ for leaf extract silver nanoparticles and 5 mL of bark extract vs. 100 µL of AgNO$_3$ for bark extract silver nanoparticles) were prepared by placing a drop of the colloidal solution on carbon-coated copper grids, allowing the films on the transmission electron microscopy grids to stand for two minutes, removing the excess solution with blotting paper, and letting the grid dry prior to measurement.

2.5. Antimicrobial activity

Antimicrobial activity of leaf extract silver nanoparticles/bark extract silver nanoparticles was performed by disc method of Bauer-Kirby (Kirby, Yoshihara, Sundsted, & Warren, 1957). Mueller Hinton Agar (M173) plates were prepared for rapid growth of *Bacillus subtilis* (G^{+ve}) and *Escherichia coli* (G^{-ve}) bacteria for this study. 10 µL of leaf extract silver nanoparticles/bark extract silver nanoparticles colloidal solution and pure extract was impregnated onto filter paper disks of 5 mm diameter, under aseptic conditions with different concentrations, then situated onto a cultured Mueller Hinton agar plates using a mechanical dispenser or sterile forceps. The plates were incubated for 18–20 h, at 37°C in the incubation chamber. Antimicrobial activity was estimated in identical by computing the zone of inhibition for the test organisms. The inhibition zone diameter that is produced will specify the susceptibility or resistance of a bacterium to the extract. Antibacterial activity can be resolute by

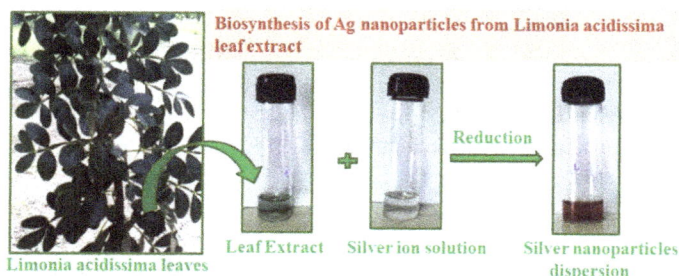

Figure 1. Schematic representation of silver nanoparticles formation.

comparing the zone diameter obtained with the known zone diameter size for susceptibility (Tetracycline). The antimicrobial activity of *L. acidissima* extracts was evaluated for both *B. subtilis* (ATCC 6633) and *E. coli* (ATCC 25922) (Sripairoj, Suwanborirux, & Tanasupawat, 2013).

3. Results and discussion

3.1. Formation of silver nanoparticles by leaf extract

The production of silver nanoparticles was confirmed by witnessing UV–vis analysis. Figure 2 shows the UV–vis spectral bands of silver nanoparticles manufactured by mixing leaf extract into Ag^+ ion solution at Lab. Conditions. The UV–vis spectral bands of formation of silver nanoparticles (5 mL of leaf extract with 300 µL of $AgNO_3$) for various intervals of time are predicted in Figure 2(a). Within 30 s the reduction was started but no clear SPR band appeared. The color of solution turned into reddish brown color, which designates the evidence of formation of silver nanoparticles. After 5 min a strong SPR band identified at 425 nm and the intensity of absorption band growths with increasing the reaction time and subsequent color fluctuations were observed, without shifting the wavelength

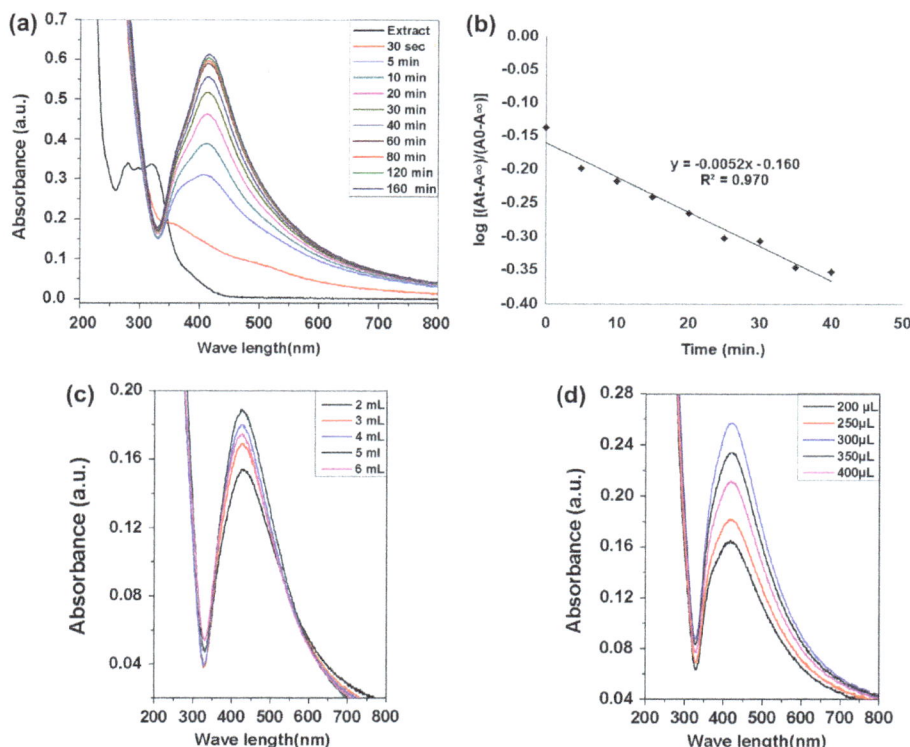

Figure 2. UV–vis spectra of Ag nanoparticles obtained using leaves extract of *L. acidissima*. (a) different time intervals; (b) rate constant determined between absorbance vs. time; (c) extract variation; and (d) $AgNO_3$ variation.

during the reaction. Mostly, the silver nanoparticles formation was started at 5 min and it might be completed within 4 h. For instance, Krishna raj et al. reported the synthesis of silver nanoparticles using the leaf extract of *Acalyphaindica* (Krishnaraj et al., 2010). The reduction of Ag^+ to silver nanoparticles during disclosure to the plant leaf extracts could be followed by color changes (Ankamwar, Chaudhary, & Sastry, 2005; Chandran & Chaudhary, 2006; Shankar, 2004). A graph plotted between absorbance vs. time, which indicates a constant absorbance value obtained after 40 min.

3.2. Effects of leaf extract concentration

To study the effect of the extract of *L. acidissima*, volume of the extract was varied from 2 to 6 mL with constant volume (Table 1) of 0.01 M $AgNO_3$ solution (Figure 2(c)). With the increase in the volume of extract, the SPR band was shifted to longer wavelength that proposed the development of enlarged size of silver nanoparticles. After addition of 5 mL extract increase in the number and size of the silver nanoparticles came to an end, may be due to the decline of the Ag^+ in the aqueous leaf extract of *L. acidissima*. The optimization of exact volume of aqueous $AgNO_3$ solution required to reduce the 5 mL of leaf extract is examined by varying both leaf and metal solutions (Table 1). 300 µL of Ag^+ was reduced by 5 mL of the extract as shown in the Figure 2(d).

3.3. Formation of silver nanoparticles by bark extract

3.3.1. Effect of contact time

Reduction of Ag^+ existing in the aqueous solution of Ag complex during the reaction with the phytochemical constituents present in the bark extract of *L. acidissima* detected by the UV–vis spectroscopy disclosed that nanoparticles in the solution may be correlated with the UV–vis analysis. A strong band noticed around 425 nm was predicted as SPR band and recognized the excitation of free electrons in the nanoparticles. The color changes were started from pale green to brown color at 5 min and its resolution increased up to 1 h reaches reddish brown color indicates the production of silver nanoparticles. Silver nanoparticles formation started between 420 nm and 590 nm whereas the strong peak was detected at 425 nm at the reaction time of 50 min due to excitation of exterior vibrations occurred in the surface of the silver nanoparticles. The intensity of SPR band primarily rises exponentially with time after that it tends to attain a persistent absorption value (Figure 3(a)) signifying the completion of the reaction. SPR band at 425 nm designates the formation of silver nanoparticles which was further confirmed by TEM analysis.

3.3.2. Effects of Ag ion concentration

The volume of the silver nitrate solution was varied (i.e. 50, 100, 150, and 200 µL) with constant volume i.e. 5 mL of *L. acidissima* bark extract for 50 min. In the UV–vis spectroscopy it was observed that the highest intense peak at 445 nm at the concentration of 100 µL of $AgNO_3$ solution which may be due to higher reduction of Ag^+ into silver nanoparticles (Figure 3(d)) and denoted that formation of silver nanoparticles was found to be at lower concentration of Ag^+ solution. In aqueous extract of *L. acidissima* bark there is a better reduction at concentration of 100 µL of $AgNO_3$ for 5 ml of bark extract (Figure 3(c)), which clearly states that very low quantity of $AgNO_3$ was required to convert Ag^+ ions to silver nanoparticles whereas in other plants higher volume of plant extract required the formation of silver nanoparticles (Dipankar & Murugan, 2012; Kotakadi et al., 2013; Philip, Philip, 2009; Raju et al., 2012; RathiSre, Reka, Poovazhagi, Arul Kumar, & Murugesan, 2015).

3.3.3. Effect of bark extracts concentration

Optimized quantity of the aqueous $AgNO_3$ (100 µL) was added to the various amounts of *L. acidissima* bark extract (3, 4, 5, and 6 mL) and their consistent SPR bands detected between 200 and 800 nm. Figure 3(c) suggests that different volumes of extract (Table 1) were monitored to synthesize the silver nanoparticles; and the production increased with increasing volume of the extract without a significant change in the size of the silver nanoparticles. Interestingly, the SPR bands noticed that the absorbance peak was substantially increased without any shift in the wavelength and sharp acute peak was observed at 446 nm with 5 mL of bark extract due to the higher reduction with 100 µL of silver nitrate (Figure 3(d)). This may be due to the presence of numerous acid and

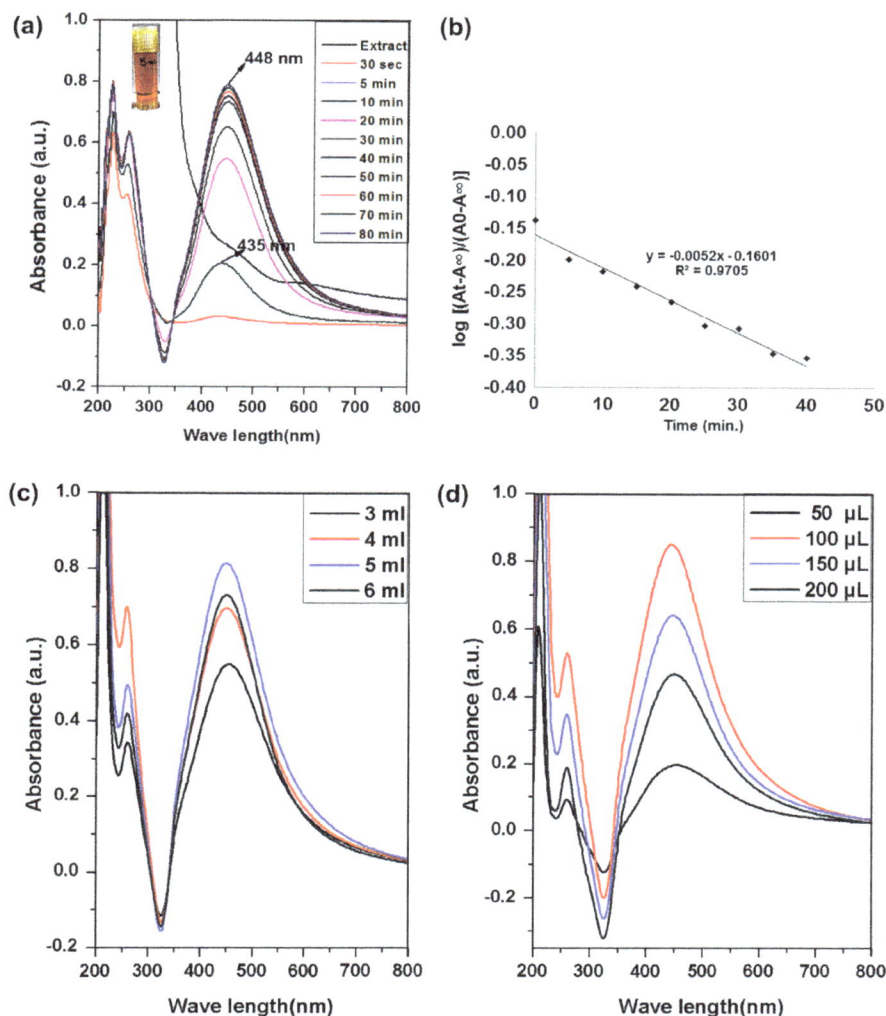

Fig. 3. UV–vis spectra of Ag nanoparticles obtained using bark extract of *L. acidissima*. (a) different time intervals; (b) rate constant determined between absorbance vs. time; (c) extract variation; and (d) AgNO₃ variation.

carboxylic groups in the bark extract of the *L. acidissima*. UV–vis spectroscopy is well known to investigate shape and size controlled of nanoparticles, frequency and width of the absorption peak depend on the size and shape of the metal nanoparticles (Srinivasan, 2011).

3.4. Kinetic studies of formation of silver nanoparticles from leaf and bark extracts

Kinetic studies of the reduction process exposed that the synthesis of nanoparticles starts within 30 s and reaches the flat terrain after 60 min showed in Figure 2(a), and 40 min for bark extract in Figure 3(a), inset. In the first 10 min there is an exponential increase in absorbance followed by slow and linear increase in the absorbance values which finally flattens out after 60 min thus, suggesting very fast reaction kinetics of silver nanoparticles synthesis. Similar control solution kept in dark for 14 days did not show much color enhancement, suggesting that light has vital role in the synthesis of silver nanoparticles. In the current study, the addition of aqueous AgNO₃ to leaf and bark aqueous extract, Ag⁺ ions were attracted by the –O⁻ groups of phytochemicals like phenols, flavonoids, and lysine. The exposure of the reaction medium to sunlight would have enabled the quick transfer of electrons from O⁻ to Ag⁺ ion resulting in the complete reduction of Ag⁺ ions. The pseudo-first-order rate constant k_{obs} was obtained from the slope of the plot of $\log(A_t - A_\infty)/(A_0 - A_\infty)$ vs. time t (Figures 2(b) and 3(b)), conferring to the equation $A_t = A_\infty + (A_0 - A_\infty) \exp(-k_{obs}t)$, where A_0 and A_∞ are the initial and final absorbance, respectively (Philip, 2009). The k_{obs} was found to be 4.79×10^{-2} min^{-1} with $R = 0.975$ for bark extract silver nanoparticles (Figure 3(b)) and 3.91×10^{-2} min^{-1} with $R = 0.970$ for leaf extract silver nanoparticles (Figure 2(b)), respectively.

$$A_t = A_\infty + (A_o - A_\infty)e^{-k_{obs}t} \tag{1}$$

3.5. Fourier transforms infrared analysis

L. acidissima leaves and bark extract contains mainly saponins, phenolic compounds, phytosterols, and quinines beta carotene, vitamin B, vitamin C, thiamin, flavanone, and riboflavin (Grotewold, Grotewold, & A. S., 2006; Malarkodi et al., 2014). To determine whether during silver nanoparticles synthesis some biomolecules particularly those with free carboxylic (–COOH) or amino (–NH$_2$) groups present in the leaf and bark extracts have been bound to surface of the Ag in the silver ion solution and formed silver nanoparticles. Fourier transforms infrared analysis spectrum of leaf extract, silver nanoparticles leaf extract and bark extract silver nanoparticles showed distinctive bands for several functional groups. Fundamental absorption peaks for phenolic (–OH), aromatic amines, aliphatic amines, carbonyl (>C=O), C–H, and C=C (benzene) functional groups were observed at 3,436, 3,401, 1,601, 1,384, 1,261, 1,040, 1,123, 1,630, 1,390, 1,178, 2,848, and 1,604 cm^{-1}, respectively in (Figure 4).

The functional group analysis performed to identify the phytochemicals responsible for the reduction of Ag$^+$ ions to Ag^o and its stabilization. Silver nanoparticles showed distinctive bands similar to those of the *L. acidissima* leaf and bark extracts (Figure 4) indicating that silver nanoparticles were coated with the biomolecules from both extracts. Peak at 1,633 cm^{-1} may be assigned to stretching

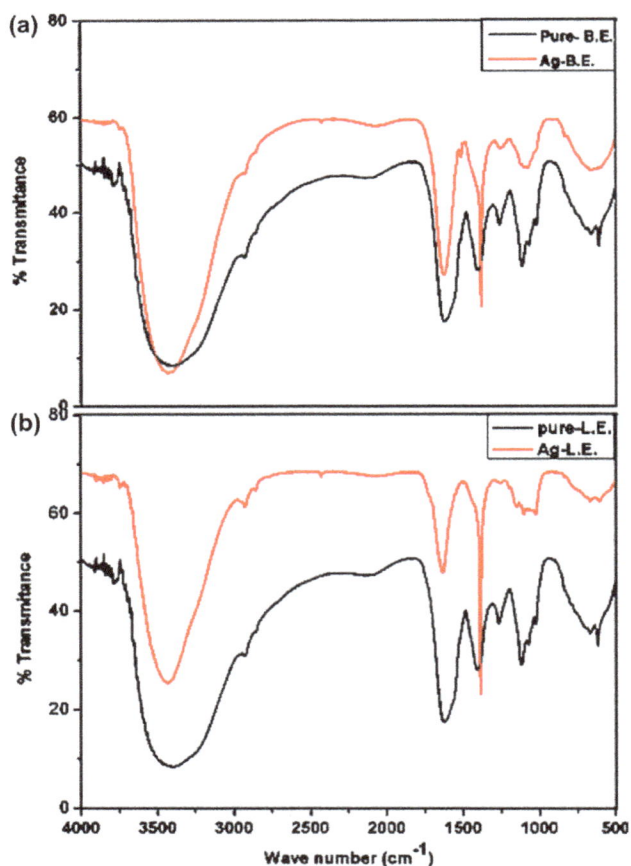

Figure 4. Fourier transforms infrared spectra of (a) *L. acidissima* bark extract and bark extract reduced silver nanoparticles and (b) *L. acidissima* leave extract and leaf extract reduced silver nanoparticles.

vibrations of C=C. There was a major shift from 3,436 to 3,401 cm^{-1} showed that the corresponding functional groups such as stretching vibration of O–H of alcohol or N–H of amines may be liable for the reduction of Ag$^+$ to stable silver nanoparticles. Very sharp intense peak was observed 1,348 cm^{-1} in both leaf extract silver nanoparticles and bark extract silver nanoparticles indicating that the phenolic carboxyl groups were affected by Ag$^+$ and to form stable Ag^0. Broad absorption band located at around 3,391 cm^{-1} may be attributed to bending vibrations of –O–H stretching and these conclusions were evaluated and the profile of phytochemicals present in leaf and bark extracts of *L. acidissima* (Banerjee, Someshwar, Subrata, & Chandra, 2011).

3.6. Transmission electron microscopic analysis and dynamic light scattering studies

Transmission electron microscopic analysis of silver nanoparticles supported to get the precise size and shape information. Figure 5 shows picture of leaf extract silver nanoparticles and bark extract silver nanoparticles, particles exhibited average mean sizes and further conformed by dynamic light scattering in Figure 6. All produced (5 mL of leaf extract vs. 300 µL of AgNO$_3$ for leaf extract silver nanoparticles and 5 mL of bark extract vs. 100 µL of AgNO$_3$ for bark extract silver nanoparticles) nanoparticles seem to be spherical in their morphology. Figure 5 shows that the particles are almost uniform within the volume of the particle, signifying the presence of silver nanoparticles supporting the probation of a single SPR band. Figure 5 shows that the silver nanoparticles obtained under normal conditions demonstrate that both leaf extract silver nanoparticles and bark extract silver nanoparticles with average sizes 25 ± 2.9 nm and 12 ± 3.7 nm, respectively.

3.7. Antibacterial activity

Antimicrobial activity of the leaf extract silver nanoparticles and bark extract silver nanoparticles was examined opposed to *B. subtilis* (ATCC 6633) and *E. coli* (ATCC 25922) using standard zone of inhibition microbiology assay. The aqueous extract effect of AgNO$_3$ has antibacterial activity against *E. coli* and *B. subtilis.* Previous studies reported that silver nanoparticles have microbial effect on micro-organisms (Mallikarjuna & John Sushma, 2014). Silver nanoparticles showed inhibition zone towards all tested bacteria (Figure 7) and maximum zone of inhibition was found to be leaf extract silver nanoparticles 1 = 14.94 ± 0.71 mm for *B. subtilis* and leaf extract silver nanoparticles 3 = 15.1 ± 0.82 mm for *E. coli* with respect to leaf extract silver nanoparticles (Figure 7) at different volumes (leaf extract = 10 µL, leaf extract silver nanoparticles 1 = 10 µL, leaf extract silver nanoparticles 2 = 11 µL, leaf extract silver nanoparticles = 12 µL, and leaf extract silver nanoparticles 4 = 13 µL of each tested bacteria) of silver nanoparticles colloidal solutions. Figure 7 suggested that bark extract silver nanoparticles (bark extract = 10 µL, bark extract silver nanoparticles 1 = 10 µL, bark extract silver nanoparticles 2 = 11 µL, bark extract silver nanoparticles 3 = 12 µL, and bark extract silver nanoparticles 4 = 13 µL of each tested bacteria) were low zone of inhibition when compared to leaf extract silver nanoparticles and reported that bark extract silver nanoparticles 2 = 13.86 ± 0.51 mm for *B. subtilis* and leaf extract silver nanoparticles 2 = 11.92 ± 1.09 mm for *E. coli* at various volumes of tested leaf extract silver nanoparticles. However, both leaf extract silver nanoparticles and bark extract silver nanoparticles showed better activity compared to plain leaf and bark extracts. Fascinatingly noticed that *E. coli* with thin cell wall is more susceptible to cell wall damage compared to *B. subtilis* with thick cell wall by leaf extract silver nanoparticles (Prasad &

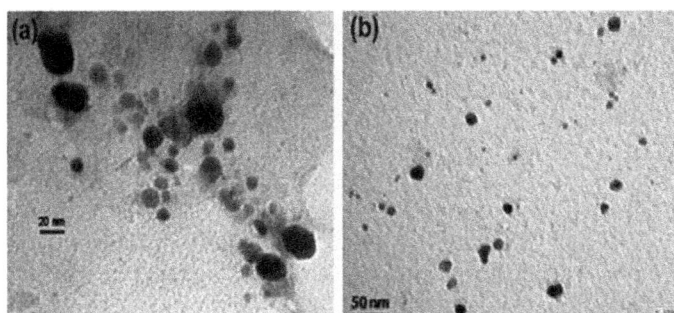

Figure 5. Transmission electron microscopic photographs of (a) leaf extract silver nanoparticles and (b) bark extract silver nanoparticles.

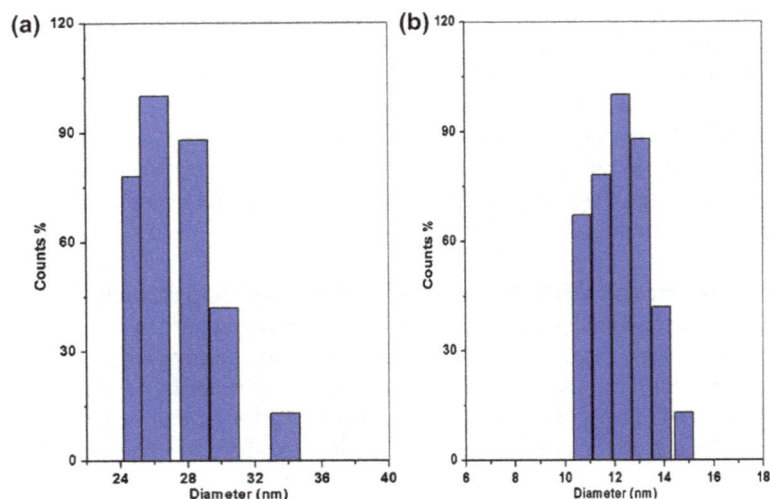

Figure 6. Particle size determination by dynamic light scattering analysis (a) leaf extracts silver nanoparticles and (b) bark extract silver nanoparticles.

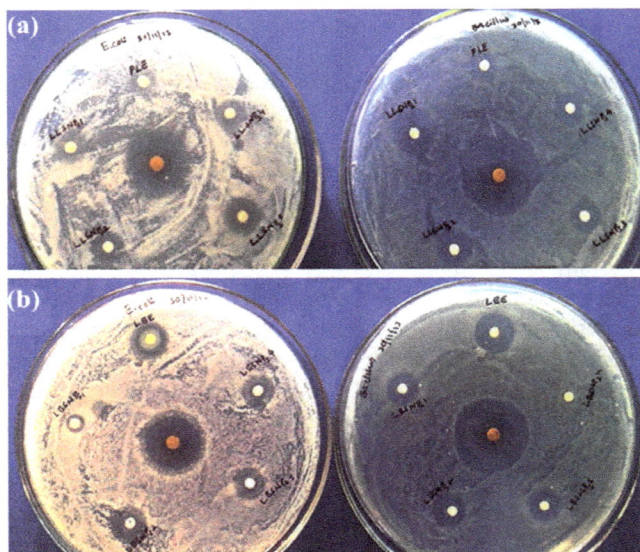

Figure 7. Antimicrobial activity of (a) leaf extract silver nanoparticles and (b) bark extract silver nanoparticles against *E. coli* and *B. subtilis* bacterial strains.

Venkateswarlu, 2014). The intensified antimicrobial effects of silver nanoparticles are analyzed and also specified that once inside the cell, silver nanoparticles would interface with the pathogenic growth indication pathway by regulating polyphenols for cell viability and the nanoparticles were not in direct contact even within the aggregates, indicating stabilization of nanoparticles by a capping agent (Duran, Marcato, Alves, De souza, & Esposito, 2005; Rahman et al., 2014).

4. Conclusion

The present study demonstrates bio-reduction synthesis of nanosized silver nanoparticles using *L. acidissima* plant materials. *L. acidissima* aqueous leaf and bark extracts appear to be environmentally eco-friendly, so that this procedure could be used for rapid production of silver nanoparticles. Silver nanoparticles synthesized by green approach in this study using *L. acidissima* leaf and bark extracts. Aqueous extract of leaf and bark extract could act not only reducing Ag$^+$ but also control the size i.e. average sizes 25 ± 26 nm for leaf extract silver nanoparticles and 12 ± 14 nm for bark extract silver nanoparticles. Further the silver nanoparticles formations conformed by UV–vis and Fourier transforms infrared analysis and size of both leaf extract silver nanoparticles and bark extract silver nanoparticles studied by dynamic light scattering and transmission electron microscopic analysis. In the future, selection of such plants may create a new platform for realizing the potential of herbal medicines in nanoscience for drug delivery and biomedical applications. The antibacterial activity of silver nanoparticles suspension showed enhancement in the activity towards *E. coli* and *B. subtilis*. By using UV–vis spectrometric analysis the rate constant k_{obs} was also calculated in this study. In future we will also study the anticancer and anti-HIV activity of silver nanoparticles colloidal suspension. In addition, this report provides a theoretical and experimental foundation for investigation of the biosynthesis of other nanoparticles.

Funding
Author Dr K.S.V. Krishna Rao thanks to University Grants Commission (UGC), New Delhi, India, for financial support under UGC-RAMAN Postdoctoral Fellowship program (No. F 5-84/ 2014 (IC)).

Author details
E. Chandra Sekhar[1]
E-mail: chandu.ac.in@gmail.com
K.S.V. Krishna Rao[2,3]
E-mail: drksvkrishna@yahoo.com
K. Madhusudana Rao[4]
E-mail: msraochem@gmail.com
S. Pradeep Kumar[5]
E-mail: sakepradeep@gmail.com
[1] Department of Chemistry, Acharya Nagarjuna University, Nagarjuna Nagar, Guntur, Andhra Pradesh, India.
[2] Polymer Biomaterial Design and Synthesis Laboratory, Department of Chemistry, Yogi Vemana University, Kadapa, Andhra Pradesh, India.
[3] Department of Chemical Engineering and Material Science, Wayne State University, Detroit, MI, USA.
[4] Nano Information Materials Laboratory, Department of Polymer Science and Engineering, Pusan National University, Busan, South Korea.
[5] Department of Microbiology, Yogi Vemana University, Kadapa. Andhra Pradesh, India.

References
Absar, A. Q., Eswar, K. K., Omer, S., & Feronia, L. (2010). A path less travelled. *International Journal of Research in Ayurveda and Pharmacy, 1*, 98–106.
Adeleh, G. R., & Hamed, A. (2012). Preparation of colloidal silver nanoparticles by laser ablation; evaluation and study on its developed applications. *Advanced Materials Research, 488–489*, 1409–1413.
Adikaram, N., Yamuna, A., Leslie, A. A., Gunatilaka, & Bandara, B. M. R. (1989). Antifungal activity, acid and sugar content in the wood apple (Limonia acidissima) and their relation to fungal development. *Plant Pathology, 38*, 258–265. http://dx.doi.org/10.1111/ppa.1989.38.issue-2
Ahmad, A., Senapati, S., Khan, M. I., Kumar, R., & Sastry, M. (2003). Extracellular biosynthesis of monodisperse gold nanoparticles by a novel extremophilic actinomycete, *Thermomonospora* sp. *Langmuir, 19*, 3550.
http://dx.doi.org/10.1021/la0267721
Ameenah, G. F. (2006). Medicinal plants: Tradition of yesterday drug of tomorrow. *Molecular Aspects of Medicine, 27*, 1–93.
Anish, B., Templeton A. K., Munshi, A., & Ramesh, R. (2013). Nanoparticle-based drug delivery for therapy of lung cancer: Progress and challenges. *Journal of Nanomaterials, 4*, 11.

Ankamwar, B., Chaudhary, M., & Sastry, M. (2005). Gold nanotriangles biologically synthesized using tamarind leaf extract and potential application in vapor sensing. *Synthesis and Reactivity in Inorganic, Metal-Organic, and Nano-Metal Chemistry, 35*, 19–26. http://dx.doi.org/10.1081/SIM-200047527
Arokiyaraj, S., Valan Arasu, M. V., Vincent, S., Oh, Y.-K., Kim, K. H., Choi, K.-C, & Choi, S. H. (2014). Rapid green synthesis of silver nanoparticles from *Chrysanthemum indicum* L. and its antibacterial and cytotoxic effects: An in vitro study. *International Journal of Nanomedicine, 9*, 379. http://dx.doi.org/10.2147/IJN
Banerjee, S. S., Someshwar, S., Subrata, L., & Chandra, G. (2011). Efficacy of *Limonia acidissima* L. (Rutaceae) leaf extract on larval immatures of Culexquinquefasciatus Say 1823. *Asian Pacific Journal of Tropical Medicine, 4*, 711–716.
Chandran, S. P., & Chaudhary, M. (2006). Synthesis of gold nanotriangles and silver nanoparticles using aloe vera plant extract. *Biotechnology Progress, 22*, 577–583. http://dx.doi.org/10.1021/bp0501423
Christensen, L., Vivekanandhan, S., Misra, M., & Mohanty, A. K. (2011). Biosynthesis of silver nanoparticles using *Murraya koenigii* (curry leaf): An investigation on the effect of broth concentration in reduction mechanism and particle size. *Advanced Materials Letters, 2*, 429. http://dx.doi.org/10.5185/amlett
Dipankar, C., & Murugan, S. (2012). The green synthesis, characterization and evaluation of the biological activities of silver nanoparticles synthesized from Iresine herbstii leaf aqueous extracts. *Colloids and Surfaces B: Biointerfaces, 98*, 112–119.
http://dx.doi.org/10.1016/j.colsurfb.2012.04.006
Duran, N., Marcato, P. D., Alves, O. L., De souza, G., & Esposito, E. (2005). Mechanistic aspects of biosynthesis of silver nanoparticles by several Fusariumoxysporum strains. *Journal of Nanbiotechnology, 3*, 1.
Gandhi, N., Sirisha, D., & Sharma, V. C. (2014). Microwave-mediated green synthesis of silver nanoparticles using *Ficus elastica* leaf extract and application in air pollution controlling studies. *International Journal of Engineering Research and Applications, 4*, 61.
Geeta Vasudevan, D. M., Kedlaya, R., Deepa, S., & Ballal, M. (2001). Activity of *Ocimum sanctum* (the traditional Indian medicinal plant) against the enteric pathogens. *Indian Journal of Medical Sciences, 55*, 434–438.
Geethalakshmi, R., & Sarada, D. V. L. (2010). Synthesis of plant-mediated silver nanoparticles using *Trianthema decandra* extract and evaluation of their anti-microbial activities. *International Journal of Engineering Science Technology, 2*, 970.
Grotewold, E. (Ed.) Erich, G., & Murphy, A. S. (2006). *The science of flavonoids* (pp. 239–267). New York, NY: Springer. http://dx.doi.org/10.1007/978-0-387-28822-2
Kasthuri, J., Veerapandian, S., & Rajendiran, N. (2009). Biological synthesis of silver and gold nanoparticles

using *Apiin* as reducing agent. *Colloids and Surfaces B: Biointerfaces, 68,* 55–60. http://dx.doi.org/10.1016/j.colsurfb.2008.09.021

Kim, J. S., Kuk, E., Yu, K. N., Kim, J. H., Park, S. J., Lee, H. J., Kim, S. H., ... Cho, M. H. (2007). Antimicrobial effects of silver nanoparticles. *Nanomedicine: Nanotechnology, Biology and Medicine., 3,* 95–101. http://dx.doi.org/10.1016/j.nano.2006.12.001

Kim, K. A., Cha, J. R., & Gong, M. S. (2013). Facile preparation of silver nanoparticles and application to silver coating using latent reductant from a silver carbamate complex. *Bulletin of the Korean Chemical Society, 34,* 505–509. http://dx.doi.org/10.5012/bkcs.2013.34.2.505

Kirby, W. M. M., Yoshihara, G. M., Sundsted, K. S., & Warren, J. H. (1957). Clinical usefulness of a single disc method for antibiotic sensitivity testing. *Antibiotics Annualc, 1956-1957,* 892.

Kithsiri, E. M., Jeratne, W. B., Bandara, B. M. R., Leslie, A. A., & Gunatilaka (1992). Chemical constituents of three rutaceae species from Sri Lanka. *Journal of Natural Products, 55,* 1261–1269.

Kotakadi, V. S., Rao, Y. S., Gaddam, S. A., Prasad, T. N. V. K. V, Reddy, A. V., & Gopal, D. V. R. S. (2013). Simple and rapid biosynthesis of stable silver nanoparticles using dried leaves of *Catharanthus roseus.* Linn. G. Donn and its anti microbial activity. *Colloids and Surfaces B: Biointerfaces, 105,* 194–198. http://dx.doi.org/10.1016/j.colsurfb.2013.01.003

Krishnaraj, C., Jagan, E. G., Rajasekar, S., Selvakumar, P., Kalaichelvan, P. T., & Mohan, N. (2010). Synthesis of silver nanoparticles using Acalypha indica leaf extracts and its antibacterial activity against water borne pathogens. *Colloids and Surfaces B: Biointerfaces, 76,* 50–56. http://dx.doi.org/10.1016/j.colsurfb.2009.10.008

Kumar, A., Vemula, P. K., Ajayan, P. M., & John, G. (2008). Silver-nanoparticle-embedded antimicrobial paints based on vegetable oil. *Nature Materials, 7,* 236. http://dx.doi.org/10.1038/nmat2099

Lim, E. K., Jang E., Lee, K., Haam, S., & Huh, Y. M. (2013). Delivery of cancer therapeutics using nanotechnology. *Pharmaceutics, 5,* 294–317. http://dx.doi.org/10.3390/pharmaceutics5020294

Malarkodi, C., Rajeshkumar, S. R., Paulkumar, K., Vanaja, M., Gnanajobitha, G., & Annadurai, G. (2014). Biosynthesis and antimicrobial activity of semiconductor nanoparticles against oral pathogens. *Bioinorganic Chemistry and Applications, 2014,* 1–10. http://dx.doi.org/10.1155/2014/347167

Mallikarjuna, K., John Sushma, N., Narasimha, G., Manoj, L., & Raju, D. P. (2014). Phytochemical fabrication and characterization of silver nanoparticles by using pepper leaf broth. *Arabian Journal of Chemistry, 7,* 1099–1103. http://dx.doi.org/10.1016/j.arabjc.2012.04.001

Manish, D., Arun N., & Ansari, S. H. (2009). PterocarpusmarsupiumRoxb—A comprehensive review. *Pharmacognosy Reviews, 3,* 359–363.

Parrotta, J. A. (2011). Healing plants of peninsular India (pp. 637–638). Oxford: CABI Publishing.

Parthasarathi, G., Asok Kumar, B., & Swapnadip, T. (1989). Acidissimin, a new tyramine derivative from the fruit of *Limonia acidissima. Journal of Natural Products, 52,* 1323–1326.

Philip, D. (2009). Biosynthesis of Au, Ag and Au-Ag nanoparticles using edible mushroom extract.

Spectrochimica Acta Part A: Molecular and Biomolecular Spectroscopy, 73, 374–381. http://dx.doi.org/10.1016/j.saa.2009.02.037

Prasad, Ch., & Venkateswarlu, P. (2014). Soybean seeds extract based green synthesis of silver nanoparticles. *Indian Journal of Advances in Chemical Science 2,* 208–211.

Praveena, V. D., & Kumar, K. V. (2014). Green synthesis of silver nanoparticles from *Achyranthes aspera* plant extract in chitosan matrix and evaluation of their antimicrobial activities. *Indian Journal of Advances in Chemical Science, 2,* 171–177.

Rahman M. M., Hattori N., Nakagawa Y., Lin X., Yagai S., Sakai M., & Yamamoto K. (2014). Japanese. Preparation and characterization of silver nanoparticles on localized surface plasmon-enhanced optical absorption. *Journal of Applied Physics, 53* 11RE01.

Rai M., Yadav A., & Gade A. (2009). Silver nanoparticles as a new generation of antimicrobials. *Biotechnology Advances, 27,* 76–83. http://dx.doi.org/10.1016/j.biotechadv.2008.09.002

Raju, V., Ramar, T., Krishnasamy, M., Palani, G., Kaveri, K., & Soundarapandian, K. (2012). Green biosynthesis of silver nanoparticles from Annonasquamosa leaf extract and it's *in vitro* cytotoxic effect on MCF-7 cells. *Process Biochemistry, 47,* 2405–2410.

RathiSre, P. R., Reka, M., Poovazhagi, R., Arul Kumar, M., & Murugesan, K. (2015). Antibacterial and cytotoxic effect of biologically synthesized silver nanoparticles using aqueous root extract of *Erythrinaindica lam. SpectrochimicaActa Part A: Molecular and Biomolecular Spectroscopy , 135,* 1137–1144.

Raveendran, P., Fu, J., & Wallen, S. L. (2006). A simple and "green" method for the synthesis of Au, Ag, and Au-Ag alloy nanoparticles. *Green Chemistry, 8,* 34. http://dx.doi.org/10.1039/B512540E

Robert, T., Morrison Robert, N., & Boyd, R. K. (1992). *Organic chemistry* (6th ed.). Prentice Hall.

Robinson, M. R., & Zhang, X. (2011). *The world medicines situation 2011 (Traditional medicines: global situation, issues and challenges)* (p. 2011). Geneva: World Health Organization.

Sathishkumar, M., Sneha, K., Won, S. W., Cho, C. W., Kim, S., & Yun, Y. S. (2009). *Cinnamon zeylanicum* bark extract and powder mediated green synthesis of nano-crystalline silver particles and its bactericidal activity. *Colloids and Surfaces B: Biointerfaces, 73,* 332. http://dx.doi.org/10.1016/j.colsurfb.2009.06.005

Shankar, S. S., Rai, A., Ahmad, A., & Sastry, M. J. (2004). Rapid synthesis of Au, Ag, and bimetallic Au core-Ag shell nanoparticles using Neem (*Azadirachta indica*) leaf broth. *Journal of Colloid and Interface Science, 275,* 496–502. http://dx.doi.org/10.1016/j.jcis.2004.03.003

Srinivasan, K. (2011). Biosynthesis of silver nanoparticles using *Citrus sinensis* peel extract and its antibacterial activity. *Spectrochimica Acta Part A: Molecular and Biomolecular Spectroscopy, 79,* 594–598.

Sripairoj, P., Suwanborirux, K., & Tanasupawat, S. (2013). Characterization and antimicrobial activity of Amycolatopsis strains isolated from Thai soils. *Journal of Applied Pharmacology Science, 3,* 011–016.

Sun, Y. (1988). *The role of traditional Chinese medicine in supportive care of cancer patients, in recent results in cancer research* (Vol. 108, pp. 327–334). Berlin: Springer-Verlag. http://dx.doi.org/10.1007/978-3-642-82932-1

Sunitha, K., & Krishna Mohan, G. (2013). Screening of *Limonia acidissima* fruit pulp for immuno modulatory activity. *Research Journal Pharmacy Biology Chemistry Science, 2,* 439–444.

Talalay, P. (2001). The importance of using scientific principles in the development of medicinal agents from plants. *Academic Medicine, 76,* 238–247. http://dx.doi.org/10.1097/00001888-200103000-00010

Valiathan, M. S. (1998). Healing plants. *Current Science, 75,* 1122–1126.

Venkateswara Rao, Ch., Viswanathan, B. (2010).

Monodispersed platinum nanoparticle supported carbon electrodes for hydrogen oxidation and oxygen reduction in proton exchange membrane fuel cells. *The Journal of Physical Chemistry C, 114*, 8661–8667.

Venkateswara Rao, Ch., & Viswanathan, B. (2012). Microemulsion synthesis and electrocatalytic properties of carbon-supported Pd–Co–Au alloy nanoparticles. *Journal of Colloid and Interface Science, 367*, 337–341. http://dx.doi.org/10.1016/j.jcis.2011.10.020

Xavier, T. F., Auxilia, A., & ISelvi, M. S. (2013). Antibacterial and phytochemical screening of Solanumerianthum D. *Journal of Natural Products and Plant Resources, 3*, 131–133.

Yang, H. W., Mu-Yi Hua, Liu H.–Li., Huang C.-Y., & Wei K.-C. (2012). Potential of magnetic nanoparticles for targeted drug delivery. *Nanotechnology, Science and Applications, 5*, 73–86.

Zhao, G. J., & Stevens, Jr., S. E. (1998). Multiple parameters for the comprehensive evaluation of the susceptibility of *Escherichia coli* to the silver ion. *Biometals, 11*, 27. http://dx.doi.org/10.1023/A:1009253223055

6

An efficient method for synthesis of bis(indolyl) methane and di-bis(indolyl)methane derivatives in environmentally benign conditions using TBAHS

Sayed Hossein Siadatifard[1], Masumeh Abdoli-Senejani[1] and Mohammad Ali Bodaghifard[2]*

*Corresponding author: Mohammad Ali Bodaghifard, Faculty of Science, Department of Chemistry, Arak University, Arak, Iran
E-mails: mbodaghi2007@yahoo.com; m-bodaghifard@araku.ac.ir
Reviewing editor: George Weaver, University of Loughborough, UK

Abstract: An efficient procedure for the synthesis of bisindolylmethanes from condensation of indole and aromatic aldehydes or ketones is described. The aromatic electrophilic substitution reactions of indole with aromatic aldehydes and ketones are achieved in the presence of tetrabutylammonium hydrogen sulfate as a mild and efficient solid acid catalyst. This methodology offers several advantages such as good yields, simple procedure, mild and environmentally benign conditions.

Subjects: Environmental Chemistry; Medicinal & Pharmaceutical Chemistry; Organic Chemistry

Keywords: bis(indolyl)methane; condensation; green synthesis; tetrabutylammonium hydrogen sulfate

1. Introduction

The indole ring system is present in many natural products, pharmaceuticals, and agrochemicals. These compounds bearing indole moiety have shown many pharmaceutical properties such as anti-microbial, antibacterial, antiviral, antifungal, antimetastatic, radical scavenging, analgesic, and anti-inflammatory activities (Bell, Carmeli, & Sar, 1994; Benabadji, Wen, Zheng, Dong, & Yuan, 2004; Fahy, Potts, Faulkner, & Smith, 1991; Irie et al., 1999; Kamal et al., 2009; Kobayashi et al., 1994; Kuethe, 2006; Shiri, Zolfigol, Kruger, & Tanbakouchian, 2010; Sivaprasad, Perumal, Prabavathy, & Mathivanan, 2006; Sujatha, Perumal, Muralidharan, & Rajendran, 2009; Valeria & Ernesto, 1998). Numerous inhibitory activities are reported for bisindolylmethanes (BIMs) derivatives in different cancer including bladder cancer, lung cancer cells, colon cancer, prostate cancer, breast tumor cells (Ge, Fares, & Yannai, 1999; Gong, Firestone, & Bjeldanes, 2006; Gong, Sohn, Xue, Firestone, & Bjeldanes, 2006;

ABOUT THE AUTHOR

Mohammad Ali Bodaghifard was born in Hamadan, Iran, in 1978. He received his BSc degree from Bu Ali Sina University, Hamedan, Iran and MSc degree from Shahid Beheshti University, Tehran, Iran, and PhD in organic chemistry in 2010 from Arak University, Arak, Iran. His doctoral thesis was on application of new catalysts in organic transformations and synthesis of heterocycles and spiro-heterocycles. He works as an assistant professor at Arak University from 2012. His current research interest is on green organic synthesis, heterogeneous catalysis, and organic–inorganic hybrid nanostructures.

PUBLIC INTEREST STATEMENT

The bisindolylmethane derivatives exhibit important biological and pharmaceutical properties such as antimicrobial, antibacterial, antiviral and antifungal, antimetastatic, analgesic, and anti-inflammatory activities. Although the reported synthetic protocols are valuable, most of them suffer from one or more drawbacks such as use of expensive and/or toxic catalysts and solvents. It is therefore clear that due to the biological activities and versatile application possibilities of BIMs, there is a continuous quest for more efficient, green, and mild methods for indole derivatives synthesis. In this research, we reported a mild and green procedure for the synthesis of mono- and di-bisindolylmethane derivatives via the reaction of indole and aldehydes or ketones in the presence of TBAHS as an efficient catalyst.

Ichite et al., 2009; Ling & Wan-Ru, 2004; Nachshon-Kedmi, Yannai, & Fares, 2004; Safe, Papineni, & Chintharlapalli, 2008; Sung et al., 2007). The oxidized forms of BIMs are utilized as dyes (Novak, Kramer, Klapper, Daasch, & Murr, 1976) and colorimetric sensors (He et al., 2006; Martínez, Espinosa, Tárraga, & Molina, 2008).

A simple and standard method for the synthesis of bis(indolyl)methanes is the Friedel–Crafts reaction between indoles and carbonyl compounds in the presence of an acid or base. Varieties of catalytic reagents used in the synthesis of BIMs have been reviewed (Shiri et al., 2010). Various Brönsted acids (Ramesh, Banerjee, Pal, & Das, 2003; Zahran, Abdin, & Salama, 2008), Lewis acids (Chen, Yu, & Wang, 1996; Kundu & Maiti, 2008; Qu et al., 2011), heterogeneous acidic catalyst (Firouzabadi, Iranpoor, & Jafari, 2006), ionic liquid (Kalantari, 2015; Veisi, Hemmati, & Veisi, 2009), and some other catalysts have been applied for this synthesis (Haghighi & Nikoofar, 2014; Hojati, Zeinali, & Nematdoust, 2013; Pore, Desai, Thopate, & Wadgaonkar, 2006; Xie, Sun, Jiang, & Le, 2014). Although these protocols are valuable, most of them suffer from one or more drawbacks including long reaction times, low yields of products, harsh reaction conditions, corrosive reagents, and use of expensive and/or toxic catalysts and solvents. Moreover, many Lewis acids are deactivated or sometimes decomposed by nitrogen-containing reactants. It is clear that due to the biological activities and versatile application possibilities of BIMs, there is a continuous quest for more efficient, mild, and clean procedures for synthesis of these compounds. This prompted us to investigate the feasibility of less hazardous, solvent-free synthesis of aryl-3,3'-bis(indolyl)methane derivatives under modified experimental conditions.

Acidic tetrabutylammonium hydrogen sulfate (TBAHS) act as a phase transfer catalyst and performs many organic transformations under mild conditions. It has been used for dehydration and cyclization step in Hantzsch reaction (Tripathi, Tewari, & Dwivedi, 2004). TBAHS has been applicated for one-pot synthesis of pyrano[2,3-c]pyrazoles via domino/Knoevenagel-hetero-Diels–Alder reaction (Parmar, Teraiya, Patel, & Talpada, 2011). It also has been used for synthesis of 3,4-dihydropyrimidin-2(1H)-ones (Shaabani, Bazgir, & Arab-Ameri, 2004), 1,8-dioxo-octahydroxanthenes (Karade, Sathe, & Kaushik, 2007), 5-substituted 1H-tetrazoles (Wang, Liu, & Cheon, 2015), and other heterocyclic compounds (Jończyk & Kuliński, 1993; Parmar, Teraiya, Barad, Sharma, & Gupta, 2013). TBAHS is an inexpensive, safe-handleable and thermally stable substance.

For unique catalyst features of TBAHS and in continuation of our efforts to develop green and mild methodologies for the synthesis of heterocycles (Bodaghifard, Solimannejad, Asadbegi, & Dolatabadifarahani, 2016; Mobinikhaledi, Foroughifar, & Bodaghifard, 2011), herein we wish to report a mild, efficient, and green procedure for the synthesis of biologically interesting mono- and di-bisindolylmethane derivatives via the reaction of indole and aldehydes or ketones in the presence of TBAHS in solvent-free condition or using a little amount of solvent (Scheme 1).

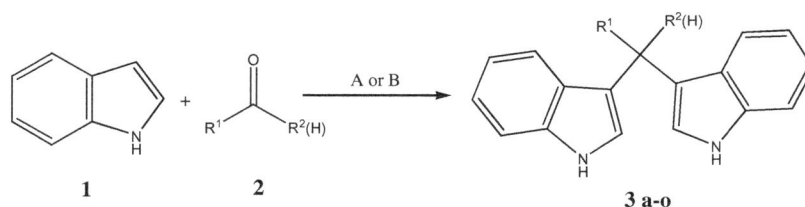

A: Bu₄NHSO₄ (0.1 mmol), 60 °C, EtOAc (1 mL).

B: Bu₄NHSO₄ (0.1 mmol), 60 °C.

Scheme 1. TBAHS efficiently catalyzed synthesis of bisindolylmethanes.

2. Results and discussion

In order to determine the optimum reaction conditions, we examined the influence of the temperature, reaction time, and the amounts of TBAHS upon a model reaction between benzaldehyde (1 mmol) and indole (2 mmol) (Scheme 2). The best result was obtained with 0.1 mmol of TBAHS at 60°C using 1 mL EtOAc or in the absence of any solvent (Table 1, entry 5 and 6). Increasing the amount of catalyst to 0.2 mmol does not affect the product yield (Table 1, entry 10). Moreover, the catalyst is essential and in the absence of the catalyst, poor yield of the corresponding BIM is produced (Table 1, entry 7, 11).

Encouraged by these results, we studied the reaction of various aldehydes and ketones under optimized conditions to better understand the scope and generality of this simple procedure (Scheme 1).

A series of aromatic aldehydes and ketones underwent an electrophilic substitution reaction with indole smoothly, to afford a range of substituted bis(indolyl)methanes in good to excellent yields (Table 2, 3**a–o**). The results showed that the reaction proceeds very efficiently in all cases. This method is equally effective for aldehydes bearing electron withdrawing or donating groups in the aromatic rings. Furthermore, reactions of ketones with indole were satisfactorily performed by the current pathway but they need longer times and afforded lower yield of products than aldehydes due to lower reactivity of ketones. The products 3**i** and 3**m** are novel compounds and have been fully characterized by their elemental analysis, FT-IR and ^{1}H and ^{13}C NMR spectra. The results indicated that the method A need less time that can be related to more and faster interaction of starting materials in the presence of little amount of ethyl acetate. In the absence of ethyl acetate and solid-state manner (method B), the interaction slowed down and in longer time produced the desired products.

2.1. Experimental

All chemicals and solvents were obtained from commercial sources and used without further purification. All known organic products were identified by comparison of their physical and spectral data with those of authentic samples. Thin layer chromatography (TLC) was performed on UV-active pre-caoted plates of silica gel (TLC Silica gel60 F254).The FTIR spectrum was recorded on a Shimadzu IR-470 spectrometer using KBr disks. The ^{1}H and ^{13}C NMR spectra were recorded on a Brucker Avance spectrometer operating at 400, 300 and 100, 75 MHz, respectively, in DMSO-d_6 or CDCl$_3$ with TMS as an internal standard. Coupling constants, J, were reported in Hertz units (Hz). Spin multiplicities are shown as s (singlet), br s (broad singlet), d (doublet), dd (doublet of doublet), t (triplet), and m (multiplet). Elemental analyses were performed by Vario EL equipment at Arak University.

2.1.1. General procedure for synthesis of bis(indolyl)methanes catalyzed by Bu$_{4NHSO4}$

Method A: To a mixture of indole (2 mmol), aldehyde, or ketone (1 mmol) ethyl acetate (1 mL) as a solvent, Bu$_4$NHSO$_4$ (0.1 mmol) was added and the mixture stirred magnetically at 60°C. After complete conversion, as indicated by TLC (hexane/ethyl acetate 4:1), the reaction mixture was cooled to room temperature and crushed ice was added. The precipitate was filtered and dried under vacuum. The product was purified by recrystallization hexane–ethyl acetate (Table 2).

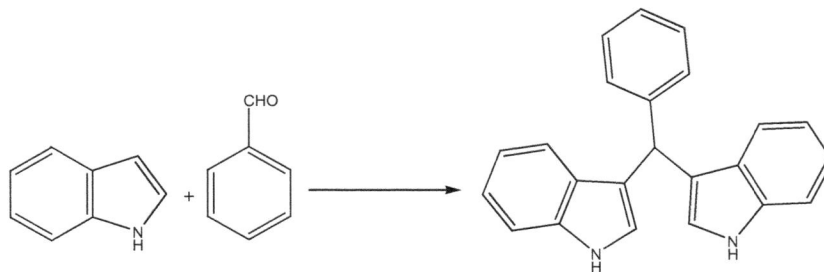

Scheme 2. Optimization of reaction condition.

Table 1. Optimization of reaction condition for synthesis of 3,3'-bis-indolyl phenylmethane[a]

Entry	Catalyst (mmol)	Solvent	Temperature (°C)	Time (h)	Yield (%)[b]
1	Bu$_4$NHSO$_4$ (0.1)	EtOH	r.t.	24	80
2	Bu$_4$NHSO$_4$ (0.1)	EtOH/H$_2$O (3:1)	r.t	24	0
3	Bu$_4$NHSO$_4$ (0.1)	CH$_3$CN	r.t	24	75
4	Bu$_4$NHSO$_4$ (0.1)	EtOAc	r.t	11	90
5	Bu$_4$NHSO$_4$ (0.1)	EtOAc	60	1	95
6	Bu$_4$NHSO$_4$ (0.1)	–	60	4	91
7	–	–	60	12	36
8	Bu$_4$NHSO$_4$ (0.1)	–	r.t	12	87
9	Bu$_4$NHSO$_4$ (0.15)	–	r.t	12	85
10	Bu$_4$NHSO$_4$ (0.2)	–	r.t	12	86
11	–	–	r.t	22	10

[a]Benzaldehyde (1 mmol), indole (2 mmol).

[b]Isolated yields.

Table 2. Synthesis of bisindolylmethanes catalyzed by TBAHS

Entry	Aldehyde or ketone	Product	A[a]		B[b]		Reference
			Time (h)	Yield (%)[c]	Time (h)	Yield (%)[c]	
1	C$_6$H$_5$CHO	3a	1	95	4	91	Hojati et al. (2013)
2	4-Me-C$_6$H$_4$CHO	3b	1.5	93	5	95	Hojati et al. (2013)
3	4-Cl-C$_6$H$_4$CHO	3c	1	94	3	95	Hojati et al. (2013)
4	4-OMe-C$_6$H$_4$CHO	3d	1	90	4	88	Hojati et al. (2013)
5	3-NO$_2$-C$_6$H$_4$CHO	3e	1	91	4	92	Hojati et al. (2013)
6	4-NO$_2$-C$_6$H$_4$CHO	3f	1	90	4	91	Hojati et al. (2013)
7	2,4-Cl$_2$-C$_6$H$_3$CHO	3g	1	93	3	94	Veisi et al. (2009)
8	2-Cl-C$_6$H$_4$CHO	3h	1.2	89	5	91	Hojati et al. (2013)
9	Isophthaldehyde	3i	1.5	88	5	91	–[d]
10	Terphthaldehyde	3j	1.5	90	5	90	Veisi et al. (2009)
11	Acetone	3k	12	45	24	40	Veisi et al. (2009)
12	Cyclohexanone	3l	10	90	24	93	Hojati et al. (2013)
13	4-Me-cyclohex-anone	3m	10	87	24	92	–[d]
14	Acetophenone	3n	12	60	24	50	Veisi et al. (2009)
15	Isatin	3o	2	93	6	96	Haghighi and Nikoofar (2014)

[a]Method A: Bu$_4$NHSO$_4$ (0.1 mmol), 60°C, EtOAc (1 mL).

[b]Method B: Bu$_4$NHSO$_4$ (0.1 mmol), 60°C.

[c]Isolated yields.

[d]Novel product.

Method B: To a mixture of indole (2 mmol) and aldehyde or ketone (1 mmol), Bu$_4$NHSO$_4$ (0.1 mmol) was added and the mixture stirred magnetically at 60°C. After complete conversion, as indicated by TLC (hexane/ethyl acetate 4:1), the reaction mixture was cooled to room temperature and crushed ice was added. The precipitate was filtered and dried under vacuum. The product was purified by recrystallization from hexane–ethyl acetate (Table 2).

2.1.2. General procedure for synthesis of di-bis(indolyl)methanes

To a mixture of indole (4.5 mmol) and dialdehyde (1 mmol) Bu_4NHSO_4 (0.1 mmol) was added and the mixture stirred magnetically at 60°C. After complete conversion, as indicated by TLC (hexane/acetone 4:1), the reaction mixture was quenched by adding ice water (10 ml). The precipitate was filtered, evaporated, and the corresponding di-bis(indolyl)methanes were obtained in excellent yields and then recrystallized from hexane–ethyl acetate to afford pure products (Table 2, entry 9–10).

3. Conclusion

In summary, we have developed a novel, efficient, and environmentally benign method for the synthesis of pharmaceutically important bis(indolyl)methanes and di-bis(indolyl)methanes using tetrabutylammonium hydrogen sulfate in excellent yield. This new protocol has advantages over previously reported procedures such as cleaner reaction profiles, use of inexpensive catalyst, simple experimental and work-up procedure, high conversions, and high yield and chemoselectivity, hence believed to be superior over many existing synthetic methods. Also two novel BIM derivatives have been synthesized and characterized by spectroscopic data.

Funding

This work is supported by Islamic Azad University, Arak Branch (IR).

Author details

Sayed Hossein Siadatifard[1]
E-mail: Siyadatifard@yahoo.com
Masumeh Abdoli-Senejani[1]
E-mail: mabdoli@iau-arak.ac.ir
Mohammad Ali Bodaghifard[2]
E-mail: mbodaghi2007@yahoo.com
[1] Department of Chemistry, Islamic Azad University-Arak Branch, Arak, Iran.
[2] Faculty of Science, Department of Chemistry, Arak University, Arak, Iran.

References

Bell, R., Carmeli, S., & Sar, N. (1994). Vibrindole A, a metabolite of the marine bacterium, *Vibrio parahaemolyticus*, isolated from the toxic mucus of the boxfish *Ostracion cubicus*. *Journal of Natural Products, 57*, 1587–1590. http://dx.doi.org/10.1021/np50113a022

Benabadji, S. H., Wen, R., Zheng, J., Dong, X., & Yuan, S. (2004). Anticarcinogenic and antioxidant activity of diindolylmethane derivatives. *Acta Pharmacologica Sinica, 25*, 666–671.

Bodaghifard, M. A., Solimannejad, M., Asadbegi, S., & Dolatabadifarahani, S. (2016). Mild and green synthesis of tetrahydrobenzopyran, pyranopyrimidinone and polyhydroquinoline derivatives and DFT study on product structures. *Research on Chemical Intermediates, 42*, 1165–1179. http://dx.doi.org/10.1007/s11164-015-2079-1

Chen, D., Yu, L., & Wang, P. G. (1996). Lewis acid-catalyzed reactions in protic media. Lanthanide-catalyzed reactions of indoles with aldehydes or ketones. *Tetrahedron Letters, 37*, 4467–4470. http://dx.doi.org/10.1016/0040-4039(96)00958-6

Fahy, E., Potts, B. C. M., Faulkner, D. J., & Smith, K. (1991). 6-bromotryptamine derivatives from the gulf of California tunicate didemnum candidum. *Journal of Natural Products, 54*, 564–569. http://dx.doi.org/10.1021/np50074a032

Firouzabadi, H., Iranpoor, N., & Jafari, A. A. J. (2006). Aluminumdodecatungstophosphate (AlPW12O40), a versatile and a highly water tolerant green Lewis acid catalyzes efficient preparation of indole derivatives. *Journal of Molecular Catalysis A: Chemical, 244*, 168–172. http://dx.doi.org/10.1016/j.molcata.2005.09.005

Ge, X., Fares, F. A., & Yannai, S. (1999). Induction of apoptosis in MCF-7 cells by indole-3-carbinol is independent of p53 and bax. *Anticancer Research, 19*, 3199–3203.

Gong, Y., Firestone, G. L., & Bjeldanes, L. F. (2006). 3,3'-diindolylmethane is a novel topoisomerase II catalytic inhibitor that induces s-phase retardation and mitotic delay in human hepatoma HepG2 cells. *Molecular Pharmacology, 69*, 1320–1327. http://dx.doi.org/10.1124/mol.105.018978

Gong, Y., Sohn, H., Xue, L., Firestone, G. L., & Bjeldanes, L. F. (2006). 3,3'-diindolylmethane is a novel mitochondrial H+-ATP synthase inhibitor that can induce p21Cip1/Waf1 expression by induction of oxidative stress in human breast cancer cells. *Cancer Research, 66*, 4880–4887. http://dx.doi.org/10.1158/0008-5472.CAN-05-4162

Haghighi, M., & Nikoofar, K. J. (2014). Nano TiO2/SiO2: An efficient and reusable catalyst for the synthesis of oxindole derivatives. *Journal of Saudi Chemical Society*, Retrieved from: https://scholar.google.com/citations?view_op=view_citation&hl=en&user=llpcGOYAAAAJ&citation_for_view=llpcGOYAAAAJ:ufrVoPGSRksC doi:10.1016/j.jscs.2014.09.002

He, X., Hu, S., Liu, K., Guo, Y., Xu, J., & Shao, S. (2006). Oxidized bis(indolyl)methane: A simple and efficient chromogenic-sensing molecule based on the proton transfer signaling mode. *Organic Letters, 8*, 333–336. http://dx.doi.org/10.1021/ol052770r

Hojati, S. F., Zeinali, T., & Nematdoust, Z. (2013). A novel method for synthesis of bis(indolyl)methanes using 1,3-dibromo-5,5-dimethylhydantoin as a highly efficient catalyst under solvent-free conditions. *Bulletin of the Korean Chemical Society, 34*, 117–120. http://dx.doi.org/10.5012/bkcs.2013.34.1.117

Ichite, N., Chougule, M. B., Jackson, T., Fulzele, S. V., Safe, S., & Singh, M. (2009). Enhancement of docetaxel anticancer activity by a novel diindolylmethane compound in human non-small cell lung cancer. *Clinical Cancer Research, 15*, 543–552. http://dx.doi.org/10.1158/1078-0432.CCR-08-1558

Irie, T., Kubushirs, K., Suzuki, K., Tsukazaki, K., Umezawa, K., & Nozawa, S. (1999). Inhibition of attachment and chemotactic invasion of uterine endometrial cancer cells by a new vinca alkaloid, conophylline. *Anticancer Research, 31*, 3061–3066.

Jończyk, A., & Kuliński, T. (1993). A simple synthesis of 2-phenylethynyl- and 2-phenylthioethynyl-2-substituted phenylacetonitriles under phase-transfer catalytic (PTC) conditions. *Synthetic Communications, 23*, 1801–1811. http://dx.doi.org/10.1080/00397919308011280

Kalantari, M. (2015). Synthesis of 1,8-dioxo-octahydroxanthenes and bis(indolyl)methanes catalyzed by [Et₃NH][H₂PO₄] as a cheap and mild acidic ionic liquid. *Arabian Journal of Chemistry, 5*, 319–323. Retrieved from: http://www.sciencedirect.com/science/article/pii/S1878535210001802

Kamal, A., Khan, M. N. A., Reddy, K. S., Srikanth, Y. V. V., Ahmed, S. K., Kumar, K. P., & Murthy, U. S. N. (2009). An efficient synthesis of bis(indolyl)methanes and evaluation of their antimicrobial activities. *Journal of Enzyme Inhibition and Medicinal Chemistry, 24*, 559–565. http://dx.doi.org/10.1080/14756360802292974

Karade, H. N., Sathe, M., & Kaushik, M. P. (2007). An efficient synthesis of 1, 8-dioxo-octahydroxanthenes using tetrabutylammonium hydrogen sulfate. *Arkivoc, xiii*, 252–258.

Kobayashi, M., Aoki, S., Gato, K., Matsunami, K., Kurosu, M., & Kitagawa, I. (1994). Trisindoline, a new antibiotic indole trimer, produced by a bacterium of Vibrio sp. separated from the marine sponge *Hyrtios altum*. *Chemical & Pharmaceutical Bulletin, 42*, 2449–2451. http://dx.doi.org/10.1248/cpb.42.2449

Kuethe, J. T. (2006). A general approach to indoles: Practical applications for the synthesis of highly functionalized pharmacophores. *CHIMIA International Journal for Chemistry, 60*, 543–553. http://dx.doi.org/10.2533/chimia.2006.543

Kundu, P., & Maiti, G. (2008). A mild and versatile synthesis of bis (indolyl)-methanes and tris (indolyl) alkanes catalyzed by antimony trichloride. *Indian Journal of Chemistry, 47B*, 1402–1406.

Ling, J., & Wan-Ru, C. (2004). U.S. Patent WO2004018475 A2.

Martínez, R., Espinosa, A., Tárraga, A., & Molina, P. (2008). Bis(indolyl)methane derivatives as highly selective colourimetric and ratiometric fluorescent molecular chemosensors for Cu2+ cations. *Tetrahedron, 64*, 2184–2191. http://dx.doi.org/10.1016/j.tet.2007.12.025

Mobinikhaledi, A., Foroughifar, N., & Bodaghifard, M. A. (2011). Simple and efficient method for three-component synthesis of spirooxindoles in aqueous and solvent-free media. *Synthetic Communications, 41*, 441–450. http://dx.doi.org/10.1080/00397911003587507

Nachshon-Kedmi, M., Yannai, S., & Fares, F. A. (2004). Induction of apoptosis in human prostate cancer cell line, PC3, by 3,3′-diindolylmethane through the mitochondrial pathway. *British Journal of Cancer, 91*, 1358–1363. http://dx.doi.org/10.1038/sj.bjc.6602145

Novak, T. J., Kramer, D. N., Klapper, H., Daasch, L. W., & Murr, B. L. (1976). Formation of dyes derived from diindolylpyridylmethanes. *The Journal of Organic Chemistry, 41*, 870–875. http://dx.doi.org/10.1021/jo00867a025

Parmar, N. J., Teraiya, S. B., Barad, H. A., Sharma, D., & Gupta, V. K. (2013). Efficient one-pot synthesis of precursors of some novel aminochromene annulated heterocycles via domino knoevenagel–hetero-diels–alder reaction. *Synthetic Communications, 43*, 1577–1586. http://dx.doi.org/10.1080/00397911.2011.652755

Parmar, N. J., Teraiya, S. B., Patel, R. A., & Talpada, N. P. (2011). Tetrabutylammonium hydrogen sulfate mediated domino reaction: Synthesis of novel benzopyran-annulated pyrano[2,3-c]pyrazoles. *Tetrahedron Letters, 52*, 2853–2856. http://dx.doi.org/10.1016/j.tetlet.2011.03.108

Pore, D. M., Desai U. V., Thopate, T. S., & Wadgaonkar, P. P. (2006). A mild, expedient, solventless synthesis of bis (indolyl) alkanes using silica sulfuric acid as a reusable catalyst. *Arkivoc, xii*, 75–80.

Qu, H.-E., Xiao, C., Wang, N., Yu, K.-H., Hu, Q.-S., & Liu, L.-X. (2011). RuCl₃·3H₂O catalyzed reactions: Facile synthesis of bis (indolyl) methanes under mild conditions. *Molecules, 16*, 3855–3868. http://dx.doi.org/10.3390/molecules16053855

Ramesh, C., Banerjee, J., Pal, R., & Das, B. (2003). Silica supported sodium hydrogen sulfate and amberlyst-15: Two efficient heterogeneous catalysts for facile synthesis of bis-and tris (1H-indol-3-yl) methanes from Indoles and carbonyl compounds [1]. *Advanced Synthesis & Catalysis, 345*, 557–559. http://dx.doi.org/10.1002/adsc.200303022

Safe, S., Papineni, S., & Chintharlapalli, S. (2008). Cancer chemotherapy with indole-3-carbinol, bis(3′-indolyl) methane and synthetic analogs. *Cancer Letters, 269*, 326–338. http://dx.doi.org/10.1016/j.canlet.2008.04.021

Shaabani, A., Bazgir, A., & Arab-Ameri, S. (2004). Tetrabutylammonium hydrogen sulfate: An efficient catalyst for the synthesis of 3,4-dihydropyrimidin-2(1H)-ones under solvent-free conditions. *Phosphorus, Sulfur, and Silicon and the Related Elements, 179*, 2169–2175. http://dx.doi.org/10.1080/10426500490474815

Shiri, M., Zolfigol, M. A., Kruger, H. G., & Tanbakouchian, Z. (2010). Bis- and trisindolylmethanes (BIMs and TIMs). *Chemical Reviews, 110*, 2250–2293. http://dx.doi.org/10.1021/cr900195a

Sivaprasad, G., Perumal, P. T., Prabavathy, V. R., & Mathivanan, N. (2006). Synthesis and anti-microbial activity of pyrazolylbisindoles—promising anti-fungal compounds. *Bioorganic & Medicinal Chemistry Letters, 16*, 6302–6305. http://dx.doi.org/10.1016/j.bmcl.2006.09.019

Sujatha, K., Perumal, P. T., Muralidharan, D., & Rajendran, M. (2009). Synthesis, analgesic and anti-inflammatory activities of bis (indolyl) methanes. *Indian Journal of Chemistry, 48B*, 267–272.

Sung, D. C., Yoon, K., Chintharlapalli, S., Abdelrahim, M., Lei, P., Hamilton, S., ... Safe, S. (2007). Nur77 Agonists induce proapoptotic genes and responses in colon cancer cells through nuclear receptor–dependent and nuclear receptor–independent pathways. *Cancer Research, 67*, 674–683.

Tripathi, R. P., Tewari, N., & Dwivedi, N. (2004). Tetrabutylammonium hydrogen sulfate catalyzed eco-friendly and efficient synthesis of glycosyl 1,4-dihydropyridines. *Tetrahedron Letters, 45*, 9011–9014.

Valeria, L., & Ernesto, M. (1998). EP0887348 A1.

Veisi, H., Hemmati, S., & Veisi, H. (2009). Highly efficient method for synthesis of bis(indolyl)methanes catalyzed by FeCl. *Journal of the Chinese Chemical Society, 56*, 240–245. http://dx.doi.org/10.1002/jccs.v56.2

Wang, Z., Liu, Z., & Cheon, S. H. (2015). Facile synthesis of 5-substituted 1H-tetrazoles catalyzed by tetrabutylammonium hydrogen sulfate in water. *Bulletin of the Korean Chemical Society, 36*, 198–202. http://dx.doi.org/10.1002/bkcs.2015.36.issue-1

Xie, Z.-B., Sun, D-Zh, Jiang, G.-F., & Le, Zh-G. (2014). Facile synthesis of bis(indolyl)methanes catalyzed by α-chymotrypsin. *Molecules, 19*, 19665–19677. http://dx.doi.org/10.3390/molecules191219665

Zahran, M., Abdin, Y., & Salama, H. (2008). Eco-friendly and efficient synthesis of bis (indolyl) methanes under microwave irradiation. *Arkivok, 11*, 256–265.

7

Synthesis, characterization and antimicrobial properties of cobalt(II) and cobalt(III) complexes derived from 1,10-phenanthroline with nitrate and azide co-ligands

Djuikom Sado Yanick Gaëlle[1], Divine Mbom Yufanyi[2], Rajamony Jagan[3] and Moise Ondoh Agwara[1]*

*Corresponding author: Moise Ondoh Agwara, Department of Inorganic Chemistry, University of Yaounde I, P. O. Box 812, Yaounde, Cameroon
E-mail: agwara29@yahoo.com
Reviewing editor: Darren Bradshaw, University of Southampton, UK

Abstract: Two complexes, a cobalt(II) complex $[Co(phen)_3(NO_3)_2]\cdot 2H_2O$ (1) and a novel Co(III) complex with mixed ligands [diazido-bis(1,10-phenanthroline-κ^2 N,N') cobalt(III)]nitrate $[Co(phen)_2(N_3)_2]NO_3$ (2), have been synthesized and were characterized by physico-chemical and spectroscopic methods. The cobalt(III) complex (**2**) crystallizes in the orthorhombic crystal system with space group *Iba*2 with four formula units. The central Co(III) atom is six coordinate with four N atoms of two 1,10-phenanthroline (1,10-phen) molecules and two terminal N atoms from two azide anions giving a distorted octahedral geometry with a CoN_6 chromophore. The molecular structure of the compound is consolidated by face-to-face π–π stacking between the aromatic rings of 1,10-phen ligand. The ligands, metal salt and the complexes were also evaluated for their antimicrobial activities *in vitro* against eight pathogens (four bacteria and four fungi species).

Subjects: Applied & Industrial Chemistry; Inorganic Chemistry; Materials Chemistry

Keywords: antimicrobial properties; azide; cobalt(III); 1,10-phenanthroline; X-ray crystal structure

ABOUT THE AUTHOR

Moise Ondoh Agwara, PhD, is Associate Professor of Chemistry at the Department of Inorganic Chemistry, University of Yaounde I in Cameroon. He obtained a PhD in chemistry from the University of Ibadan, Nigeria in 1986. Research activities within his research group are focused on the development of the chemistry of transition metal complexes with heterocyclic N-, O- and N,O-donor ligands and some co-ligands. Such interest derives from the fascinating structural chemistry of the complexes obtained, their interesting physico-chemical properties and their diverse applications such as antimicrobials, in photoluminescence and as precursors for the development of nanostructured functional materials.

PUBLIC INTEREST STATEMENT

The frequent, inappropriate and abusive use of drugs to fight infections caused by micro-organisms has led to resistant pathogens which are resistant to many drugs. Efforts made to develop drugs to fight against these resistant pathogens include protection of the efficacy and appropriate use of existing drugs as well as research and development of new drugs. Metal complexes have found application as novel diagnostic and antimicrobial agents. This study presents the synthesis, characterization and structure elucidation of cobalt(II) and cobalt(III) complexes with 1,10-phen, nitrate and azide co-ligands. The complexes were screened for their activities against four pathogenic bacteria and four fungi species. The results indicate that the complexes have activities higher than that of the metal salt but lower than that of 1,10-phen. The most sensitive strains are *E. coli, P. aeruginosa, S. typhi, C. albicans ATCC P37037* and *C. neoformans*. The positive results suggest that the complexes have a broad spectrum of activity.

1. Introduction

The design and synthesis of new metal complexes based on transition metal ions and diimine ligands, such as 1,10-phenanthroline (1,10-phen) and 2,2′-bipyridine, is an interesting field for the development of new functional materials with intriguing structures and potential applications (Amani, Safari, Khavasi, & Mirzaei, 2007). The use of mixed functional ligands in this process can enable the modification of the physical and chemical properties of these metal complexes. Mixed ligand complexes with 1,10-phen is also interesting due to their potential role as models for biological systems such as binding of small molecules to DNA (Jennifer & Muthiah, 2014; Yesilel, Olmez, Yılan, Pasaoglu, & Buyukgungor, 2006). Since the aromatic rings have an extended π-system and can give the ligands various non-covalent π-interactions which mimic various biological processes, the study of these complexes has gained importance (Jennifer & Muthiah, 2014; Pook, Hentrich, & Gjikaj, 2015). The geometries of the complexes are greatly influenced by factors such as different coordination abilities of the ligands and the counterion. These nitrogen-containing heterocycles are metal-coordinating, electron-deficient aromatic systems which can undergo π–π stacking interactions as π-acceptors (Pook et al., 2015).

The chemistry of transition metal complexes has received much attention in recent years on account of their applications, amongst others, in biological systems as antimicrobial agents (McCann et al., 2000; Mishra, Kaushik, Verma, & Gupta, 2008; Shaabani, Khandar, Dusek, Pojarova, & Mahmoudi, 2013; Yenikaya et al., 2009), DNA studies (Arounaguiri, Easwaramoorthy, Ashokkumar, Dattagupta, & Maiya, 2000a, 2000b; Gopinathan, Komathi, & Arumugham, 2014; Kou et al., 2009; Kumar et al., 2008), larvicidal activity (Gopinathan & Arumugham, 2015), and cytotoxicity studies (Deegan, McCann, Devereux, Coyle, & Egan, 2007; Pivetta et al., 2012; Silva et al., 2011). The metal complexes of diimine ligands such as 1,10-phen and bipyridine have received much research interest because of their versatile roles as building blocks for the synthesis of metallo-dendrimers as well as their bioinorganic and biomedicinal applications (Ahmed & Khaled, 2015; Gopinathan & Arumugham, 2015; Molphy, Slator, Chatgilialoglu, & Kellett, 2015; Rajarajeswari, Ganeshpandian, Palaniandavar, Riyasdeen, & Akbarsha, 2014; Wang et al., 2014).

1,10-phen is a heterocyclic bidentate chelating ligand that bonds to metal atoms (ions) using the lone pairs of electrons on the nitrogen atoms resulting in a five-membered ring structure (Bencini & Lippolis, 2010). In this ligand, the σ-donation is complemented by the π-acceptor ability giving the complex formed greater stability (Janiak, 2000; Pook et al., 2015).

Cobalt is a component of vitamin B_{12} complex that is useful in the prevention of anaemia and the production of erythrocytes. Interest in cobalt complexes has also increased due to their therapeutic and biological applications (Mishra et al., 2008).

The azide anion is a versatile ligand that has been intensively studied because of its ability to coordinate to transition metals with different coordination modes generating a wide variety of fascinating structures (discrete molecules to 3D arrays) and their promising applications in functional materials (Chen, Jiang, Yan, Liang, & Batten, 2009; Lazari et al., 2009). The coordination modes of the azide anion range from monodentate to bridging bi-, tri- and tetra-dentate (Adhikary & Koner, 2010; Goher & Mautner, 1995; Lazari et al., 2009; Ye, Cheng, Li, & Hu, 2004). The most common bridging coordination modes of the azide ligand are μ–1,1 (end-on, EO), μ–1,3 (end-to-end, EE), μ–1,1,3 and others (Batten & Murray, 2003). The azide ion easily coordinates to transition metal ions and due to the π-donor properties of this ligand, its occupied π-orbitals can also overlap with the d-orbitals of the corresponding metal ion.

The nitrate group can function as a bidentate ligand, bridging group, monodentate ligand, and ionic species in various inorganic systems (Biagetti & Haendler, 1966). The functional characteristics probably depend on the nature and number of molecules of other ligands present.

In order to fight against antimicrobial resistance, our research team has recently focused on the synthesis and antimicrobial screening of some transition metal complexes of the ligands hexamethyl-enetetramine (Tabong et al., 2016), pyridine-2-carboxylic acid (Amah, Ondoh, Yufanyi, & Gaelle, 2015), 2-aminopyridine (Yuoh et al., 2015), mixture of ligands 1,10-phen and 2,2'-bipyridine (Ndosiri et al., 2013; Sado, Agwara, Yufanyi, Nenwa, & Jagan, 2016). The upsurge of resistant pathogens impedes the effective prevention and treatment of an ever-increasing variety of infections caused by bacteria, parasites, viruses and fungi (WHO, 2014). This situation poses a public health and economic burden to countries worldwide (Hawkey & Jones, 2009; Tanwar, Das, Fatima, & Hameed, 2014; WHO, 2014). Several efforts have been made to develop antimicrobial agents to fight against these resistant pathogens amongst which are the protection of the efficacy and appropriate use of existing drugs as well as research and development of new antimicrobial agents that are not affected by the currently known, predicted, or unknown mechanisms of resistance (Spellberg et al., 2008; Weinstein & Fridkin, 2003). Metal complexes of biologically active ligands are a target for the development of new active agents.

In view of the varied applications of cobalt mixed ligand complexes and exploring the good biological properties of cobalt and 1,10-bipy as well as the structure-directing properties of N_3^-, we report herein the synthesis and structure elucidation of cobalt(II) and cobalt(III) complexes of 1,10-phen, N_3^- and NO_3^-. The effects of the co-ligands on the biological activities of the complexes towards some resistant pathogens, evaluated using *in vitro* assays, are also presented.

2. Experimental

2.1. Materials and methods
$Co(NO_3)_2 \cdot 6H_2O$, sodium azide, 1,10-phen and anhydrous methanol 99.8% were obtained from Sigma–Aldrich. All chemicals (reagent grade) and solvents were used as received.

2.2. Synthesis of the complexes

2.2.1. Synthesis of [Co(phen)$_3$(NO$_3$)$_2$]·2H$_2$O (1)
A 25 mL methanol solution of 1,10-phen (0.60 g, 3.0 mmol) was added drop wise to a 25 mL methanol solution of $Co(NO_3)_2 \cdot 6H_2O$ (0.29 g, 1.0 mmol). The mixture was refluxed at 90°C for 4 h. The pink precipitate obtained was filtered, washed with methanol and dried *in vacuo*. Dark brown block-shaped crystals suitable for single crystal X-ray diffraction obtained from the filtrate within two weeks were washed with acetone and dried *in vacuo*.

2.2.2. Synthesis of [Co(phen)$_2$(N$_3$)$_2$]NO$_3$ (2)
A 25 mL of methanol solution of 1,10-phen (0.39 g, 2.0 mmol) was added drop wise to a 25 mL methanol solution of $Co(NO_3)_2 \cdot 6H_2O$ (0.29 g, 1.0 mmol). The mixture was refluxed at 90°C for an hour. Sodium azide (0.13 g, 2 mmol) dissolved in 2 mL of water and 8 mL of methanol was added drop wise to the reaction mixture and it was further refluxed for 4 h. The yellow precipitate obtained was filtered, washed with ethanol and dried *in vacuo*. Dark red block-shaped crystals suitable for single crystal X-ray diffraction obtained from the filtrate within two months were washed with acetone and then dried *in vacuo*.

2.3. Characterization
The melting point/decomposition temperatures were recorded using a STUART SCIENTIFIC melting point apparatus. The conductivity of the complex was measured in distilled water using a HANNA multimeter type H19811–5, pH/°C/EC/TDS meter at room temperature. The infrared spectrum was recorded using a Bruker ALPHA-P spectrophotometer directly on a small sample of the complex in the range 400–4,000 cm^{-1}, while the UV–vis spectrum of an aqueous solution of the complex was recorded using a Bruker HACH DR 3,900 UV–vis spectrophotometer at room temperature. Thermogravimetric measurements were obtained using a Pyris 6 PerkinElmer TGA 4,000 thermal analyser. TGA analysis was conducted between 35 and 930°C under nitrogen atmosphere at a flow rate of 20 mL/min and a temperature ramp of 10°C/min.

2.4. Single crystal X-ray structure determination of [Co(phen)$_2$(N$_3$)$_2$]NO$_3$ (2)

Intensity data for the compound was collected using a Bruker AXS Kappa APEX II single crystal CCD Diffractometer, equipped with graphite-monochromated MoKα radiation ($\lambda = 0.71073$ Å) at room temperature. The selected crystal for the diffraction experiment had a dimension of $0.25 \times 0.25 \times 0.20$ mm^3. Accurate unit cell parameters were determined from the reflections of 36 frames measured in three different crystallographic zones by the method of difference vectors. The data collection, data reduction and absorption correction were performed by APEX2, SAINT-plus and SADABS program (Bruker AXS, 2004). The structure was solved by direct methods procedure using SHELXS-97 program (Sheldrick, 2008) and the non-hydrogen atoms were subjected to anisotropic refinement by full-matrix least squares on F^2 using SHELXL-97 program (Sheldrick, 2008). The positions of all the hydrogen atoms were identified from difference electron density map and were fixed accordingly. All the aromatic hydrogen atoms were constrained to ride on the corresponding non-hydrogen atoms with a distance of C-H = 0.93 Å and U$_{iso}$(H) = 1.2U$_{eq}$(C). Whereas the hydrogen atoms associated with all the N atoms were restrained to a distance of N-H = 0.88(2) Å.

2.5. Antimicrobial tests

The antimicrobial tests were carried out in the laboratory of Phytobiochemical and Medicinal Plant Study, University of Yaounde I. The tests were done on eight pathogenic micro-organisms, four bacterial strains, *Staphylococcus aureus* CIP 7625 (Gram-positive) and *Pseudomonas aeruginosa* CIP 76110, *Salmonella typhi* and *Escherichia coli* ATCC 259224 (Gram-negative) and four yeasts, *Candida albicans* ATCC P37039, *C. albicans* 194B, *Candida glabrata* 44B, *Cryptococcus neoformans* obtained from Centre Pasteur Yaoundé, Cameroon. Reference antibacterial drug gentamycin and antifungal drug nystatin were evaluated for their antibacterial and antifungal activities and their results were compared to those of the free ligands and the complex.

The disc diffusion method, using Muller Hinton Agar, from the protocol described by the National Committee for Clinical Laboratory Standard was used for preliminary screening.

Mueller-Hinton agar was prepared from a commercially available dehydrated base according to the manufacturer's instructions. Several colonies of each micro-organism was collected and suspended in saline (0.9% NaCl). Then, the turbidity of the test suspension was standardized to match that of a 0.5 McFarland standard (corresponds to approximately 1.5×10^8 CFU/mL for bacteria or 1×10^6 to 5×10^6 cells/mL for yeast). Each compound or reference was accurately weighed and dissolved in the appropriate diluents (DMSO at 10%, Methanol at 10% or distilled water) to yield the required concentration (2 mg/mL for compound or 1 mg/mL for reference drug), using sterile glassware.

Whatman filter paper No. 1 was used to prepare discs approximately 6 mm in diameter, which were packed up with aluminium paper and sterilized by autoclaving. Then, 25 µL of stock solutions of compound or positive control were delivered to each disc, leading to 50 µg of compound or 25 µg of reference drug.

The dried surface of a Müeller-Hinton agar plate was inoculated by flooding over the entire sterile agar surface with 500 µL of inoculum suspensions. The lid was left ajar for 3–5 min to allow for any excess surface moisture to be absorbed before applying the drug impregnated discs. Discs containing the compounds or antimicrobial agents were applied within 15 min of inoculating the MHA plate. Six discs per petri dish were plated. The plates were inverted and placed in an incubator set to 35°C. After 18 h (for bacteria) and 24 h (for yeasts) of incubation, each plate was examined. The disc diameter and the diameter of the zones of complete inhibition (as judged by the unaided eye) were measured. Zones were measured to the nearest whole millimetre, using sliding callipers, which was held on the back of the inverted petri plate. All experiments were carried out in duplicate. The compound was considered active against a microbe if the inhibition zone was 6 mm and above.

Table 1. Physical data of the complexes

Complex	Nature	Colour	Yield (%)	Melting point (°C)	Molar conductivity (Ω^{-1} cm^2 mol^{-1})	Elemental analyses: % Found (% Calc.)		
						% C	% H	% N
[Co(phen)$_3$](NO$_3$)$_2$·2H$_2$O (1)	Crystals	Dark brown	92	314	68.4	50.52 (49.75)	3.28 (3.45)	14.66 (14.51)
[Co(phen)$_2$(N$_3$)$_2$]NO$_3$ (2)	Crystals	Dark red	75	/	120	49.28 (50.94)	2.52 (2.83)	27.71 (27.24)

2.6. Minimum inhibitory concentration (MIC) of the complexes

The microbroth dilution method was used to determine the MIC of the compounds and the reference antibiotic on a given micro-organism. A polystyrene tray containing 80 wells is filled with small volumes of serial twofold dilutions of the complex and reference antibiotics. The inoculum suspension and standardization is done according to McFarland standard. The bacterial inoculum is then inoculated into the wells and incubated at 37°C overnight. The lowest concentration of antibiotic that completely inhibits visual growth of bacteria (no turbidity) is recorded as MIC.

3. Results and discussion

3.1. Synthesis of the complex

The reaction of Co(NO$_3$)$_2$·6H$_2$O and 1,10-phen with and without azide in methanol yielded two complexes whose physicochemical properties are summarized in Table 1. The complexes which are coloured and air stable were obtained in good yields (>75%). Complex **1**, a Co(II) complex, had a sharp melting point (314°C) indicating its purity, while the melting point of complex **2**, a Co(III) complex, could not be determined due to the explosive nature of the azide. The molar conductivity values of 68.4 and 120 Ωcm^{-2} mol^{-1} in water, for complexes **1** and **2**, respectively, indicate that complex **1** is a non-electrolyte, while complex **2** is a 1:1 electrolyte type.

The chemical reaction leading to the formation of complex **2** is a redox process where Co(II) becomes oxidized to Co(III). The oxidation of Co(II) to Co(III) generates a diamagnetic complex and indicates an octahedral structure for complex **2**. Most Co^{3+} complexes are diamagnetic; the mixed ligand under the investigation behaves as a strong field.

3.2. Crystal structure of complex 2

The crystals obtained for complex **1** were not suitable for single crystal X-ray structure determination. Thus, only the crystal structure of complex **2** was determined.

The ORTEP representation of the crystal structure of [Diazido-bis(1,10-phen-κ^2 N,N′)cobalt(III)]nitrate, [Co(C$_{12}$H$_8$N$_2$)$_2$(N$_3$)$_2$]NO$_3$ with the atom numbering scheme used in the corresponding tables is shown in Figure 1. The crystal packing diagram for [Co(C$_{12}$H$_8$N$_2$)$_2$(N$_3$)$_2$]NO$_3$ seen along the crystallographic c-axes is shown in Figure 2. The crystal data and structure refinement are presented in Table 2, while the selected bond lengths and bond angles are shown in Table 3.

Complex **2** crystallizes in the orthorhombic crystal system with space group *Iba*2 with four formula units. The asymmetric unit consists of one molecule of 1,10-phen, one azide anion, one nitrate anion and one Co(III) ion. The molecular structure of complex **2** contains one monomeric cation, [Co(phen)$_2$(N$_3$)$_2$]$^+$ and one nitrate ion, as revealed in Figure 1. Chemically, each Co atom is six coordinate; it is bonded to four N atoms of two phen molecules [Co1—N1 1.955 (2) Å, Co1—N1^i 1.955 (2) Å, Co1—N2 1.9356 (17) Å, Co1—N2^i 1.9356 (17) Å] and two terminal N atoms from two azide anions [Co1—N3 1.933 (3) Å, Co1—N3^i 1.933 (3) Å] giving a slightly distorted octahedral geometry around the Co atom with CoN$_6$ chromophore. The axial positions are occupied by the N2 and N2^i atoms of two different phen molecules [N2—Co1—N2^i 179.98 (16)°] while the equatorial plane is formed by

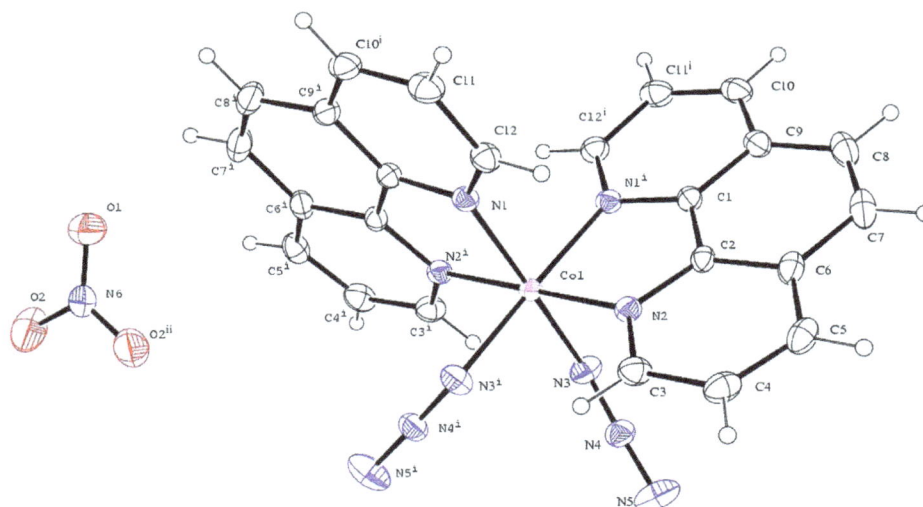

Figure 1. ORTEP view of crystal structure showing atom numbering scheme.

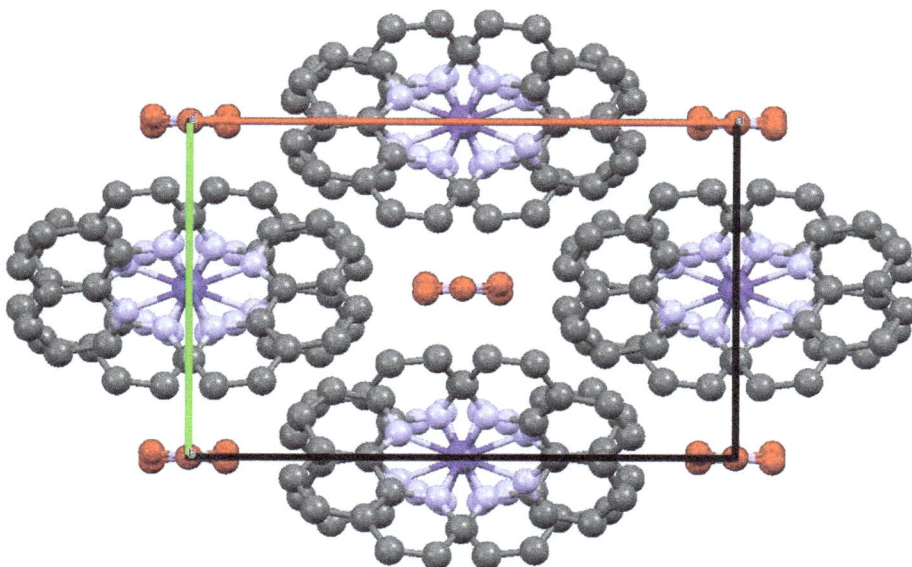

Figure 2. Packing diagram of the complex seen along the crystallographic c-axis.

the coordinating atoms N1, N1*i* [N1—Co1—N1^i 89.01 (13)°] from phen and N3, N3*i* [N3—Co1—N3^i 93.86 (17)°] from the two azide anions. The Co-N(phen) bond lengths are shorter than Co-N(phen) bonds reported in the literature (Guo, Shi, Si, Duan, & Shi, 2013; Pook et al., 2015). The structure of (2) contains one crystallographically independent Co(III) atom with the angles around Co1 slightly distorted from the ideal 90° and 180° of a perfect octahedron (Richers, Bertke, & Rauchfuss, 2015). The bond angles N3—Co1—N2 89.55 (11)°, N3^i—Co1—N2 90.44 (10)°, N3—Co1—N2^i 90.44 (10)° and N3^i—Co1—N2^i 89.55 (11)° indicate that the Co1—N2 and N3—Co1 bonds are in two different molecular planes perpendicular to each other. The two phen ligands which form five-membered chelate rings with Co are oriented in two different molecular planes (Liang et al., 2003; Meundaeng, Prior, & Rujiwatra, 2013). These bond angles slightly deviate from 90° indicating a distorted square planar arrangement in the equatorial plane. The N5—N4—N3 bond angle of 175.3° (3) indicates that the azide anion is non-linear. The presence of nitrate O and azide N atoms in the molecular structure act as acceptor centres for the formation of C-H ... O and C-H ... N intermolecular interactions. These

Table 2. Crystal data and structure refinement for [Co(C$_{12}$H$_8$N$_2$)$_2$(N$_3$)$_2$]NO$_3$ (2)

Crystal data

Chemical formula	C$_{24}$H$_{16}$CoN$_{11}$O$_3$
M$_r$	565.41
Crystal system, space group	Orthorhombic, *Iba*2
Temperature (K)	296
a, b, c (Å)	15.5018 (16), 9.8379 (8), 14.7376 (11)
V (Å^3)	2247.6 (3)
Z	4
Radiation type	Mo Kα
μ (mm^{-1})	0.82
Crystal size (mm)	0.25 × 0.25 × 0.20
Data collection	
Diffractometer	Bruker kappa apex2 CCD Diffractometer
Absorption correction	Multi-scan *SADABS*
T$_{min}$, T$_{max}$	0.801, 0.882
No. of measured, independent and observed [I > 2σ(I)] reflections	18,221, 3,100, 2,655
R$_{int}$	0.024
(sin θ/λ)$_{max}$ (Å^{-1})	0.708
Refinement	
R[F^2 > 2σ(F^2)], wR(F^2), S	0.025, 0.070, 1.07
No. of reflections	3100
No. of parameters	178
No. of restraints	1
H-atom treatment	H-atom parameters constrained
Δρ$_{max}$, Δρ$_{min}$ (e Å^{-3})	0.23, −0.29
Absolute structure	Flack x determined using 1,168 quotients [(I+)(I)]/[(I+)+(I-)]
Absolute structure parameter	−0.008 (5)

Table 3. Selected bond lengths (Å) and angles (°) for complex (2)

Co1—N3	1.933 (3)	N3—Co1—N3^i	93.86 (17)
Co1—N3^i	1.933 (3)	N3—Co1—N2	89.55 (11)
Co1—N2	1.9356 (17)	N3^i—Co1—N2	90.44 (10)
Co1—N2^i	1.9356 (17)	N3—Co1—N2^i	90.44 (10)
Co1—N1	1.955 (2)	N3^i—Co1—N2^i	89.55 (11)
Co1—N1^i	1.955 (2)	N2—Co1—N2^i	179.98 (16)
O1—N6	1.212 (9)	N3—Co1—N1	173.75 (11)
O2—N6	1.230 (4)	N3^i—Co1—N1	88.85 (9)
N4—N3—Co1	120.6 (2)	N2—Co1—N1	96.07 (9)
N5—N4—N3	175.3 (3)	N2^i—Co1—N1	83.95 (9)
O1—N6—O2	120.8 (3)	N3—Co1—N1^i	88.85 (9)
O1—N6—O2^{ii}	120.8 (3)	N3^i—Co1—N1^i	173.75 (12)
O2—N6—O2^{ii}	118.5 (5)	N2—Co1—N1^i	83.94 (9)
N2^i—Co1—N1^i	96.07 (9)	N1—Co1—N1^i	89.01 (13)

Table 4. Characteristic IR bands of the ligands and the complexes					
	C = N	C-H	C = C	N = N = N	M-N
Phen	1,586	1,638	1,492	/	/
N$_3$	/	/	/	2,103	/
[Co(phen)$_3$](NO$_3$)$_2$·2H$_2$O (1)	1,518	1,625	1,480	/	497
[Co(phen)$_2$(N$_3$)$_2$]NO$_3$ (2)	1,578	1,622	1,512	2,073	419

weak, C5-H5 ... O2, C11-H11 ... O2, C8-H8 ... N5, intermolecular interactions, result in the formation of two dimensional supramolecular sheets in the *ab* plane. The adjacent two-dimensional supramolecular sheets exchange π–π interactions experienced between the 1,10-phen moiety forming an extended three-dimensional network in the solid. The cobalt atoms between the stacked sheets running along the *c*-direction have an interlayer Co–Co distance of 7.369 Å.

3.3. Infrared spectroscopy
The characteristic absorption bands in the IR spectra of the ligands and the complexes (**1** and **2**) are summarized in Table 4.

In the spectrum of the phen ligand, the absorption bands at 1,586 and 1,492 cm^{-1} assigned to $\nu_{(C=N)}$ and $\nu_{(C=C)}$ stretching vibrations, respectively, are shifted in the complexes to 1,518 and 1,480 cm^{-1} (1) and 1,578 and 1,512 cm^{-1} (2), respectively, indicating the participation of the C = N of phen in bonding (Liang et al., 2003; Lu, Zhu, & Yang, 2003). The strong absorption band at 2,103 cm^{-1} in the spectrum of the azide ligand, assigned to the asymmetric stretching vibration $\nu_{as}(N_3)$ of the azide is shifted to 2,073 cm^{-1} in complex **2** indicating terminal coordination of the azide to the metal ion (Goher et al., 2004; Liang et al., 2003). The new bands at 497 and 419 cm^{-1} which were not found in the spectra of the ligands, indicates the presence of Co-N bonding between the metal and the nitrogen atoms of both phen and the azide.

3.4. UV–vis spectroscopy
The UV–vis spectra of the complexes were measured in aqueous solution at ambient temperature.

The ground term symbol for d^6 Co(III) is ^{5}D and the octahedral high spin spectroscopic ground state is ^{5}T$_{2g}$. Thus, a single spin-allowed transition ^{5}E$_g$ ← ^{5}T$_{2g}$ is predicted but this will generally be split because of Jahn-Teller distortion. Most of the six coordinate complexes of cobalt(III) are low spin and have the spectroscopic ground term symbol ^{1}A$_{1g}$. Two transitions: ^{1}T$_{1g}$ ← ^{1}A$_{1g}$ and ^{1}T$_{2g}$ ← ^{1}A$_{1g}$ are expected for six coordinate low spin Co(III) complexes (Lever, 1984). The ground term symbol of d^7 Co(II) is ^{4}F. In octahedral Co(II) complexes, three transitions ^{4}T$_{2g}$ ← ^{4}T$_{1g}$(F), ^{4}A$_{2g}$ ← ^{4}T$_{1g}$(F) and ^{4}T$_{1g}$(P) ← ^{4}T$_{1g}$(F) are expected. The transition to ^{4}A$_{2g}$ is usually very weak and often appears as a shoulder.

The electronic absorption spectrum of Co(III) (complex **2**) shows a single broad band centred at 18,518.5 cm^{-1} (530 nm). This d–d absorption band in has been assigned to ^{5}E$_g$ ← ^{5}T$_{2g}$ transition (Abdelhak et al., 2014). The electronic spectrum of Co(II) (complex **1**) shows a split absorption band centred at 34,482 cm^{-1} (290 nm) and 31,746 cm^{-1} (315 nm) probably due to intraligand transitions. A shoulder is observed at 28,571 cm^{-1} (350 nm) assigned to ligand-to-metal charge transfer transition.

3.5. Thermogravimetric analysis of [Co(phen)$_2$(N$_3$)$_2$]NO$_3$
To examine the thermal stability of the complex **2**, thermal gravimetric (TG) analysis was carried out. This analysis (Figure 3) shows that the complex is thermally stable up to 250°C and upon further heating, it decomposes in a single step from 300 to 420°C. The major weight loss (~83%) takes place rapidly at 300°C, attributed to the complete decomposition of the sample (calc. 89.6%). A stable mass is reached at 460°C.

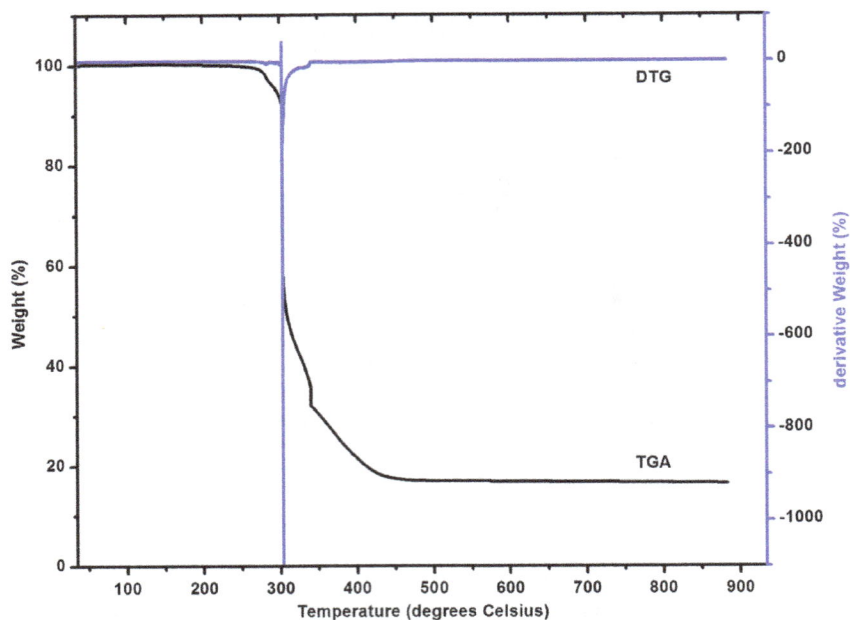

Figure 3. Thermogravimetric and differential thermogravimetric analyses of complex 2.

Caution! Azide complexes of metals with organic ligands are potentially explosive and should be handled with care.

3.6. Antimicrobial tests

The potency of the metal salt, 1,10-phen, N_3^- and the complexes together with the reference anti-bacterial drug (Gentamycin) and antifungal drug (Nystatin) were evaluated against four bacteria and four fungi strains. The diameter of the zone of inhibition (mm) was used to compare the antimicrobial activity of the test compound with that of the reference drug. Results of the antimicrobial activity are presented in Table 5 and histograms shown in Figures 4–6.

The metal salt, 1,10-phen and azide showed considerable antimicrobial activity against the micro-organisms. The ligand 1,10-phen was very active against all the fungi and bacteria species tested. The metal salt and azide were more active against bacteria species while 1,10-phen showed higher activity against the fungi. Coordination of the ligands to the metal ion impacted the antimicrobial activity against all fungi and bacteria species tested with inhibition zone diameters in the range 13–38 mm. Complex **1** showed higher activity against the bacteria species while complex **2**

Table 5. Diameter of zone of inhibition of the complex, ligands and the metal salt								
	Micro-organisms	**Zone of inhibition (mm)**						
		N_3	**Phen.**	**$Co(NO_3)_2 \cdot 6H_2O$**	**Complex 1**	**Complex 2**	**Gentamycin**	**Nystatin**
Bacteria	*E. coli*	26 ± 1.4	24.5 ± 0.7	17 ± 0	20.5 ± 0.7	20.5 ± 0.7	22 ± 0	6.0 ± 0
	P. aeruginosa	37 ± 1.4	13.5 ± 0.7	17.5 ± 0.7	23.5 ± 0.7	16 ± 1.4	25 ± 0	7.0 ± 0
	S. typhi	0 ± 0	14 ± 0	14.5 ± 0.7	17.5 ± 0.7	13 ± 1.2	30 ± 0	6.0 ± 0
	S. aureus	13.5 ± 0.7	17 ± 1.4	16 ± 1.2	18.5 ± 1.2	16 ± 0	30 ± 0	6.0 ± 0
Yeast	*C. albicans* ATCC 12C	35 ± 0	30 ± 1.4	11.5 ± 0.7	17.5 ± 1.2	22 ± 0	6.0 ± 0	6.0 ± 0
	C. albicans ATCC P37037	0 ± 0	40 ± 0	10.5 ± 0.7	16 ± 0	36 ± 0	7.5 ± 0	7.0 ± 0
	C. albicans ATCC P37039	9 ± 1.2	30 ± 1.4	13 ± 0	38 ± 0	17.5 ± 0.7	6.0 ± 0	6.0 ± 0
	C. neoformans	6 ± 0	26 ± 0	24 ± 1.4	0 ± 0	20 ± 0	6.5 ± 0	9.5 ± 0

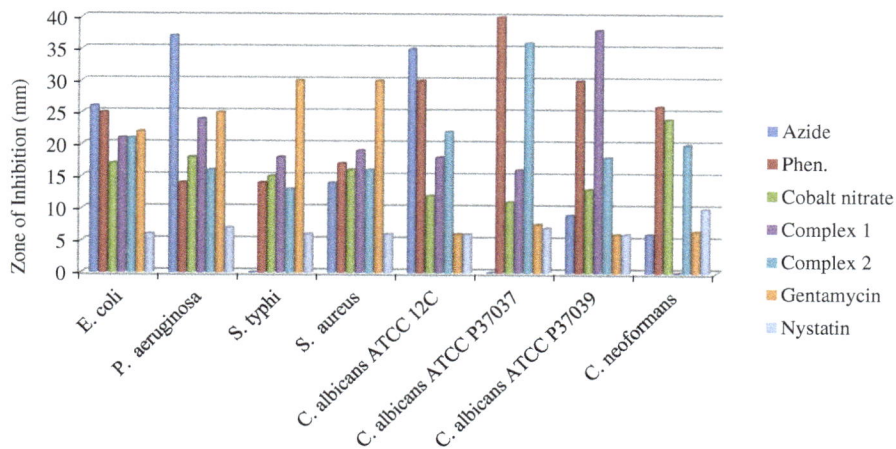

Figure 4. Histogram of inhibition zones.

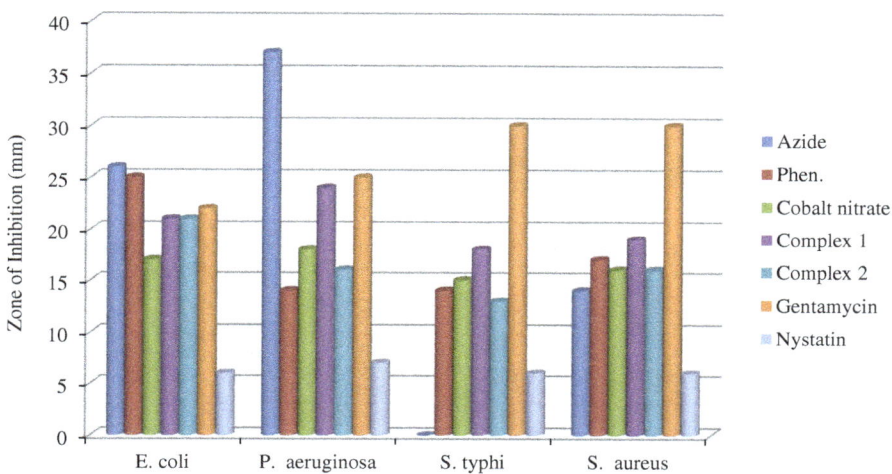

Figure 5. Histogram of inhibition zones against bacteria species.

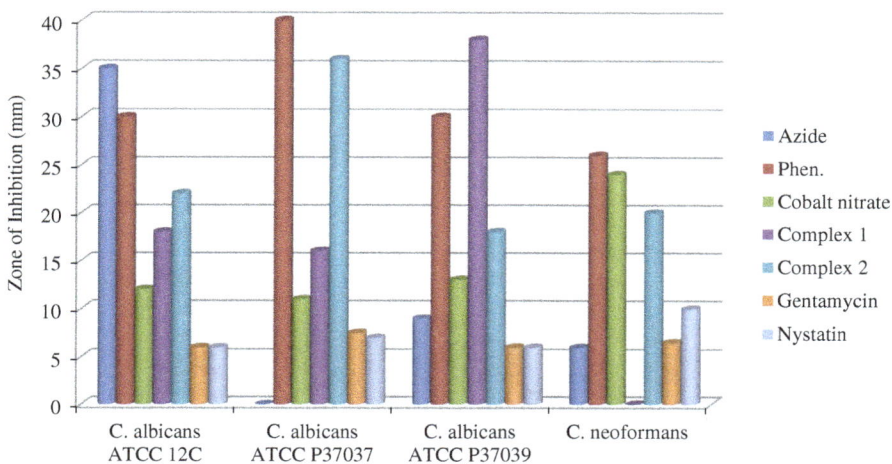

Figure 6. Histogram of inhibition zones against fungi species.

Essential Concepts of Chemistry

exhibited higher activities against the fungi species. The complexes were generally more potent towards the fungi species than the reference antifungal (nystatin). The complexes exhibited highest
activity against the fungus *C. albicans ATCC P37039*, which can lead to the development of vagina
candidiasis. Compared to the reference antibiotic gentamicin, the complexes showed comparable
antibacterial activity against the bacteria species *E. coli* and *P. aeruginosa*. These results are comparable to those of the complex [azido-bis(1,10-phen-κ^2 N,N')copper(II)]nitrate hydrate, [Cu(C$_{12}$H$_8$N$_2$)$_2$N$_3$]
NO$_3$·$\frac{1}{3}$H$_2$O (Sado et al., 2016).

The complexes are also more active than the reference drug nystatin towards the fungi species,
with complex **2** being most active. This indicates that reaction of metal ions with the ligands plays
an important role in enhancing its antimicrobial activity. This increase in activity could be due to the
reduction of the polarity of the metal ion by partial sharing of the positive charge with the ligand's
donor atoms so that there is electron delocalization within the metal complex. This may increase the
hydrophobic and lipophilic character of the metal complex, enabling it to permeate the lipid layer of
the organism killing them more effectively (Amah et al., 2015; Chohan, Munawar, & Supuran, 2001).

The antimicrobial activities of the complexes were quantified by the determination of their MICs.
The MIC indicates the lowest drug concentration at which a visible inhibition growth of the micro-
organisms is noticed. The values for the different pathogens were determined by using the microbroth dilution test. The MICs values for the complexes are summarized in Table 6 and the histograms
on Figure 7.

Table 6. MIC (mg/mL) of the complexes

	Species	MIC (mg/mL)						
		Co(NO$_3$)$_2$·6H$_2$O	O-Phen	N$_3^-$	Complex 1	Complex 2	Gentamycin	Nystatin
Bacteria	E. coli	0.125	0.039	0.156	1.25	0.313	0.156	>2.5
	P. aeruginosa	0.625	0.039	0.078	0.078	0.313	1.25	>2.5
	S.typhi	1.25	0.039	0.156	0.156	0.625	1.25	>2.5
	S. aureus	1.25	0.078	0.625	0.625	0.625	1.25	>2.5
Fungi	C. albicans ATCC 12C	0.625	0.039	1.25	0.156	1.25	>2.5	>2.5
	C. albicans ATCC P37037	0.625	0.078	0.625	0.156	0.313	>2.5	>2.5
	C. albicans ATCC P37039	0.625	0.156	1.25	0.156	1.25	>2.5	>2.5
	C. neoformans	1.25	0.039	2.5	0.078	1.25	>2.5	>2.5

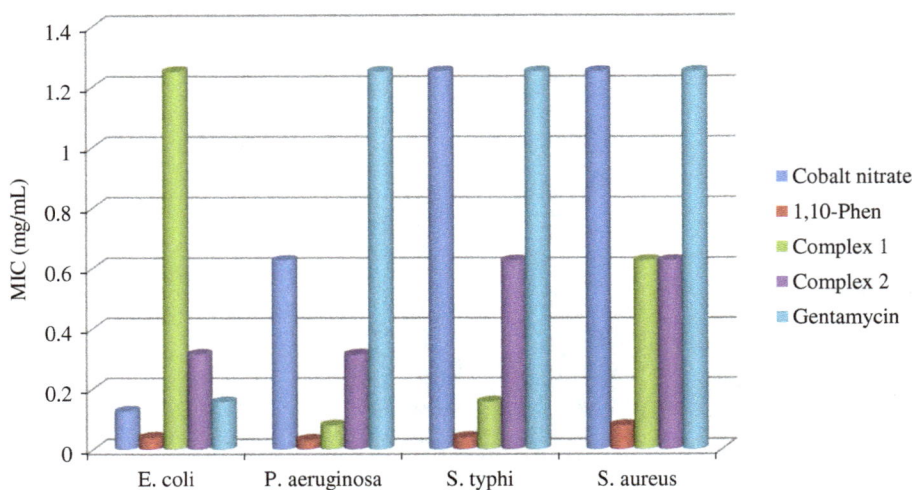

Figure 7. Histogram of MIC against bacteria species.

The MIC values indicate that the ligands (1,10-phen and azide) have higher activity against the bacteria species than the fungi, with 1,10-phen being more active than azide. The complexes have activities higher than that of the metal salt but lower than that of 1,10-phen, towards the micro-organisms. The most sensitive strains are the bacteria species *E. coli, P. aeruginosa, S. typhi* and the fungi *C. albicans ATCC P37037* and *C. neoformans.* The complexes have a broad spectrum of activity, since they showed high activity towards all the tested strains. The MIC values obtained are lower than those of a copper(II) complex against the same micro-organisms, indicating that the cobalt complexes synthesized are more potent (Yenikaya et al., 2009). Comparison of the MIC values of the complexes with that of the reference drugs indicates that the complexes are potentially more active than the reference drugs. The disparity in the effectiveness of the complexes against the tested pathogens depends either on the impermeability of the cells of the microbes or the difference in the ribosomes of the microbial cells (Yenikaya et al., 2009).

4. Conclusion

A novel mixed ligand Co(III) complex with 1,10-phen and azide co-ligands has been reported. The equatorial and terminal azide ligands coordinate in a non-linear manner to the central metal ion while two 1,10-phen ligands coordinate to the Co(III) ion through four N-atoms, two axial and two equatorial. The Co(III) ion in the complex adopts a slightly distorted octahedral environment comprising four phenanthroline N-atoms and two nitrile N-atoms from the azides. The molecular structure of the compound is consolidated by weak C-H ... O and C-H ... N intermolecular interactions and face-to face π–π stacking between the aromatic rings of 1,10-phen ligand. The results of the preliminary antimicrobial screening against four pathogenic bacteria and four fungi species indicates that the complexes are very active and could be further screened *in vitro* against a wide range of pathogens. The MIC values indicate that the ligands (1,10-phen and azide) have higher activity against the bacteria species than the fungi, with 1,10-phen being more active than azide. The complexes have activities higher than that of the metal salt but lower than that of 1,10-phen, towards the micro-organisms. The most sensitive strains are the bacteria species *E. coli, P. aeruginosa, S. typhi* and the fungi *C. albicans ATCC P37037* and *C. neoformans.* The generally positive results of the antimicrobial screening against four bacteria and four fungi species suggest that complex **1** is more active towards the bacteria, while complex **2** is more active towards the fungi. Both complexes (**1** and **2**) may represent good candidates as an antibacterial and antifungal agent, respectively. However, additional and profound *in vitro* antimicrobial studies mainly in relation to elucidation of the mechanism of growth inhibition and toxicity of the complexes is ongoing.

Acknowledgement
The authors thank the Laboratory of Phytobiochemical and Medicinal Plant Study, University of Yaounde I for the antimicrobial tests.

Funding
AMO and DMY acknowledge the Government of Cameroon for financial support through the Fonds d'Appuis à la Recherche.

Author details
Djuikom Sado Yanick Gaëlle[1]
E-mail: yanicksado@yahoo.fr
Divine Mbom Yufanyi[2]
E-mail: dyufanyi@yahoo.com
ORCID ID: http://orcid.org/0000-0001-8889-611X
Rajamony Jagan[3]
E-mail: phyjagan@gmail.com
Moise Ondoh Agwara[1]
E-mail: agwara29@yahoo.com

ORCID ID: http://orcid.org/0000-0001-9112-7637

[1] Department of Inorganic Chemistry, University of Yaounde I, P. O. Box 812, Yaounde, Cameroon.
[2] Department of Chemistry, The University of Bamenda, P.O. Box 39 Bambili, Bamenda, Cameroon.
[3] Sophisticated Analytical Instruments Facility, Indian Institute of Technology, Chennai 600036, India.

References

Abdelhak, J., Namouchi Cherni, S., Amami, M., El Kébir, H., Zid, M. F., & Driss, A. (2014). Iron(III) and cobalt(III) complexes with oxalate and phenanthroline: synthesis, crystal structure, spectroscopy properties and magnetic properties. *Journal of Superconductivity and Novel Magnetism, 27,* 1693–1700. doi:10.1007/s10948-014-2479-2

Adhikary, C., & Koner, S. (2010). Structural and magnetic studies on copper(II) azido complexes. *Coordination Chemistry Reviews, 254,* 2933–2958. doi:10.1016/j.ccr.2010.06.001

Ahmed, S. K., & Khaled, S. (2015). Syntheses, spectral characterization, thermal properties and DNA cleavage studies of a series of Co(II), Ni(II) and Cu(II) polypyridine complexes with some new imidazole derivatives of 1,10-phenanthroline. *Arabian Journal of Chemistry.* doi:10.1016/j.arabjc.2015.04.025

Amah, C., Ondoh, A. M., Yufanyi, D. M., & Gaelle, D. S. Y. (2015). Synthesis, crystal structure and antimicrobial properties of an anhydrous copper(ii) complex of pyridine-2-carboxylic acid. *International Journal of Chemistry, 7,* 10–20. doi:10.5539/ijc.v7n1p10

Amani, V., Safari, N., Khavasi, H. R., & Mirzaei, P. (2007). Iron(III) mixed-ligand complexes: Synthesis, characterization and crystal structure determination of iron(III) hetero-ligand complexes containing 1,10-phenanthroline,

2,2'-bipyridine, chloride and dimethyl sulfoxide, [Fe(phen) Cl₃(DMSO)] and [Fe(bipy)Cl₃(DMSO)]. *Polyhedron, 26,* 4908–4914. doi:10.1016/j.poly.2007.06.038

Arounaguiri, S., Easwaramoorthy, D., Ashokkumar, A., Dattagupta, A., & Maiya, B. G. (2000a). Cobalt(III), nickel(II) and ruthenium(II) complexes of 1,10-phenanthroline family of ligands: DNA binding and photocleavage studies. *Journal of Chemical Sciences, 112*(1), 1–17. doi:10.1007/bf02704295

Arounaguiri, S., Easwaramoorthy, D., Ashokkumar, A., Dattagupta, A., & Maiya, B. G. (2000b). Cobalt(III), nickel(II) and ruthenium(II) complexes of 1,10-phenanthroline family of ligands: DNA binding and photocleavage studies. *Proceedings of the Indian Academy of Sciences (Chemical Sciences), 112*(1), 1–17. http://dx.doi.org/10.1007/BF02704295

Batten, S. R., & Murray, K. S. (2003). Structure and magnetism of coordination polymers containing dicyanamide and tricyanomethanide. *Coordination Chemistry Reviews, 246,* 103–130. http://dx.doi.org/10.1016/S0010-8545(03)00119-X

Bencini, A., & Lippolis, V. (2010). 1,10-Phenanthroline: A versatile building block for the construction of ligands for various purposes. *Coordination Chemistry Reviews, 254,* 2096–2180. doi:10.1016/j.ccr.2010.04.008

Biagetti, R. V., & Haendler, H. M. (1966). Pyridine complexes of cobalt(II) and nickel(II) nitrates. *Inorganic Chemistry, 5,* 383–386. http://dx.doi.org/10.1021/ic50037a012

Bruker AXS. (2004). *APEX2, SAINT-Plus and XPREP.* Madison, WI: Author.

Chen, Z.-L., Jiang, C.-F., Yan, W.-H., Liang, F.-P., & Batten, S. R. (2009). Three-dimensional metal azide coordination polymers with amino carboxylate coligands: Synthesis, structure, and magnetic properties. *Inorganic Chemistry, 48,* 4674–4684. doi:10.1021/ic802026n

Chohan, Z. H., Munawar, A., & Supuran, C. T. (2001). Transition metal ion complexes of schiff-bases. Synthesis, characterization and antibacterial properties. *Metal Based Drugs, 8,* 137–143. doi:10.1155/MBD.2001.137

Deegan, C., McCann, M., Devereux, M., Coyle, B., & Egan, D. A. (2007). In vitro cancer chemotherapeutic activity of 1,10-phenanthroline (phen), [Ag₂(phen)₃(mal)]·2H₂O, [Cu(phen)₂(mal)]·2H₂O and [Mn(phen)₂(mal)]·2H₂O (malH2=malonic acid) using human cancer cells. *Cancer Letters, 247,* 224–233. doi:10.1016/j.canlet.2006.04.006

Goher, M. A. S., Hafez, A. K., Abu-Youssef, M. A. M., Badr, A. M. A., Gspan, C., & Mautner, F. A. (2004). New metal(II) complexes containing monodentate and bridging 3-aminopyridine and azido ligands. *Polyhedron, 23,* 2349–2356. doi:10.1016/j.poly.2004.06.011

Goher, M. A. S., & Mautner, F. A. (1995). New unexpected coordination modes of azide and picolinato anions acting as bridging ligands between copper(II) and sodium or potassium ions. Synthesis, crystal structures and spectral characterizations of [MCu(picolinato)(N₃)₂]ₙ (M = Na or K) complexes. *Polyhedron, 14,* 1439–1446. http://dx.doi.org/10.1016/0277-5387(94)00415-B

Gopinathan, H., & Arumugham, M. N. (2015). Larvicidal activity of synthesized copper(II) complexes against *Culex quinquefasciatus* and *Anopheles subpictus. Journal of Taibah University for Science, 9,* 27–33. doi:10.1016/j.jtusci.2014.04.008

Gopinathan, H., Komathi, N., & Arumugham, M. N. (2014). Synthesis, structure, DNA binding, cleavage and biological activity of cobalt (III) complexes derived from triethylenetetramine and 1,10 phenanthroline ligands. *Inorganica Chimica Acta, 416,* 93–101. doi:10.1016/j.ica.2014.03.015

Guo, Y., Shi, T., Si, Z., Duan, Q., & Shi, L. (2013). Novel magnetic CoII complexes: Synthesis and characterization. *Inorganic*

Chemistry Communications, 34, 15–18. doi:10.1016/j.inoche.2013.04.039

Hawkey, P. M., & Jones, A. M. (2009). The changing epidemiology of resistance. *Journal of Antimicrobial Chemotherapy, 64*(Supplement 1), i3–i10. doi:10.1093/jac/dkp256

Janiak, C. (2000). A critical account on π–π stacking in metal complexes with aromatic nitrogen-containing ligands†. *Journal of the Chemical Society, Dalton Transactions,* 3885–3896. http://dx.doi.org/10.1039/b003010o

Jennifer, S. J., & Muthiah, P. T. (2014). Synthesis, crystal structures and supramolecular architectures of square pyramidal Cu(II) complexes containing aromatic chelating N,N'-donor ligands. *Chemistry Central Journal, 8,* 42. Retrieved from http://journal.chemistrycentral.com/content/8/1/42 http://dx.doi.org/10.1186/1752-153X-8-42

Kou, Y.-Y., Tian, J.-L., Li, D.-D., Liu, H., Gu, W., & Yan, S.-P. (2009). Oxidative DNA cleavage by Cu(II) complexes of 1,10-phenanthroline-5,6-dione. *Journal of Coordination Chemistry, 62,* 2182–2192. doi:10.1080/00958970902763271

Kumar, R. S., Arunachalam, S., Periasamy, V. S., Preethy, C. P., Riyasdeen, A., & Akbarsha, M. A. (2008). Synthesis, DNA binding and antitumor activities of some novel polymer-cobalt(III) complexes containing 1,10-phenanthroline ligand. *Polyhedron, 27,* 1111–1120. doi:10.1016/j.poly.2007.12.008

Lazari, G., Stamatatos, T. C., Raptopoulou, C. P., Psycharis, V., Pissas, M., Perlepes, S. P., & Boudalis, A. K. (2009). A metamagnetic 2D copper(II)-azide complex with 1D ferromagnetism and a hysteretic spin-flop transition. *Dalton Transactions, 3215–3221,* doi:10.1039/b823423j

Lever, A. B. P. (1984). *Inorganic electronic spectroscopy* (2nd ed.). Netherlands: Elsevier Science.

Liang, M., Wang, W.-Z., Liu, Z.-Q., Liao, D.-Z., Jiang, Z.-H., Yan, S.-P., & Cheng, P. (2003). A new mixed-ligand copper(II) complex containing azide and 1,10-phenanthroline: crystal structure and properties. *Journal of Coordination Chemistry, 56,* 1473–1480. doi:10.1080/00958970310001617058

Lu, L.-P., Zhu, M.-L., & Yang, P. (2003). Crystal structure and nuclease activity of mono (1,10-phenanthroline) copper complex. *Journal of Inorganic Biochemistry, 95,* 31–36. doi:10.1016/S0162-0134(03)00049-7

McCann, M., Geraghty, M., Devereux, M., O'Shea, D., Mason, J., & O'Sullivan, L. (2000). Insights into the mode of action of the anti-candida activity of 1,10-phenanthroline and its metal chelates. *Metal Based Drugs, 7,* 185–193. http://dx.doi.org/10.1155/MBD.2000.185

Meundaeng, N., Prior, T. J., & Rujiwatra, A. (2013). Bis(1,10-phenanthroline-κ 2 N, N')(sulfato-κ O)copper(II) ethanol monosolvate. *Acta Crystallographica Section E: Structure Reports Online, 69,* m568–m569. doi:10.1107/s1600536813026093

Mishra, A., Kaushik, N. K., Verma, A. K., & Gupta, R. (2008). Synthesis, characterization and antibacterial activity of cobalt(III) complexes with pyridine-amide ligands. *European Journal of Medicinal Chemistry, 43,* 2189–2196. doi:10.1016/j.ejmech.2007.08.015

Molphy, Z., Slator, C., Chatgilialoglu, C., & Kellett, A. (2015). DNA oxidation profiles of copper phenanthrene chemical nucleases. *Frontiers in Chemistry, 3,* 28. doi:10.3389/fchem.2015.00028

Ndosiri, N., Agwara, M., Paboudam, A., Ndifon, P., Yufanyi, D., & Amah, C. (2013). Synthesis, charaterization and antifungal activities of Mn(II), Co(II), Cu(II) and Zn(II) mixed-ligand complexes containing 1,10-phenanthroline and 2,2-bipyridine. *Research Journal of Pharmaceutical, Biological and Chemical Sciences* 4, 386–397.

Pivetta, T., Isaia, F., Verani, G., Cannas, C., Serra, L., Castellano, C., & Demartin, F. (2012). Mixed-1,10-phenanthroline Cu(II) complexes: Synthesis, cytotoxic activity versus hematological and solid tumor cells and complex formation equilibria with glutathione. *Journal of Inorganic Biochemistry, 114*, 28–37. doi:10.1016/j.jinorgbio.2012.04.017

Pook, N.-P., Hentrich, P., & Gjikaj, M. (2015). Crystal structure of bis[tris(1,10-phenanthroline k^2 N,N')cobalt(II)] tetranitrate N,N'-(1,4-phenylenedicarbonyl)diglycine solvate octahydrate. *Acta Cryst, E71*, 910–914. doi:10.1107/S2056989015013006

Rajarajeswari, C., Ganeshpandian, M., Palaniandavar, M., Riyasdeen, A., & Akbarsha, M. A. (2014). Mixed ligand copper(II) complexes of 1,10-phenanthroline with tridentate phenolate/pyridyl/(benz)imidazolyl Schiff base ligands: Covalent vs non-covalent DNA binding, DNA cleavage and cytotoxicity. *Journal of Inorganic Biochemistry, 140*, 255–268. doi:10.1016/j.jinorgbio.2014.07.016

Richers, C. P., Bertke, J. A., & Rauchfuss, T. B. (2015). Crystal structure of di-m-hydroxido-k^{4}O:O-bis-[bis(acetylacetonato-k^{2}O,O')cobalt(III)]. *Acta Crystallographica, E71*, 983–985. doi:10.1107/S2056989015013663

Sado, D. Y. G., Agwara, M. O., Yufanyi, M. D., Nenwa, J., & Jagan, R. (2016). Crystal structure and antimicrobial properties of a copper(II) complex with 1,10-phenanthroline and azide co-ligand. *Synthesis and Reactivity in Inorganic, Metal-Organic, and Nano-Metal Chemistry, 00–00*, doi:10.1080/15533174.2016.1212220

Shaabani, B., Khandar, A. A., Dusek, M., Pojarova, M., & Mahmoudi, F. (2013). Synthesis, crystal structure, antimicrobial activity and electrochemistry study of chromium(III) and copper(II) complexes based on semicarbazone Schiff base and azide ligands. *Inorganica Chimica Acta, 394*, 563–568. doi:10.1016/j.ica.2012.08.027

Sheldrick, G. M. (2008). A short history of SHELX. *Acta Crystallographica Section A, 64*, 112–122.

Silva, P. P., Guerra, W., Silveira, J. N., Ferreira, A. M. D. C., Bortolotto, T., Fischer, F. L., & Terenzi, H. (2011). Two New Ternary Complexes of Copper(II) with Tetracycline or Doxycycline and 1,10-Phenanthroline and Their Potential as Antitumoral: Cytotoxicity and DNA Cleavage. *Inorganic Chemistry, 50*, 6414–6424. doi:10.1021/ic101791r

Spellberg, B., Guidos, R., Gilbert, D., Bradley, J., Boucher, H. W., Scheld, W. M., & Infectious Diseases Society of America (2008). The epidemic of antibiotic-resistant infections: A call to action for the medical community from the infectious diseases society of America. *Clinical Infectious Diseases 46*, 155–164. doi:10.1086/524891

Tabong, C. D., Yufanyi, D. M., Paboudam, A. G., Nono, K. N., Eni, D. B., & Agwara, M. O. (2016). Synthesis, Crystal structure, and antimicrobial properties of [diaquabis(hexamethylen etetramine)diisothiocyanato-kn]nickel(ii) complex. *Advances in Chemistry, 2016*, 8. Article ID 5049718. doi:10.1155/2016/5049718

Tanwar, J., Das, S., Fatima, Z., & Hameed, S. (2014). Multidrug resistance: An emerging crisis. *Interdisciplinary Perspectives on Infectious Diseases, 2014*, 7. Article ID 541340. doi:10.1155/2014/541340

Wang, Q., Huang, M., Huang, Y., Zhang, J.-S., Zhou, G.-F., Zeng, R.-Q., & Yang, X.-B. (2014). Synthesis, characterization, DNA interaction, and antitumor activities of mixed-ligand metal complexes of kaempferol and 1,10-phenanthroline/2,2'-bipyridine. *Medicinal Chemistry Research, 23*, 2659–2666. doi:10.1007/s00044-013-0863-2

Weinstein, R. A., & Fridkin, S. K. (2003). Routine cycling of antimicrobial agents as an infection-control measure. *Clinical Infectious Diseases, 36*, 1438–1444. doi:10.1086/375082

WHO. (2014). *Antimicrobial resistance: Global report on surveillance*. Geneva: Author.

Ye, M.-D., Cheng, Y.-Q., Li, X.-H., & Hu, M.-L. (2004). Crystal structure of bis[di(1,10-phenanthroline)azidocopper(II)] biphenyl-4,4'-dicarboxylate pentahydrate, [Cu(C$_{12}$H$_8$N$_2$)$_2$N$_3$]$_2$(C$_{14}$H$_8$O$_4$)·5H$_2$O. *Zeitschrift für Kristallographie, NCS219*, 165–167.

Yenikaya, C., Poyraz, M., Sarı, M., Demirci, F., İlkimen, H., & Büyükgüngör, O. (2009). Synthesis, characterization and biological evaluation of a novel Cu(II) complex with the mixed ligands 2,6-pyridinedicarboxylic acid and 2-aminopyridine. *Polyhedron, 28*, 3526–3532. doi:10.1016/j.poly.2009.05.079

Yesilel, O. Z., Olmez, H., Yılan, O. O., Pasaoglu, H., & Buyukgungor, O. (2006). Syntheses, spectral and thermal studies, and crystal structure of 1,10-phenanthroline and picolinamide complexes of cobalt(II) squarate. *Z. Naturforsch, 61b*, 1094–1100. Retrieved from http://znaturforsch.com

Yuoh, A. C. B., Agwara, M. O., Yufanyi, D. M., Conde, M. A., Jagan, R., & Eyong, K. O. (2015). Synthesis, crystal structure, and antimicrobial properties of a novel 1-d cobalt coordination polymer with dicyanamide and 2-aminopyridine. *International Journal of Inorganic Chemistry, 2015*, 8. Article ID 106838. doi:10.1155/2015/106838

8

Effect of surfactant-free addition and γ-irradiation on the synthesis of CdO quantum dots by thermal decomposition of γ-irradiated anhydrous cadmium acetate[‡]

Refaat M. Mahfouz[1]*, Gamal A-W Ahmed[1] and Tahani Al-Rashidi[2]

*Corresponding author: Refaat M. Mahfouz, Department of Chemistry, Faculty of Science, Assiut University, P. O. Box 71516 Assiut, Egypt.
E-Mail: rmhfouz@science.au.edu.eg
Reviewing editor: Ananda Kumar Kanduluru, On Target Laboratories, USA

Abstract: Pure phase of cubic (FCC) CdO quantum dots were successfully synthesized by thermal oxidation of γ-irradiated anhydrous cadmium acetate at 400°C for three hours in the presence and absence of benzyl alcohol as surfactant-free. Morphological and structural characteristic of the as-synthesized quantum dots were performed with X-ray diffraction (XRD), transmission electron microscopy (TEM), scanning electron microscopy (SEM), and Fourier transform infrared spectroscopy (FT-IR). SEM image of CdO quantum dots synthesized using γ-irradiated anhydrous cadmium acetate with 10^2 kGy absorbed dose shows the formation of the mesoporous nanostructure. In the presence of benzyl alcohol surfactant, the calcination process afforded cauliflower-like structure. TEM image of CdO quantum dots synthesized using γ-irradiated anhydrous cadmium acetate in presence of benzyl alcohol surfactant shows formation of nanoribbon of CdO quantum dots.

Subjects: Chemistry; Material Science; Physical Chemistry

Keywords: γ-irradiation; CdO quantum dots; thermal decomposition

1. Introduction
Cadmium oxide nanoparticles can potentially be applied to optoelectronics and other applications including solar cells, phototransistors, gas sensors, photodiodes, and transparent electrodes

ABOUT THE AUTHORS
Refaat M. Mahfouz is now a professor of Materials Science and Nuclear Chemistry at the Faculty of Science, Assiut University, Egypt. He got his BSc, MSc, and PhD degrees from Assiut University, Egypt. He got more than one fellowship to work as guest scientist at a nuclear research centre, Juelich, Germany in the period from 1086 to 1990. He has more than 110 publications in ISI scientific journals in the field of radiochemistry and materials sciences and one patent.

Gamal A-W Ahmed is now an associate professor in Physical Chemistry at the Faculty of Science, Assiut University. He has more than 20 publications in the field of kinetic reactions in solutions and nanotechnology.

Tahani Al-Rashidi is now a PhD student. She got her MSc from King Saud University, Saudi Arabia in the field of nanotechnology.

PUBLIC INTEREST STATEMENT
This paper describes the utilization of gamma rays to produce nanostructured materials of uniform morphology. Irradiating the precursor prior to the calcination process could lead to significants changes in the size, shape, and morphology of the metal oxide nanoparticles. The addition of surfactant-free has also another effect on the morphology of nanoparticles. We hope this study opens new approach in the synthesis of metal oxides nanoparticles with controlled size , shape , and morphology.

(Giribabu, Suresh, Manigandan, Stephen, & Narayanan, 2013; Kondawar, Mahore, Dahegaonkar, & Agrawal, 2011; Shiori, 1997; Tadjarodi & Imani, 2011; Zhang, Wang, Lin, & Huang, 2010). Bulk CdO is an *n*-type semiconductor with a band gap of 2.2–2.7 eV and an indirect band gap of 1.36–1.98 eV. Different values of band gap have been reported in the literature that can be the results of lattice imperfections originated from different preparation conditions (Grado-Caffaro & Grado-Caffaro, 2008; Ye, Zhong, Zheng, Li, & Li, 2007).

Different chemical methods have been reported for the synthesis of nanostructured CdO, like sol–gel (Dong & Zhu, 2003), spray pyrolysis (Manickathai, Viswanathan, & Alagar, 2008), pulsed laser deposition (Askarinejad & Morsali, 2009), hydrothermal (Ghosh & Rao, 2014), and solvothermal methods (Yang, Siu, Zhang, & Yang, 2004).

Radiation-induced synthesis of nanostructured materials plays an important role in the investigation and production of well-shaped and mono-dispersed nanoparticles. Methods based on the interaction of high energy-charged particles, and γ-ray is widely used in making ion-track membranes, polymeric nanocomposites, and metal oxide nanoparticles. Irradiation effects on the preparation of CdO nanoparticles were scanty reported (Veeraputhiran, Gomathinayagam, Udhaya, Francy, & Kathrunnisa, 2015).

In the present, work, we describe the synthesis of CdO nanoparticles by thermal oxidation of anhydrous cadmium acetate. Factors including a surfactant-free addition and γ-irradiation of cadmium acetate precursor were thoroughly investigated in order to shed more light on the role of γ-irradiation and surfactant-free addition on the morphology, shape, and size of the as-synthesized nanoparticles.

2. Experimental

2.1. Synthesis of CdO nanoparticles
Anhydrous cadmium acetate (CdAc), 99.93% (metal basis), and anhydrous benzyl alcohol, 99.8% were purchased from Sigma-Aldrich and used without any further purification.

Two samples of anhydrous cadmium acetate were prepared for the experiment. The first sample contains 0.l mole of γ-irradiated cadmium acetate, the second one contains 0.l mole of γ-irradiated cadmium acetate and 2 mL of anhydrous benzyl alcohol. The two samples were encapsulated into two separate glass Pyrex cells and allowed to stand in a muffle furnace. The temperature was raised at a heating rate of 10°C min^{-1}–400°C and kept constant for three hours. The reaction took place under the autogenic pressure of the encapsulated materials. At the end of the reaction, the containers were gradually cooled (5 h) to room temperature, and after opening, the obtained nanoparticles were collected in clean and dry containers and subjected to characterization.

We will refer to the calcination using γ-irradiated precursor only as a method (a), and the calcination using γ-irradiated precursor in the presence of benzyl alcohol surfactant as a method (b). It should be mentioned that a trial to prepare CdO nanoparticles by thermal decomposition of un-irradiated cadmium acetate under the experimental conditions mentioned above afford microcrystalline plates of cadmium oxide. Further investigation of this product by TEM investigation indicated that the obtained material is not in nano-scale range.

2.2. Characterization
X-ray powder diffraction patterns (XRD) were recorded on Siemens D 5,000 *X*-ray diffractometer with CuKα radiation ($\lambda = 1.54$ Å). TG measurements were recorded on Perkin-Elmer TG A7 thermogravimetric analyzer in the temperature range of 30–1,000°C. The sample weight was 10.0 ± 0.1 mg with a heating rate of 10°C min^{-1}. FT-IR measurements were recorded as KBr pellets in the range of 200–4,000 cm^{-1} on Perkin-Elmer FT-IR spectrophotometer (spectrum 1,000). SEM and TEM images were captured using the models (SEM, JSM-6360 ASEM, JEOL, Japan) and (TEM, JEM-2100F, JEOL, Japan) electron microscopes.

For irradiation, samples were encapsulated under vacuum in glass vials and exposed to successively increasing doses of radiation at a constant intensity. A Co-60 γ-ray source model gamma cell 220 from MDS (Nordion, Canada) was used for irradiation of the samples. The source was calibrated against a Fricke ferrous sulfate dosimeter and the absorbed doses in the irradiated samples were calculated by applying appropriate corrections on the basis of photon mass attenuation and the energy absorption coefficients for the sample and the dosimeter solutions (Spinks & Woods, 1990). The transient dose was estimated to be 12.07 Gy and the dose rate was 9.83 kGy h^{-1}. All of the irradiations were conducted at 25°C. After irradiation, the samples were stored at room temperature for 24 h before analysis.

3. Results and discussion

Figures 1 and 2 show XRD patterns of CdO nanoparticles synthesized by methods (a) and (b), respectively. The sharp and well-defined peaks indicate the crystalline nature of the as-synthesized CdO. The 2θ values of 32.90°, 38.20°, 55.20°, 65.30°, and 69.30° are indexed, respectively, as the (1 1 1), (2 0 0), (3 1 1), (2 2 0), and (2 2 2) crystal planes and correspond to the (FCC) cubic structure of CdO with a lattice parameter $a = 4.693$ Å (JCPDSO5-0640). No foreign lines from impurities were detected indicating that the obtained CdO was pure phase. The (1 1 1) plane was also selected to calculate size of the obtained CdO nanoparticles using the following Debye–Sherrer's formula

$$D = \frac{0.89\lambda}{\beta \cos \theta_\beta}$$

where D is the average size of the crystallite, assuming that the grains are spherical, λ is the wavelength (in Å), β is the broadening of the diffraction peak (in radians) of full width at the half maximum (FWHM), and θ_β is the Bragg diffraction angle. The calculated average particle size of CdO nanoparticles was found in the range of 2–3 nm indicating quantum dots morphology of the as-synthesized CdO product.

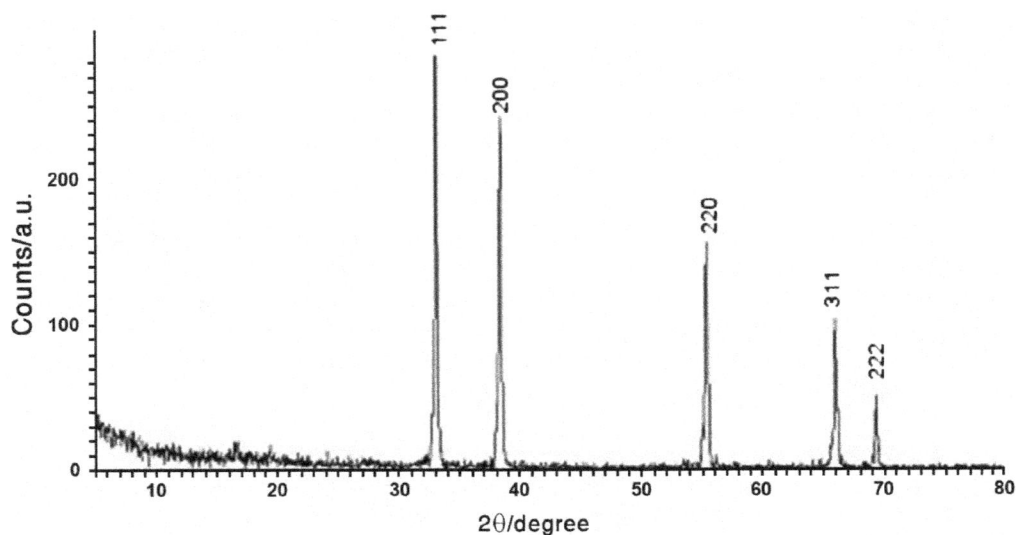

Figure 1. XRD patterns of as- synthesized CdO nanoparticles obtained by calcination of γ-irradiated cadmium acetate for three hours at 400°C.

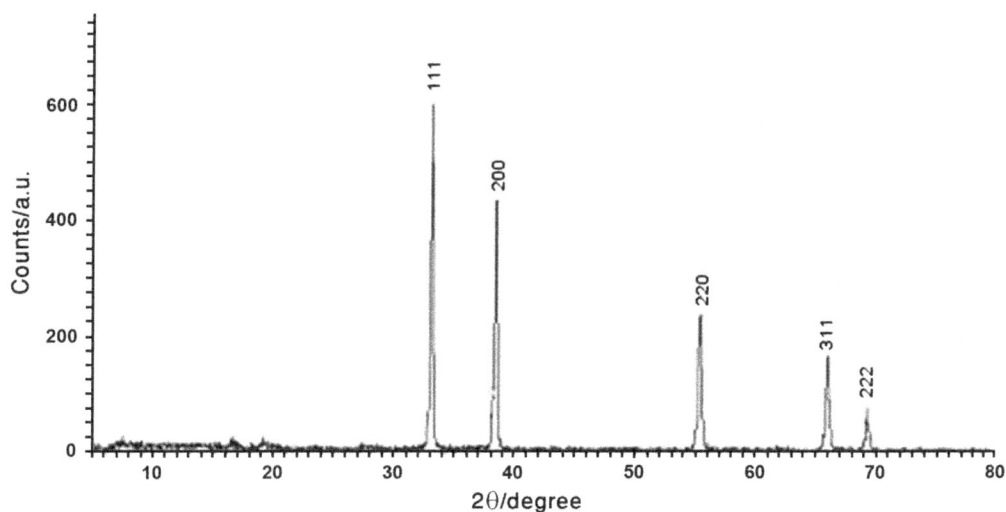

Figure 2. XRD patterns of as- synthesized CdO nanoparticles obtained by calcination of γ-irradiated cadmium acetate for three hours at 400°C in presence of benzyl alcohol as surfactant-free

Figure 3. SEM images of CdO nanoparticles synthesied by calcination of γ-irradiated cadmium acetate for three hours at 400°C.

Figure 3 shows the SEM images of CdO nanoparticles synthesized by method (a). The image displays the formation of the mesoporous structure of CdO nanoparticles.

For the method (b) The SEM image, Figure 4 displays the formation of cauliflower-like mesoporous structure of CdO.

The TEM image, Figure 5 shows the formation of dark spots of tiny nanostructured grains of CdO quantum dots obtained by method (a). The nanocrystalline grains were single crystalline in nature with an average size in the range of 2.5 nm.

The TEM image, Figure 6 shows the formation of nanoribbon of CdO quantum dots of average size 2–3 nm synthesized by method (b).

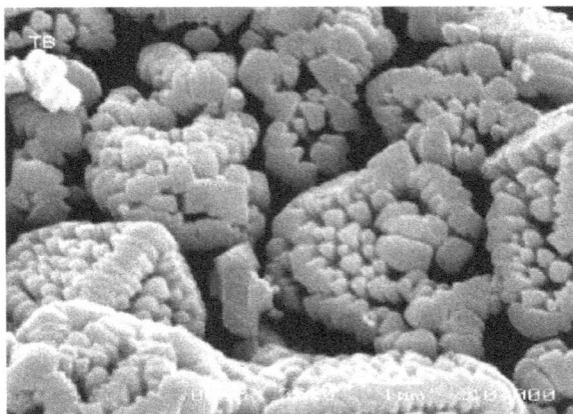

Figure 4. SEM images of CdO nanoparticles synthesized by calcination of γ-irradiated cadmium acetate for three hours at 400°C in presence of benzyl alcohol as surfactant-free.

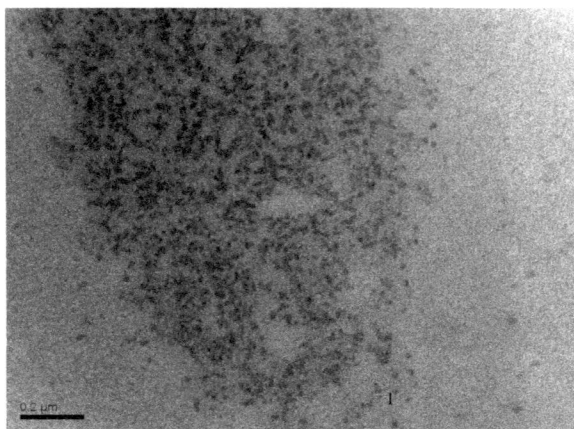

Figure 5. TEM image of CdO nanoparticles synthesized by calcination of γ-irradiated cadmium acetate for three hours at 400°C.

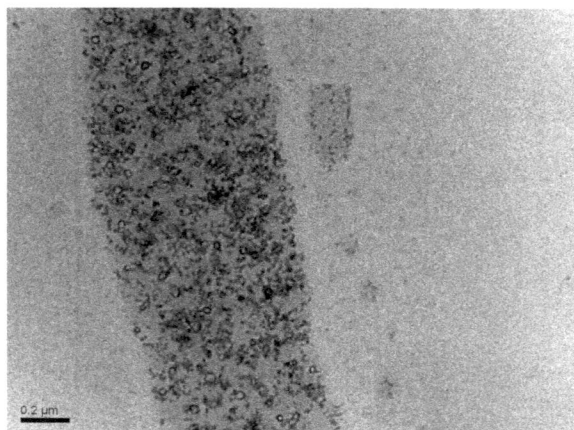

Figure 6. TEM image of CdO nanoparticles synthesized by calcination of γ-irradiated cadmium acetate for three hours at 400°C in presence of benzyl alcohol as surfactant-free

FT-IR spectrum of CdO quantum dots synthesized by method (a) is shown in Figure 7. The spectrum exhibits stretching broad band around 3,422 cm^{-1} attributed to adsorbed water (v_{OH}). The formation of CdO phase was characterized by a very broad IR band in the range of 400–1,000 cm^{-1}. This result gave further evidence for the formation of CdO quantum dots (Askarinejad & Morsali 2008). No

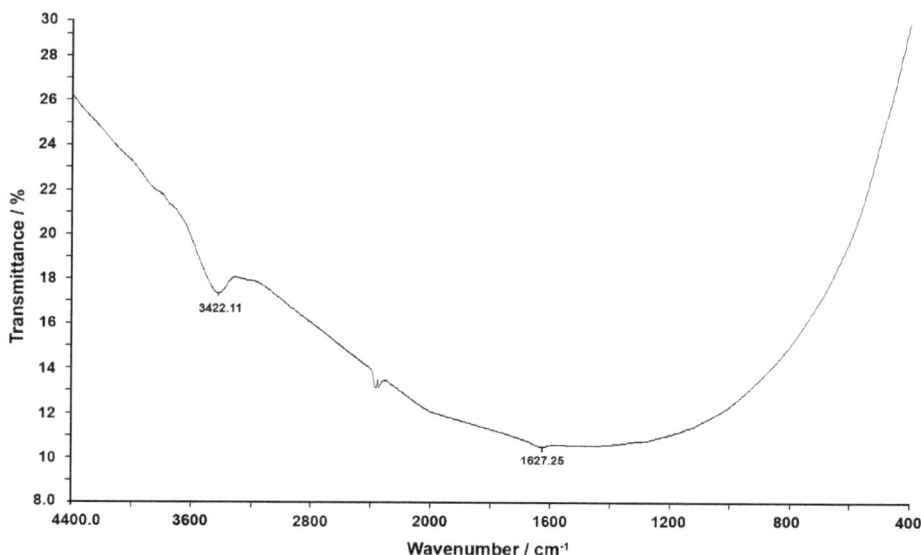

Figure 7. FT-IR spectrum of as- synthesized CdO nanoparticles obtained by calcination of γ-irradiated cadmium acetate for three hours at 400°C.

significant change in the IR spectra of the as-synthesized CdO nanoparticles was obtained as a result of the surfactant-free addition. Therefore, only one FT-IR spectrum is displayed for representation.

4. Role of irradiation

In solid sample the radiation effect is dominated by direct ionization of the material, where's for aqueous solutions the reaction with radical species, such as OH· or solvated electrons is dominate mechanism for damage to a solute.

Upon irradiation with Co-60 γ-ray source, Compton scattering is the main mode of interaction of the γ-ray with Cd and O atoms and multiple indirect ionization occurs. These events create electron–hole pairs, lattice imperfection, and extended defects. These defects lead to the creation of specific damage in the host lattice of CdO crystal and may be responsible for the formation of CdO quantum dots.

In the presence of benzyl alcohol, the trapped electrons in the host lattice of cadmium acetate may react with surfactant molecule to create another source of point and line defects in cadmium acetate crystal due to the damage of benzyl alcohol on the surface of new phase formed by CdO nanoparticles as a result of λ-irradiation. This damage may increase the rate of nucleation of the nanoparticles to give another impact on the morphologies and shapes of the as-synthesized quantum dots (West, 1999).

5. Conclusion

This study reports the synthesis of CdO quantum dots of high purity by thermal decomposition of γ-irradiated anhydrous cadmium acetate precursor. Calcination of precursor irradiated with 10^2 kGy absorbed dose at 400°C for three hours led to the formation of the mesoporous structure of CdO quantum dots. In the presence of surfactant-free as benzyl alcohol, the calcination afforded nanoribbon of CdO quantum dots and cauliflower-like structure. The interaction of γ-ray with matter lead to the formation of the lattice imperfections (ion pairs, ion defects, and extended defects) responsible for the formation of CdO quantum dots as result of the radiation damage induced by these defects in host lattice of cadmium acetate precursor.

Funding

The authors received no direct funding for this research.

Author details

Refaat M. Mahfouz[1]

E-mail: rmhfouz@science.au.edu.eg

Gamal A-W Ahmed[1]

E-mail: gamala@aun.edu.eg

Tahani Al-Rashidi[2]

E-mail: rmhfouz@hotmail.com

[1] Department of Chemistry, Faculty of Science, Assiut University, P. O. Box 71516 Assiut, Egypt.

[2] Department of Chemistry, College of Science, King Saud University, P. O. Box 2455, Riyadh, 11451, Kingdom of Saudi Arabia.

References

Askarinejad, K., & Morsali, A. (2009). Synthesis of cadmium(II) hydroxide, cadmium(II) carbonate and cadmium(II) oxide nanoparticles; investigation of intermediate products. *Chemical Engineering Journal, 150*, 569–571. http://dx.doi.org/10.1016/j.cej.2009.03.005

Dong, W., & Zhu, C. (2003). Optical properties of surface-modified CdO nanoparticles. *Optical Materials, 22*, 227–233. http://dx.doi.org/10.1016/S0925-3467(02)00269-0

Ghosh, M., & Rao, C. N. (2014). Solvothermal synthesis of CdO and CuO nanocrystals. *Chem. Phys. Lett, 393*, 493–497.

Giribabu, K., Suresh, R., Manigandan, R., Stephen, A., & Narayanan, V. J. (2013). Cadmium oxide nanoplatelets: synthesis, characterization and their electrochemical sensing property of catechol. *Journal of the Iranian Chemical Society, 10*, 771–776. http://dx.doi.org/10.1007/s13738-012-0211-3

Grado-Caffaro, M. A., & Grado-Caffaro, M. (2008). A quantitative discussion on band-gap energy and carrier density of CdO in terms of temperature and oxygen partial pressure. *Physics Letters A, 372*, 4858–4860. http://dx.doi.org/10.1016/j.physleta.2008.04.068

Kondawar, S., Mahore, R., Dahegaonkar, A., & Agrawal, S. (2011). Electrical conductivity of Cadmium Oxide nanoparticles embedded Polyaniline nanocomposites.

Advances in Applied Science Research, 2, 401–406.

Manickathai, K., Viswanathan, S. K., & Alagar, J. (2008). Synthesis and characterization of CdO and CdS nanoparticles. *Indian Journal of Pure & Applied Physics, 46*, 561–564.

Shiori, A. (1997). *Jpn. Patent No., 7, 909.*

Spinks, J. W. T., & Woods, R. J. (1990). *An Introduction to radiation chemistry* (3rd ed.). Toronto, Canada: John-Wiley and Sons Inc.

Tadjarodi, A., & Imani, M. (2011). Synthesis and characterization of CdO nanocrystalline structure by mechanochemical method. *Materials Letters, 65*, 1025–1027. http://dx.doi.org/10.1016/j.matlet.2010.12.054

Veeraputhiran, V., Gomathinayagam, V., Udhaya, A., Francy, K., & Kathrunnisa, B. (2015). Microwave Mediated Synthesis and Characterizations of CdO Nanoparticles. *Journal of Advanced Chemical Sciences, 1*, 17–19.

West, A .R. (1999). *Basic solid state chemistry* (2nd ed.). Chichester:John-Wiley and Sons, LTD: 1999, translated into Arabic by Mahfouz R. M., Al-Farhan, Kh, A., king Saud university press: Riyadh, KSA, 2008.

Yang, H., Siu, G., Zhang, X., & Yang, W. J. (2004). Preparation of CdO nanoparticles by mechanochemical reaction. *Journal of Nanoparticle Research, 6*, 539–542. http://dx.doi.org/10.1007/s11051-004-3327-2

Ye, M., Zhong, H., Zheng, W., Li, R., & Li, Y. (2007). Ultralong Cadmium Hydroxide Nanowires: Synthesis, Characterization, and Transformation into CdO Nanostrands. *Langmuir, 23*, 9064–9068. http://dx.doi.org/10.1021/la070111c

Zhang, J., Wang, Y., Lin, Z., & Huang, F. (2010). Formation and self-assembly of cadmium hydroxide nanoplates in molten composite-hydroxide solution. *Crystal Growth & Design, 10*, 4285–4291. http://dx.doi.org/10.1021/cg901559y

9

A route to simple nonionic surfactants

Sindija Brica[1], Maris Klavins[1] and Andris Zicmanis[1]*

*Corresponding author: Andris Zicmanis, Faculty of Chemistry, University of Latvia, 19, Rainis Boulevard, Riga LV- 1586, Latvia
E-mail: zicmanis@latnet.lv
Reviewing editor: Chris Smith, University of Reading, UK

Abstract: A method for the synthesis of nonionic surfactants – *N*-alkyl-*O*-(2-hydroxyethyl) carbamates is proposed by acylation of fatty amines with ethylene carbonate without any solvent or catalyst. The surface tension of the prepared surfactants was measured, toxicity and biodegradability were determined for the surfactant with *n*-dodecyl as a hydrophobic group and *N*-monosubstituted amide and hydroxyl groups for their hydrophilic part.

Subjects: Chemistry; Materials Science; Physical Sciences

Keywords: surfactants; nonionic surfactants; synthesis; surface tension; toxicity; biodegradability

1. Introduction

Surfactants can be found among the most widely spread man-manufactured substances in the world, and they are produced in millions of tons per year. Very strict regulations are in place for their utilization/exploitation and production. They should not only be highly efficient in washing and cleaning processes but also practically nontoxic, sufficiently biodegradable, and reasonably cheap to make their application favorable to people. Various methods are elaborated for production of modern surfactants, both chemical and enzymatic (Friedli, 2001; Holmberg, 2003; Rosen & Kunjappu, 2013; Rosenholm, 2014; Somasundaran, 2006) allowing the preparation of anionic, cationic, zwitterionic, and nonionic surfactants, useful for different applications. Syntheses of these surfactants usually include several stages and demand time. Prices of reagents and the degree of complexity of their preparations together with the biological properties of these surfactants quite frequently limit exploitation of such surfactants. Therefore, the elaboration of a straightforward method for the synthesis of nonionic surfactants is discussed in this communication.

2. Results and discussion

Recently we have found that a simple acylation reaction of primary alkyl amines (**2**) with ethylene carbonate (**1**) at reasonably high temperatures (~150°C) resulted in the formation of promising nonionic surfactants—*N*-alkyl carbamates with hydrophobic alkyl group at nitrogen atom and N–H and O–H bonds in the hydrophilic parts of the surfactants (**3**) (Scheme 1).

ABOUT THE AUTHORS

Sindija Brica, Maris Klavins, and Andris Zicmanis are interested in the development of modern surfactants with better biodegradability than existing ones useful for different applications. Their research is focused on the elaboration of such materials that might be produced in a large scale, including industrial production.

PUBLIC INTEREST STATEMENT

This paper describes a simple method for the preparation of nonionic surfactants that are useful in different exploitation areas. These surfactants form by heating fatty amines with ethylene carbonate at 150°C without any other solvent or catalyst. Obtained surfactants have surface activity comparable to other nonionic surfactants, and they are highly biodegradable materials. Therefore, they may be used when the simplicity of their preparation is important.

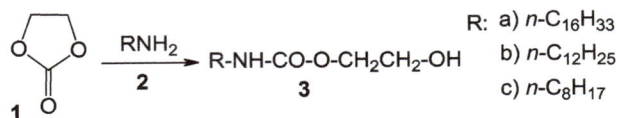

Scheme 1. Reaction scheme showing the acylation reaction of primary amines with ethylene carbonate.

Preparation of these surfactants is extremely simple—just heating both reagents **1** and **2** in equimolar ratio without any solvent, any catalyst and any inert gas protection. The simplicity of the proposed method differs it from approaches described in literature for similar purposes (Fujita, Bhanage, Kanamaru, & Arai, 2005; Hazard, Cheymol, Charbrier, Sekera, & Eche-Fialaire, 1961; Sekera, 1962). Acylation reactions of alcohols with freshly prepared carbamoyl chlorides (Hazard et al., 1961; Sekera, 1962) or transesterification reactions of corresponding carbamates with alcohols proposed by various authors (Fujita et al., 2005; Hazard et al., 1961) are very much more complicated and time-consuming methods for fabrication of carbamates. Even the recently shown acylation of octylamine with ethylene carbonate has required the use of a poisonous catalyst—$Bu_2Sn(OCH_3)_2$, and the necessity to remove the excess of amine under vacuum at the end of the process (Hellawell et al., 2014). Traces of poisonous and dangerous catalyst have been acceptable in compositions of greases—the subject of a patent (Hellawell et al., 2014) but their presence is certainly not imaginable in materials foreseen for washing and cleaning.

Structures of obtained carbamates (**3**) were confirmed by their ^{1}H-NMR and IR spectra; acceptable data of elemental analyses were also obtained. Good correspondence of spectral and analytical data was achieved (Table 1).

Sufficiently long linear carbon atom chains ($C_8 - C_{16}$) form the lipophilic part of the proposed surfactants, and N–H bond of monosubstituted amide group together with O–H bond provide acceptable surface tension (32–43 mN/cm) and critical micelle concentration (cmc, 8.4–11.9 mg/L) for these novel surfactants (**3**) (Table 2). The surface tension of obtained N-alkyl carbamates (**3**) measured by the du Noüy ring method at 25°C differs a little depending on the length of alkyl chain but remains in range typical for nonionic surfactants (Friedli, 2001; Holmberg, 2003; Rosen & Kunjappu, 2013; Somasundaran, 2006).

Table 1. Yields and analyses of surfactants–carbamates

Surfactant	Yield, %	M.p.,°C	Analyses, %*			^{1}H-NMR spectra, δ, ppm**			IR spectra, cm^{-1}***		
			C	H	N	CH$_2$OC	CH$_2$OH	CH$_2$N	$\nu_{C=O}$	ν_{N-H}	δ_{N-H}
3a	79.8	78–80	69.46 / 69.25	12.16 / 11.93	4.28 / 4.25	4.20	3.80	3.17	1,685	3,353	1,541
3b	79.1	58–60	66.01 / 65.89	11.60 / 11.43	5.14 / 5.12	4.20	3.80	3.17	1,685	3,298	1,568
3c	79.0	45–46	60.75 / 60.80	10.84 / 10.67	6.26 / 6.45	4.19	3.79	3.16	1,699	3,311	1,551

*Found results vs. calculated data.

**Determined in CDCl$_3$ solution on *Varian 400 NMR* spectrometer, internal standard TMS.

***Registered in KBr tablet on the instrument *Perkin Elmer Spectrum One*.

Table 2. The behavior of surfactants–carbamates in water solutions

Surfactant	Surface tension, mN/cm*	cmc, mg/L
3a	42.1	11.9
3b	38.5	9.8
3c	32.5	8.4

*Determined by the du Noüy ring method at 25°C at concentrations of surfactants 100 mg/L.

It should be mentioned that these surfactants are also comparatively nontoxic substances. *Daphnia's* are the most commonly used crustacean test species for determination of the effects of xenobiotics on primary consumers in freshwater aquatic ecosystems (*Daphtoxkit F Magna*, 2000). EC_{50} (half maximum effective concentration) for the carbamate **3b** after 24 and 48 h was found to be 2 and 0.5 mg L^{-1}, respectively. The comparison of our results obtained for *Daphnia magna* with earlier literature in the context of ecotoxicity (EC_{50}) of carbamates showed that surfactant **3b** displayed similar or lower toxicity than other surfactants (Rosen & Kunjappu, 2013; Somasundaran, 2006). Germination tests with *Lepidium sativum L.*, *Raphanus sativus L.* and *Secale cereale L.* in the concentration range of the surfactant **3b** from 0.5 to 100 L^{-1} did not reveal any inhibition effect.

The biodegradability of the surfactant **3b** was determined by CO_2 evolution in sealed vessels (*OECD Guidelines for the Testing of Chemicals*, 2014). BOD_3 test has revealed some difference between sets amended with surfactant **3b** (10 mg L^{-1}) and control without any chemical. After 72 h incubation period, oxygen consumption and without it was found to be 18.4 ± 2 mg L^{-1} and 8.4 ± 0.0 mg L^{-1}, respectively. The results indicate potentially high biodegradability of the surfactant **3b** by microorganisms (Holmberg, 2003; Rosen & Kunjappu, 2013; Somasundaran, 2006).

3. Experimental

All reagents were used as purchased from commercial suppliers without further purification. All melting points were determined in open capillaries, using Thomas Hoover melting point apparatus, expressed in °C, and are uncorrected. ^{1}H-NMR spectra of the compounds were recorded on Varian 400 MR NMR spectrometer using TMS as an internal standard. Infrared spectra (IR) were recorded on Perkin Elmer Spectrum One Infrared Spectrometer, and samples were screened in potassium bromide (KBr) pellets. Elemental analyses were performed on a Flash EA1112 CHNS analyzer (Thermo Electron Corporation).

N-Dodecyl-*O*-(2-hydroxyethyl)-carbamate (**3b**) (a typical experiment). Dodecyl amine (5.55 g; 30 mmol) and ethylene carbonate (2.64 g; 30 mmol) were stirred in a round bottom flask at 150°C (±5°C) for 3 h. No inert gas protection was required. *tert*-butyl methyl ether (40 mL) was added to the cooled to ~90°C reaction mixture, and the obtained solution was carefully mixed. The precipitate was separated from the mother liquid by suction using a Büchner funnel after keeping the reaction mixture at room temperature during 16 h. The precipitate was dried in open air. *N*-Dodecyl-*O*-(2-hydroxyethyl)-carbamate (**3b**, 6.48 g, 79.1%) was obtained in the form of white crystals with m.p. 58–60°C. Found, %: C, 66.01; H, 11.60; N, 5.14. $C_{15}H_{31}NO_3$. Calculated, %: C 65.89; H, 11.43; N, 5.12. ^{1}H-NMR spectrum (CDCl$_3$, δ-ppm): 4.78 (m, 1H); 4.20 (t, 2H); 3.80 (t, 2H); 3.18 (m, 2H); 2.42 (m, 1H); 1.49 (t, 2H); 1.27 (m, 18H); 0.88 (t, 3H). IR spectrum (KBr ν_{max} cm^{-1}): 3,298 (ν, N–H); 2,951–2,853 (ν, C–H); 1,685 (ν, C=O); 1,568 (δ, N–H).

Other carbamates (**3a** and **3c**) were prepared and characterized in a similar way.

N-Hexadecyl-*O*-(2-hydroxyethyl)-carbamate (**3a**), yield 79.8%, white crystals, m.p. 78–80°C. Found, %: C, 69.46; H, 12.16; N, 4.28. $C_{19}H_{39}NO_3$. Calculated, %: C 69.25; H, 11.93; N, 4.25. ^{1}H NMR spectrum (CDCl$_3$, δ-ppm): 4.77 (m, 1H); 4.20 (t, 2H); 3.80 (t, 2H); 3.17 (m, 2H); 2.39 (m, 1H); 1.49 (t, 2H); 1.25 (m, 26H); 0.88 (t, 3H). IR spectrum (KBr ν_{max} cm^{-1}): 3,353 (ν, N–H); 2,951–2,853 (ν, C–H); 1,685 (ν, C=O); 1,541 (δ, N–H).

N-Octyl-*O*-(2-hydroxyethyl)-carbamate (**3c**), yield 79.0%, white crystals, m.p. 45–46°C. Found, %: C, 60.75; H, 10.84; N, 6.26. $C_{11}H_{23}NO_3$. Calculated, %: C 60.80; H, 10.67; N, 6.45. ^{1}H NMR spectrum (CDCl$_3$, δ-ppm): 4.87 (m, 1H); 4.19 (t, 2H); 3.79 (t, 2H); 3.16 (m, 2H); 2.42 (m, 1H); 1.48 (t, 2H); 1.28 (m, 8H); 0.87 (t, 3H). IR spectrum (KBr ν_{max} cm^{-1}): 3,311 (ν, N–H); 2,951–2,853 (ν, C–H); 1,699 (ν, C=O); 1,551 (δ, N–H).

4. Conclusion

Accordingly, our results confirm that primary fatty alkyl amines can be very easily converted into corresponding surfactants—alkyl carbamates. Conditions of their syntheses are truly simple—just stirring and heating reagents in an equimolar ratio without any inert gas protection in a common laboratory flask. Obtained surfactants have surface activity comparable to other nonionic surfactants but they are considerably less toxic and highly biodegradable materials. Hence, these surfactants may have a wide exploitation when the simplicity of their preparation is important.

Acknowledgments

Authors are grateful to Dr. O. Muter for determination of toxicity and biodegradability of investigated surfactants and Mr. L. Arbidans for measuring the surface tension.

Funding

Financial support from the European Regional Development Foundation [grant number 2DP/2.1.1.1.0/14/APIA/VIAA/016] is gratefully acknowledged.

Author details

Sindija Brica[1]
E-mail: sindijabrica@gmail.com
Maris Klavins[1]
E-mail: maris.klavins@lu.lv
Andris Zicmanis[1]
E-mail: zicmanis@latnet.lv
[1] Faculty of Chemistry, University of Latvia, 19, Rainis Boulevard, Riga LV-1586, Latvia.

References

Daphtoxkit F Magna. Crustacean toxicity screening test for freshwater. Standard operational procedure. (2000). Mariakerke: MicroBioTests. Retrieved from http://www.microbiotests.be/SOPs/Daphtoxkit%20magna%20F%20SOP%20-%20A5.pdf

Friedli, F. E. (Ed.). (2001). Detergency of specialty surfactants. New York, NY: Marcel Dekker. ISBN 978-8793102040.

Fujita, S., Bhanage, B., Kanamaru, H., & Arai, M. (2005). Synthesis of 1,3-dialkylureas from ethylene carbonate and amines using calcium oxide. Journal of Molecular Catalysis A: Chemical, 230, 43–48. doi:10.1016/j.molcata.2004.12.014

Hazard, R., Cheymol, J., Charbrier, P., Sekera, A., & Eche-Fialaire, R. (1961). Chemistry and pharmacology of the amino esters of carbamic acids and their quaternary ammonium salts. Bulletin de la Société chimique de France, 11, 2087–2091.

Hellawell, A., Kone, E., Massey, A., Mead, H. B., Van Zwieten, D., Visser, T., & Wilkinson, R. J. (2014). Process for preparing a urea grease [WO 2014/122273 A1]. New York, NY: Shell Oil Company.

Holmberg, K. (Ed.). (2003). Novel surfactants: Preparation, applications and biodegradability (2nd ed.). New York, NY: Marcel Dekker. ISBN 978082474300.

OECD Guidelines for the Testing of Chemicals. (2014). Test No. 310: Ready biodegradability – CO_2 in sealed vessels (Headspace Test). doi:10.1787/2074577x

Rosen, M. J., & Kunjappu, J. T. (2013). Surfactants and interfacial phenomena (4th ed.). New York, NY: Wiley. ISBN 978-0-470-54194-4.

Rosenholm, J. B. (2014). Phase equilibriums, self-assembly and interactions in two-, three- and four medium-chain length component systems. Advances in Colloid and Interface Science, 205, 9–47. doi:10.1016/j.cis.2013.08.009

Sekera, A. (1962). Syntheses and pharmacological studies of a series of basic esters of substituted carbamic acids. Journal mondial de pharmacie, 5, 5–18.

Somasundaran, P. (Ed.). (2006). Encyclopedia of surface and colloid science (2nd ed.). New York, NY: Taylor & Francis. ISBN 9780849396151.

10

Varying the flexibility of the aromatic backbone in half sandwich rhodium(III) dithiolato complexes: A synthetic, spectroscopic and structural investigation

Phillip S. Nejman[1], Alexandra M.Z. Slawin[1], Petr Kilian[1] and J. Derek Woollins[1]*

*Corresponding author: J. Derek Woollins, EaStChem School of Chemistry, University of St Andrews, St Andrews, Fife KY16 9ST, UK

E-mail: jdw3@st-andrews.ac.uk

Reviewing editor: Massimiliano Arca, University of Cagliari, Italy

Abstract: A series of rhodium(III) half sandwich complexes of the type [Cp*Rh(PMe$_3$) (S–R–S)]; S–R–S = naphthalene-1,8-dithiolate, acenaphthene-5,6-dithiolate, [1,1'-biphenyl]-2,2'-dithiolate and [2,2'-binaphthalene]-1,1'-dithiolate are reported. In the case of [2,2'-binaphthalene]-1,1'-dithiolate, this represents an infrequent example of a metal complex containing this ligand. All the complexes have been fully characterised using multinuclear NMR spectroscopy and single-crystal X-ray diffraction. The single-crystal X-ray structure of the starting material [Cp*Rh(PMe$_3$)Br$_2$] (**1**) is also reported for the first time.

Subjects: Chemistry; Chemical Spectroscopy; Inorganic Chemistry

Keywords: dithiolates; half sandwich; complexation; rhodium

1. Introduction

The coordination of S,S'-bidentate ligands remains an important area of chemistry. Complexes bearing this type of ligand have a number of industrial applications including vulcanisation (Bond & Martin, 1984; Burns & McAuliffe, 1979; Burns, McCullough, & McAuliffe, 1980; Eisenberg, 2007), lubricant additives (Phillips et al., 1995) and catalysis (Bond & Martin, 1984; Burns & McAuliffe, 1979; Burns et al., 1980; Eisenberg, 2007). In addition, S,S'-donors can support unusual magnetic properties (Tuna et al., 2012; Zhou, Wang, Wang, & Gao, 2011) and are important in biological systems (Woollins, 1996). As part of our interest in the properties of sulfur donor systems, we have investigated a series of dithiolate ligands bound to aromatic backbones of varying flexibility (Figure 1).

There has been little study on the coordination chemistry of these types of ligands compared to dithiolates such as benzene-1,2-dithiolate or ethane-1,2-dithiolate. One of the most notable exceptions to this was a series of publications by Teo and co-workers in the late 1970s and early 1980s (Teo, Bakirtzis, & Snyder-Robinson, 1983; Teo & Snyder-Robinson, 1978, 1979a, 1979b, 1981, 1984; Teo, Wudl, Hauser, & Kruger, 1977; Teo, Wudl, Marshall, & Kruger, 1977). They investigated the oxidative addition of the structurally related compounds tetrathionaphthalene (TTN), tetrachlorotetrathionaphthalene (TCTTN) and tetrathiotetracene (TTT) (Figure 2) to a variety of low-valent metal

ABOUT THE AUTHOR

J. Derek Woollins has a long-standing interest in the synthetic and structural chemistry of group 16 elements and their coordination compounds. Recent work has investigated peri substituted naphthalenes and related systems, and the work described here examines the coordination of some peri and related ligand systems.

PUBLIC INTEREST STATEMENT

S,S'-bidentate ligands remain an important area of chemistry. Complexes bearing this type of ligand have a number of industrial applications including vulcanisation, lubricant additives and catalysis. In addition, S,S'-donors can support unusual magnetic properties and are important in biological systems This work describes the coordination of some simple bidentate S,S ligands to Iridium

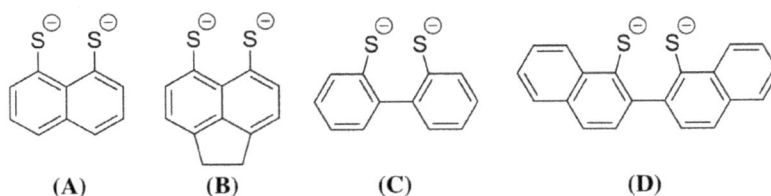

Figure 1. Dithiolate ligands studied in this work (charges omitted for clarity).

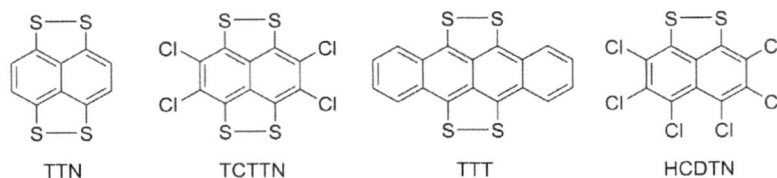

Figure 2. Structurally related aromatic sulfur donating ligands.

centres. With the work focusing on the extensive redox chemistry associated with these "non-inno-cent" ligands and their potential use as organic solid-state conductors (Teo & Snyder-Robinson, 1978, 1979a, 1979b, 1981, 1984; Teo, Wudl, Hauser, et al., 1977; Teo, Wudl, Marshall, 1977; Teo et al., 1983). Another interesting system bearing the related hexachlorodithionaphthalene (HCDTN) (Figure 2) resulted in an unusual trinuclear nickel complex [Ni$_3$(PPh$_3$)$_3$(S$_2$C$_{10}$Cl$_6$)$_3$] with the HCDNT acting as a bridging ligand (Bosman & van der Linden, 1977).

Some of the most recent work involving derivatives of naphthalene-1,8-dithiolate (**A**) has been in designing [FeFe]-hydrogenase mimics for the production of hydrogen, by both electrochemical and photochemical processes (Figure 3) (Figliola, Male, Horswell, & Grainger, 2015; Figliola et al., 2014; Samuel, Co, Stern, & Wasielewski, 2010; Wright, Lim, & Tilley, 2009). Complexes involving [1,1'-biphenyl]-2,2'-dithiolate (**C**) bound to iron have also been investigated as electron transfer catalysts designed to mimic iron hydrogenases (Figure 3) (Albers et al., 2014; Ballmann, Dechert, Demeshko, & Meyer, 2009; Charreteur et al., 2010). The coordination chemistry of the structurally related ligand acenaphthene-5,6-dithiolate (**B**) has received very little investigation out with our own research. (Topf, Monkowius, and Knör (2012) used the acenaphthene backbone as a linker be-tween a 1,2-diimine unit and a dithiolate binding site. The iron carbonyl complex formed using this ligand showed potential as a multielectron transfer photosensitiser for artificial photosynthesis and as a bio-inspired photoredox catalyst (Figure 3).

Beyond the electron transfer mimics, there are few examples of complexes incorporating ligand **C**. Two molybdenum complexes have been reported with one containing an Mo oxygen triple bond (Conry & Tipton, 2001; McNaughton, Tipton, Rubie, Conry, & Kirk, 2000). In addition, two methods to synthesise titanocene-2,2'-dithiolatobiphenyl have been published (Aucott et al., 2005; Stafford, Rauchfuss, Verma, & Wilson, 1996). The 2,2'-binaphthalene-based ligand, **D**, has been largely

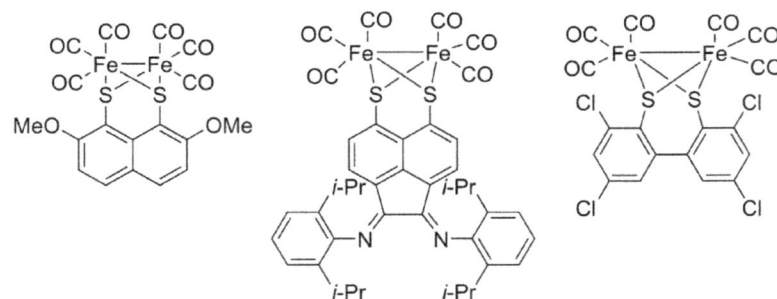

Figure 3. Iron-based catalysts based on A (left), B (centre) and C (right).

overlooked with regards to its complexation chemistry compared to the 1,1'-binaphthalene derivative. The only work reported involving this ligand thus far has been its preparation by Armarego (1960), the single-crystal X-ray structure (Kempe, Sieler, Hintzsche, & Schroth, 1993), one titanium complex (Aucott et al., 2005) and a platinum complex (Aucott, Kilian, Robertson, Slawin, & Woollins, 2006).

We have recently published two papers investigating the use of ligands **A–C** in the formation of half sandwich rhodium and iridium complexes (Nejman, Morton-Fernandez, Black, et al., 2015; Nejman, Morton-Fernandez, Moulding, et al., 2015). This paper describes the synthesis of a further four half sandwich rhodium (III) dithiolato complexes bearing a neutral phosphine donor. In this contribution, a different phosphine donor has been used compared to the previous work (trimethylphosphine instead of triethylphosphine) (Nejman, Morton-Fernandez, Black, et al., 2015). Additionally, a new ligand has been investigated [2,2'-binaphthalene]-1,1'-dithiolate (**D**). Two synthetic methods were employed due to the varying difficulty in preparing the ligand precursors. Both of these methods were different to the procedures used within our previous work (Nejman, Morton-Fernandez, Black, et al., 2015; Nejman, Morton-Fernandez, Moulding, et al., 2015). Furthermore, the single-crystal X-ray structure of the complex precursor, [Cp*Rh(PMe$_3$)Br$_2$] (**1**), is reported for the first time. All the complexes have been fully characterised, principally by multinuclear NMR spectroscopy, mass spectrometry and single-crystal X-ray diffraction.

2. Results and discussion

2.1. Synthetic methods

The dithiol pro-ligands [Naphth(SH)$_2$] (**H$_2$a**), [Acenap(SH)$_2$] (**H$_2$b**) and [Biphen(SH)$_2$] (**H$_2$c**) were prepared from their respective disulphides, naphtho[1,8-*cd*]-1,2-dithiole (Ashe, Kampf, & Savla, 1994), 5,6-dihydroacenaphtho[5,6-*cd*]-1,2-dithiole (Benson et al., 2013) and dibenzo[*c,e*]-1,2-dithiine (Cossu, Delogu, Fabbri, & Maglioli, 1991). The reduction of the disulphides was performed using NaBH$_4$ followed by an acidic work up which afforded the three pro-ligands (Figure 6) (Yui, Aso, Otsubo, & Ogura, 1988). The disulphide precursor to **D**, [2,2'-BinapS$_2$], was prepared according to the literature procedure by Armarego (1960). The reduction to the dithiol was not attempted as the amount of disulphide prepared was not sufficient. For this reason, the reduction to the reactive dithiolate was performed *in situ* using lithium triethylborohydride.

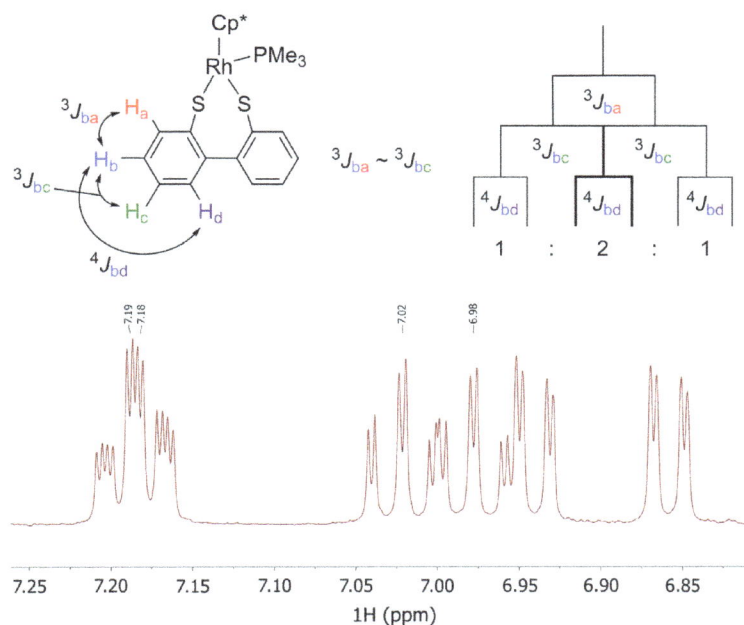

Figure 4. The section of the ¹H NMR spectrum of 2c showing the four *pseudo* triplet of doublets with a splitting diagram showing how these signals are formed.

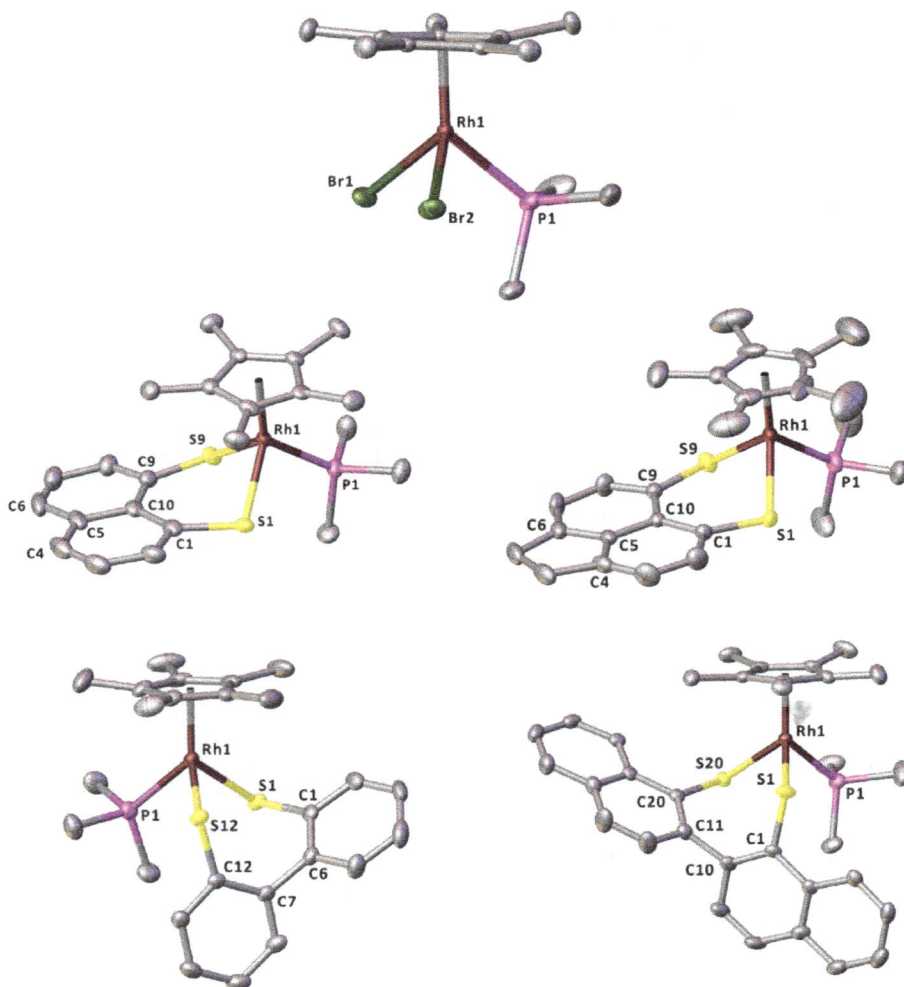

Figure 5. Crystal structures of 1 (Top), 2a (Middle left), 2b (Middle right), 2c (Bottom left) and 2d (Bottom right). Hydrogen atoms are omitted from all structures for clarity. Ellipsoids are plotted at the 50% probability level.

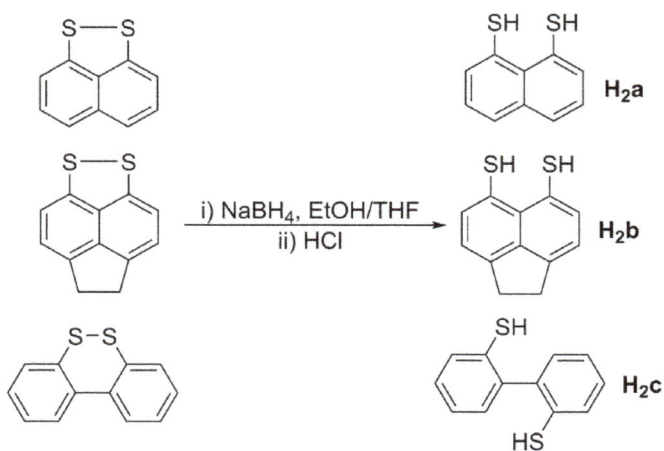

Figure 6. Synthesis of the pro-ligands H₂a-c.

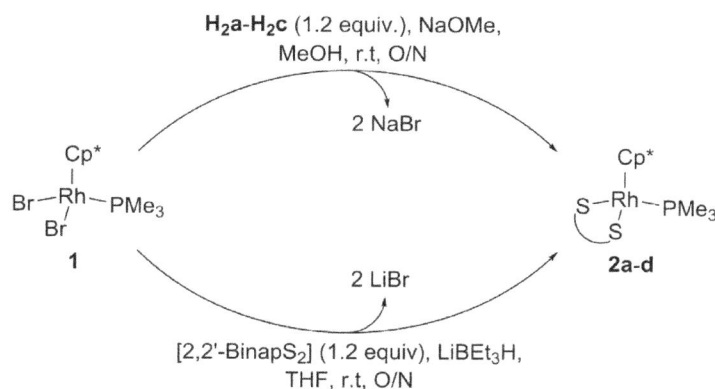

Figure 7. Reaction conditions for the preparation of 2a-d.

The synthesis of [Cp*Rh(PMe$_3$)(NaphthS$_2$)] (**2a**), [Cp*Rh(PMe$_3$)(AcenapS$_2$)] (**2b**), [Cp*Rh(PMe$_3$)(BiphenS$_2$)] (**2c**) and [Cp*Rh(PMe$_3$)(2,2'-BinapS$_2$)] (**2d**) is shown in Figure 7. The metathesis of the bromide ligands with dithiolates **A–D** proceeds smoothly at room temperature. The presence of the phosphine group allowed the reaction progress to be monitored by ^{31}P NMR spectroscopy. In the case of **2a–c**, this meant the loss of the signal from **1**, whilst for **2d**, the product and **1** are soluble in THF so the conversion could be observed. Upon completion of the reaction for **2a–c** filtration of the precipitate followed by washing with methanol was sufficient to afford pure compound. For **2d**, purification by column chromatography was required (silica/CH$_2$Cl$_2$). Excellent isolated yields of between 82 and 91% were obtained after purification. Compared to the similar complexes incorporating triethylphosphine, this represents an increase in the isolated yield of approximately 10% (Nejman, Morton-Fernandez, Black, et al., 2015). The synthesis of **2a–d** all proceeded via a reactive dithiolate intermediate, whereas the triethylphosphine derivatives were prepared by reacting **H$_2$a–c** directly with the dichloro rhodium precursor. The dithiols represent a less reactive sulfur centre which could explain the higher yields obtained for the complexes reported here.

2.2. Data analysis

The ^{1}H NMR spectra (CDCl$_3$) for **2a** and **2b** show the expected signals, with splitting, from the aromatic backbones in the range of 7.88–6.91 ppm. In the case of **2c** and **2d**, we observe 8 and 10 signals, respectively, as the two joined aryl ring systems are inequivalent. This is due to the inability of the ligand backbones to rotate around the aryl–aryl bond. Four of the aromatic signals observed for **2c** appear as a *pseudo* triplet of doublets instead of the expected doublet of doublet of doublets. This is due to the $^3J_{HH}$ coupling constants observed between H$_b$ and H$_a$ as well as H$_b$ and H$_c$ (Figure 4) being almost identical. Two of these signals overlap closely (δ_H 7.19 and 7.18 ppm, Figure 4); however, both can be distinctly observed and the coupling constants easily extracted. These observations mirror those made for other similar rhodium complexes we have prepared incorporating ligands **A–C** (Nejman, Morton-Fernandez, Black, et al., 2015). For **2d**, this signal is not observed as the equivalent positions from each of the two naphthalene ring systems overlap resulting in multiplets. The η^5–Cp* methyl signals range from 1.58 to 1.42 ppm and are split into doublets by long range phosphorus coupling ($^4J_{Hp}$ = 3.0–3.2 Hz). The signals from the methyl groups attached to the phosphorus atom appear as a doublet of doublets for **2a–c**, with $^3J_{Hp}$ coupling ($^3J_{Hp}$ = 10.3–10.5 Hz) and long range $^4J_{HRh}$ coupling ($^4J_{HRh}$ = 0.6–0.7 Hz). Only a doublet is observed for this signal in **2d** with a similar $^3J_{Hp}$ coupling to that seen in complexes **2a–c**.

The ^{31}P{^{1}H} NMR spectra (CDCl$_3$) for **1** and **2a–d** are shown in Table 1. Complexes **2a–d** all display an upfield shift in the ^{31}P{^{1}H} NMR spectra compared to the starting material **1** ($\Delta\delta$ = 0.7–3.3 ppm). The coordination of the dithiolate ligand is also accompanied by a small increase in the $^1J_{PRh}$ coupling (Δ^1J_{PRh} = 11–15 Hz) in **2a–d** when compared to **1**. Both of these observations match those made in the ^{31}P{^{1}H} NMR spectra of the triethylphosphine derivatives of **2a–c** (Nejman, Morton-Fernandez, Black, et al., 2015).

Table 1. $^{31}P\{^1H\}$ NMR data (CDCl$_3$) for 1 and 2a-d. All δ values are in ppm and J values are in hertz

	1[†]	2a	2b	2c	2d
δ_p	3.6	1.7	2.9	0.3	2.2
$^1J_{PRh}$	137	150	148	153	152

[†]Values obtained from a sample run on a Bruker Avance II 400 NMR spectrometer (162 Hz).

Table 2. Selected bond lengths [Å] and angles [°] for 1

	1
Rh1–P1	2.284(2)
Rh1–Br1	2.529(1)
Rh1–Br2	2.550(1)
P1–Rh1–Br1	87.14(3)
P1–Rh1–Br2	86.91(3)
Br1–Rh1–Br2	94.82(2)

As expected, the $^{13}C\{^1H\}$ NMR spectra (CDCl$_3$) of **2a–d** mirrors that of the ^1H NMR spectra, with distinct signals for all carbons in the biphenyl and binaphthyl examples. Interestingly, only one of the quaternary carbons bound to the sulfur atoms is split into a doublet ($^3J_{Cp}$ = 6.5 Hz (**2c**) and 6.7 Hz (**2d**)) by the phosphorus atom in **2c** and **2d**. This provides further support of the difference between the two sides of the aryl–aryl bond. The mass spectra of **2a–d** each showed a peak corresponding to [M–PMe$_3$ + H]$^+$ ions at m/z 429, 455, 455 and 555, respectively. Homogeneity of the complexes **2a–d** was confirmed by means of accurate elemental analysis.

2.3. Single-crystal X-ray diffraction

Despite the previously reported synthesis and spectroscopic characterisation of precursor complex **1** (Jones & Feher, 1984), its single-crystal X-ray data have not been published. Therefore, for completeness we include it here. The crystal structures of **1** and **2a–d** are shown below in Figure 5 with selected structural parameters in Tables 2 and 3.

All of the complexes, **1** and **2a–d**, adopt the piano stool geometry around the rhodium centre we have seen previously (Nejman, Morton-Fernandez, Black, et al., 2015; Nejman, Morton-Fernandez, Moulding, et al., 2015). The η^5-Cp* ring in **1** is slightly tilted as the Rh–C bond lengths vary from 2.141(4) to 2.252(4) Å. The Rh–Br bond lengths (2.550(1) and 2.529(1) Å) and Rh–P bond length (2.284(2) Å) are similar to those previously reported within the Cambridge Structural Database for compounds of a similar type (Rh–Br; 2.543 Å, Rh–P; 2.288 Å) (Bruno et al., 2002; Macrae et al., 2008; Thomas et al., 2010). The angles around the rhodium centre vary with the two P–Rh–Br angles being below the idealised 90°. This is accompanied by a widening of the Br–Rh–Br angle to 94.82(2)° as the two larger atoms try and sit further apart.

The Rh–S bond lengths of **2a** and **2b** were almost identical (**2a**; 2.331(1) and 2.332(1) Å, **2b**; 2.330(2) Å), whilst in **2c** and **2d** there was more variation and they were slightly longer (**2c**; 2.3691(8) and 2.3705(8) Å, **2d**; 2.3677(7) and 2.3994(8) Å). These are comparable to other half sandwich complexes with Rh–S bonds reported by ourselves and Jin and co-workers ranging from 2.340 to 2.386 Å (Nejman, Morton-Fernandez, Black, et al., 2015; Wang, Lin, Blacque, Berke, & Jin, 2008; Xiao & Jin, 2008; Yao, Xu, Huo, & Jin, 2013). The Rh–P bond lengths show no appreciable change compared to **1** with little variation across the series of **2a–d**.

All of the non-Cp* angles around the rhodium centre are reduced to less than 90° for **2a** and **2b**. This is a consequence of the rigid ligand backbone preventing the sulfur atoms from adopting a more ideal geometry. The effect is most obvious for the naphthalene system as the *peri* positions are restricted to a slightly shorter distance than those in the acenaphthene system. For both **2a** and **2b**,

Table 3. Selected bond lengths [Å], angles [°] and displacements [Å] for 2a-d

	2a	2b	2c	2d
Rh1–P1	2.284(1)	2.272(2)	2.2773(9)	2.2851(8)
Rh1–S1	2.331(1)	2.330(2)	2.3691(8)	2.3677(7)
Rh1–S9	2.332(1)	2.330(2)		
Rh1–S12			2.3705(8)	
Rh1–S20				2.3994(8)
P1–Rh1–S1	89.85(5)	87.02(3)	88.79(3)	96.29(3)
P1–Rh1–S9	84.87(5)	87.37(4)		
P1–Rh1–S12			91.27(3)	
P1–Rh1–S20				84.11(3)
S1–Rh1–S9	85.41(5)	87.78(3)		
S1–Rh1–S12			93.94(3)	
S1–Rh1–S20				95.00(3)
Splay angle[a]	19.2(5)	21.0(3)		
Torsion angles				
S1–C1···C9–S9	8.9(3)	5.2(2)		
C1–C10–C5–C6	177.5(6)	178.5(3)		
C9–C10–C5–C4	178.1(6)	177.3(3)		
C1–C6–C7–C12			68.0(4)	
C1–C10–C11–C20				79.0(4)
Out of plane displacements				
S1	0.213	0.117	0.184	0.189
S9	0.149	0.121		
S12			0.003	
S20				0.086

[a]Calculated as [(S1–C1–C10)+(C1–C10–C9)+(C10–C9–S9)−360].

the splay angles are large and positive (**2a**; 19.2(5)°, **2b**; 21.0(3)°) as the rhodium centre forces the sulfur atoms apart. The S1–C1···C9–S9 torsion angle is larger in **2a** than **2b**, again as a consequence of the more limited movement of the sulfur atoms imposed by the backbone. The central C–C–C–C torsion angles are similar in both complexes showing limited buckling of the ring system. The out of plane displacement of the sulfur atoms is slightly greater in **2a** than **2b**.

In complexes **2c** and **2d** the non-Cp* angles show a broader range than in **2a** and **2b** ranging from 88.79(3)–93.94(3)° to 84.11(3)–96.29(3)°, respectively. The ability of the backbone to twist around the aryl–aryl bond allows the sulfur atoms to adopt a more idealised geometry. The torsion angle between the two aryl rings is larger in **2d** (79.0(4)°) than **2c** (68.0(4)°) most likely due to the added steric bulk of having a binaphthyl instead of biphenyl-based system. In both **2c** and **2d**, the out of plane displacement of the sulfur atoms are similar.

3. Conclusions

We have prepared and fully characterised a series of new rhodium(III) η⁵–e have prepared and fully characterised a series of new rhodium(III) ηad ₃)Br₂] with a series of dithiolates attached to aromatic backbones. Similar features were seen in both the NMR spectra and single-crystal X-ray structures compared to previous rhodium complexes incorporating ligands **A–C** we have reported (Nejman, Morton-Fernandez, Black, et al., 2015; Nejman, Morton-Fernandez, Moulding, et al., 2015). The work herein clearly demonstrates the utility of these sulfur ligands in organometallic complexes, with complexes of this type having potential uses in the formation of multimetallic systems.

4. Experimental

4.1. General

Unless otherwise stated all manipulations were performed under an oxygen-free nitrogen atmosphere using standard Schlenk techniques and glassware. Solvents were collected from an MBraun Solvent Purification System or dried and stored according to common procedures (Armarego & Chai, 2009). [Cp*Rh(PMe$_3$)Br$_2$] was prepared following literature procedures (Ojima, Vu, & Bonafoux, 2002). The disulphide ligand precursors were made according to literature methods (Armarego, 1960; Ashe et al., 1994; Benson et al., 2013; Cossu et al., 1991). The pro-ligand H$_2$a was prepared following the literature procedure (Yui et al., 1988), with H$_2$c prepared following an identical procedure. H$_2$b was prepared according to literature (Nejman, Morton-Fernandez, Black, et al., 2015). ^{1}H, ^{13}C{^{1}H}, ^{31}P and ^{31}P{^{1}H} NMR spectra were obtained on either a Bruker Avance II 400 or Bruker Avance III 500 spectrometer. Full assignments of the ^{1}H and ^{13}C{^{1}H} NMR spectra were made with the aid of H–H DQF COSY, H–C HSQC and H–C HMBC experiments. For ^{1}H and ^{13}C{^{1}H} spectra δ_H and δ_C are reported relative to TMS, residual solvent peaks (CDCl$_3$; δ_H 7.26, δ_C 77.2 ppm) were used for calibration. For ^{31}P and ^{31}P{^{1}H} spectra δ_P are reported relative to external 85% H$_3$PO$_4$. All measurements were performed at 21°C with shifts reported in ppm. p-td has been used to denote a pseudo-triplet of doublet. IR spectra were collected on a Perkin Elmer 2000 NIR/Raman Fourier transform spectrometer with a dipole pumped NdYAG near-IR excitation laser. Mass spectra were acquired by the EPSRC UK National Mass Spectrometry Facility at Swansea University. Elemental analysis was performed by Stephen Boyer at the London Metropolitan University.

4.2. Dithiolato complexes

4.2.1. [Cp*Rh(PMe$_3$)(NaphthS2)] (2a)

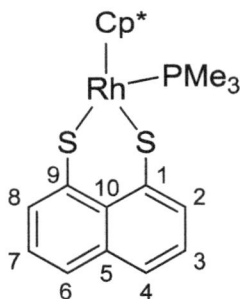

A methanol (20 mL) solution of [Cp*Rh(PMe$_3$)Br$_2$] (120 mg, 0.25 mmol), [Naphth(SH)$_2$] (60 mg, 0.31 mmol) and NaOMe (17 mg, 0.31 mmol) was stirred at room temperature overnight. The red precipitate was filtered, washed with MeOH then dried under vacuum for 3 h. The product was obtained as a red solid (110 mg, 0.21 mmol, 83%). Crystals suitable for X-ray work were obtained by slow evaporation from CH$_2$Cl$_2$. **Anal. calcd.** for C$_{23}$H$_{30}$PRhS$_2$ (504.06 g mol^{-1}): C, 54.76; H, 5.99. Found: C, 54.69; H, 5.96. **^{1}H NMR (400 MHz, CDCl$_3$):** δ 7.88 (dd, $^3J_{HH}$ = 7.3, $^4J_{HH}$ = 1.3 Hz, 2 H, H2,8), 7.47 (dd, $^3J_{HH}$ = 8.1, $^4J_{HH}$ = 1.1 Hz, 2 H, H4,6), 7.05 (dd, $^3J_{HH}$ = 8.1 & 7.3 Hz, 2 H, H3,7), 1.54 (dd, $^2J_{Hp}$ = 10.3, $^3J_{HRh}$ = 0.7 Hz, 9 H, PMe$_3$), 1.49 (d, $^4J_{Hp}$ = 3.0 Hz, 15 H, Cp*-Me). **^{13}C{^{1}H} NMR (100 MHz, CDCl$_3$):** δ 139.5 (d, $^3J_{Cp}$ = 5.4 Hz, C$_q$, C1,9), 136.3 (C$_q$, C5), 133.9 (C$_q$, C10), 128.1 (CH, C2,8), 124.9 (CH, C4,6), 123.8 (CH, C3,7), 99.7 (dd, $^1J_{CRh}$ = 4.5, $^2J_{Cp}$ = 2.9 Hz, C$_q$, Cp*), 14.9 (d, $^1J_{Cp}$ = 32.7 Hz, CH$_3$, PMe$_3$), 8.9 (CH$_3$, Cp*). **^{31}P NMR (162 MHz, CDCl$_3$):** δ 1.7 (br d, $^1J_{PRh}$ = 149 Hz). **^{31}P{^{1}H} NMR (162 MHz, CDCl$_3$):** δ 1.7 (d, $^1J_{PRh}$ = 150 Hz). **HRMS (APCI+):** m/z (%) Calcd. for C$_{20}$H$_{22}$RhS$_2$: 429.0212, found 429.0209 (100) [M–PMe$_3$ + H]. **IR (KBr):** ν_{max}/cm^{-1} 3040w (ν_{Ar-H}), 2907 m (ν_{C-H}), 1536s, 1195 m, 952s, 810 m, 761 m.

4.2.2. [Cp*Rh(PMe₃)(AcenapS2)] (2b)

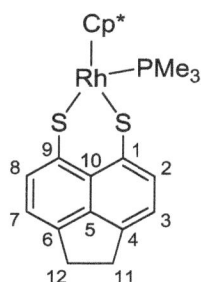

A methanol (20 mL) solution of [Cp*Rh(PMe$_3$)Br$_2$] (120 mg, 0.25 mmol), [Acenap(SH)$_2$] (68 mg, 0.31 mmol) and NaOMe (17 mg, 0.31 mmol) was stirred at room temperature overnight. The red precipitate was filtered, washed with MeOH then dried under vacuum for 3 h. The product was obtained as a red solid (121 mg, 0.22 mmol, 91%). Crystals suitable for X-ray work were obtained by slow evaporation from CH$_2$Cl$_2$. **Anal. calcd.** for C$_{25}$H$_{32}$PRhS$_2$ (530.07 g mol^{-1}): C, 56.60; H, 6.08. Found: C, 56.49; H, 6.11. **^{1}H NMR (400 MHz, CDCl$_3$):** δ 7.77 (d, $^3J_{HH}$ = 7.2 Hz, 2 H, H2,8), 6.91 (d, $^3J_{HH}$ = 7.2 Hz, 2 H, H3,7), 3.17 (s, 4 H, H11,12), 1.54 (dd, $^2J_{Hp}$ = 10.3, $^3J_{HRh}$ = 0.7 Hz, 9 H, PMe$_3$), 1.50 (d, $^4J_{Hp}$ = 3.0 Hz, 15 H, Cp*–Me). **^{13}C{^{1}H} NMR (100 MHz, CDCl$_3$):** δ 141.7 (C$_q$, C4,6), 141.1 (C$_q$, C5), 134.8 (d, $^2J_{CRh}$ = 5.5 Hz, C$_q$, C1,9), 132.3 (C$_q$, C10), 128.6 (CH, C2,8), 117.9 (CH, C3,7), 99.5 (dd, $^1J_{CRh}$ = 4.9, $^2J_{Cp}$ = 2.9 Hz, C$_q$, Cp*), 30.1 (CH$_2$, C11,12), 15.0 (d, $^1J_{Cp}$ = 33.0 Hz, CH$_3$, PMe$_3$), 8.9 (CH$_3$, Cp*). **^{31}P NMR (162 MHz, CDCl$_3$):** δ 2.9 (br d, $^1J_{PRh}$ = 148 Hz). **^{31}P{^{1}H} NMR (162 MHz, CDCl$_3$):** δ 2.9 (d, $^1J_{PRh}$ = 148 Hz). **HRMS (APCI+):** m/z (%) Calcd. for C$_{22}$H$_{24}$RhS$_2$: 455.0369, found 455.0362 (80) [M–PMe$_3$ + H], 216.0060 (95) [C$_{12}$H$_8$S$_2$], 184.0339 (55) [C$_{12}$H$_8$S], 152.0618 (100) [C$_{12}$H$_8$]. **IR (KBr):** ν_{max}/cm^{-1} 3037w (ν_{Ar-H}), 2907 m (ν_{C-H}), 1552 m, 1404 m, 1027 m, 952s, 837 m, 734 m.

4.2.3. [Cp*Rh(PMe₃)(BiphenS2)] (2c)

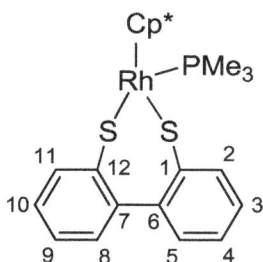

A methanol (20 mL) solution of [Cp*Rh(PMe$_3$)Br$_2$] (120 mg, 0.25 mmol), [Biphen(SH)$_2$] (71 mg, 0.33 mmol) and NaOMe (19 mg, 0.33 mmol) was stirred ct room temperature overnight. The red precipitate was filtered, washed with MeOH then dried under vacuum for 3 h. The product was obtained as a red solid (110 mg, 0.21 mmol, 83%). Crystals suitable for X-ray work were obtained by slow evaporation from CH$_2$Cl$_2$. **Anal. calcd.** for C$_{25}$H$_{32}$PRhS$_2$ (530.07 g mol^{-1}): C, 56.60; H, 6.08. Found: C, 56.49; H, 6.15. **^{1}H NMR (400 MHz, CDCl$_3$):** δ 7.66 (dd, $^3J_{HH}$ = 7.6, $^4J_{HH}$ = 1.3 Hz, 1 H, H2), 7.64 (dd, $^3J_{HH}$ = 7.7, $^4J_{HH}$ = 1.3, 1 H, H11), 7.19 (p-td, $^3J_{HH}$ = 7.5, $^4J_{HH}$ = 1.4 Hz, 1 H, H4), 7.18 (p-td, $^3J_{HH}$ = 7.6, $^4J_{HH}$ = 1.4 Hz, 1 H, H9), 7.03 (p-td, $^3J_{HH}$ = 7.6, $^4J_{HH}$ = 1.6 Hz, 1 H, H3), 6.98 (p-td, $^3J_{HH}$ = 7.6, $^4J_{HH}$ = 1.6 Hz, 1 H, H10), 6.94 (dd, $^3J_{HH}$ = 7.5, $^4J_{HH}$ = 1.5 Hz, 1 H, H5), 6.86 (dd, $^3J_{HH}$ = 7.5, $^4J_{HH}$ = 1.5 Hz, 1 H, H8), 1.58 (d, $^4J_{Hp}$ = 3.2 Hz, 15 H, Cp*–Me), 1.39 (dd, $^2J_{Hp}$ = 10.5, $^3J_{HRh}$ = 0.6 Hz, 9 H, PMe$_3$). **^{13}C{^{1}H} NMR (100 MHz, CDCl$_3$):** δ 151.0 (C$_q$, C6), 150.0 (C$_q$, C7), 143.0 (d, $^3J_{Cp}$ = 6.5 Hz, C$_q$, C1), 140.2 (C$_q$, C12), 137.2 (CH, C2), 135.2 (CH, C11), 130.9 (CH, C8), 130.6 (CH, C5), 126.2 (CH, C3,9), 125.7 (CH, C4), 125.5 (CH, C10), 99.3 (dd, $^1J_{CRh}$ = 5.3, $^2J_{Cp}$ = 3.2 Hz, C$_q$, Cp*), 15.9 (d, $^1J_{Cp}$ = 31.6 Hz, CH$_3$, PMe$_3$), 8.8 (CH$_3$, Cp*). **^{31}P NMR (162 MHz, CDCl$_3$):** δ 0.3 (br d, $^1J_{PRh}$ = 153 Hz). **^{31}P{^{1}H} NMR (162 MHz, CDCl$_3$):** δ 0.3 (d, $^1J_{PRh}$ = 152 Hz). **MS (APCI+):** m/z (%) Calcd. for C$_{22}$H$_{24}$RhS$_2$: 455.0369, found 455.0365 (100) [M–PMe$_3$ + H]. **IR (KBr):** ν_{max}/cm^{-1} 3037w (ν_{Ar-H}), 2905 m (ν_{C-H}), 1451 m, 1404 m, 1280w, 959s, 750s.

4.2.4. [Cp*Rh(PMe$_3$)(2,2'-BinapS2)] (2d)

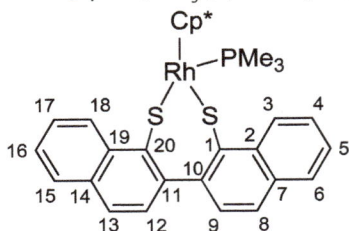

To a THF (15 mL) solution of [2,2'-BinapS$_2$] (117 mg, 0.37 mmol) was added lithium triethylborohydride (0.80 mL, 0.80 mmol, 1 M sol^n in THF) at room temperature. The reaction was left to stir for 0.5 h during which time the solution turned from yellow to almost colourless. To this was added [Cp*Rh(PMe$_3$)Br$_2$] (148 mg, 0.31 mmol) and the solution turned instantly to a very dark red/purple colour. The reaction was left to stir at room temperature overnight. The solvent was removed under vacuum and the crude product purified by column chromatography (silica/CH$_2$Cl$_2$). The solvent was removed to afford the product as a dark purple solid (147 mg, 0.23 mmol, 82%). Crystals suitable for X-ray work were obtained by slow evaporation from CH$_2$Cl$_2$. **Anal. calcd.** for C$_{33}$H$_{36}$PRhS$_2$ (630.11 g mol^{-1}): C, 62.84; H, 5.75. Found: C, 62.73; H, 5.66. **^{1}H NMR (500 MHz, CDCl$_3$):** δ 9.20 (d, $^3J_{HH}$ = 8.6 Hz, 1 H, H3), 9.17 (d, $^3J_{HH}$ = 8.5 Hz, 1 H, H18), 7.80 (d, $^3J_{HH}$ = 8.3 Hz, 1 H, H6), 7.77 (d, $^3J_{HH}$ = 8.2 Hz, 1 H, H15), 7.65 (d, $^3J_{HH}$ = 8.3 Hz, 1 H, H13), 7.64 (d, $^3J_{HH}$ = 8.2 Hz, 1 H, H8), 7.54–7.48 (m, 2 H, H4,17), 7.45–7.39 (m, 2 H, H5,16), 7.01 (d, $^3J_{HH}$ = 8.3 Hz, 1 H, H9), 6.86 (d, $^3J_{HH}$ = 8.3 Hz, 1 H, H12), 1.42 (d, $^4J_{Hp}$ = 3.2 Hz, 15 H, Cp*–Me$_3$), 1.31 (d, $^2J_{Hp}$ = 10.6 Hz, 9 H, PMe$_3$). **^{13}C{^{1}H} NMR (125 MHz, CDCl$_3$):** δ 150.1 (C$_q$, C11), 148.1 (C$_q$, C10), 140.8 (d, $^3J_{Cp}$ = 6.7 Hz, C$_q$, C1), 138.3 (C$_q$, C2), 138.0 (C$_q$, C19), 137.5 (C$_q$, C20), 132.9 (C$_q$, C14), 132.1 (C$_q$, C7), 129.5 (CH, C12), 129.2 (CH, C3,9), 128.0 (CH, C15), 127.9 (CH, C18), 127.7 (CH, C6), 125.4 (CH, C17), 125.0 (CH, C13), 124.8 (CH, C8,16), 124.7 (CH, C5), 124.3 (CH, C4), 99.7 (dd, $^1J_{CRh}$ = 5.5, $^2J_{Cp}$ = 3.5 Hz, C$_q$, Cp*), 15.7 (d, $^1J_{Cp}$ = 31.5 Hz, CH$_3$, PMe$_3$), 9.2 (CH$_3$, Cp*). **^{31}P NMR (202 MHz, CDCl$_3$):** δ 2.2 (br d, $^1J_{PRh}$ = 152 Hz). **^{31}P{^{1}H} NMR (202 MHz, CDCl$_3$):** δ 2.2 (d, $^1J_{PRh}$ = 152 Hz). **HRMS (APCI+):** m/z (%) Calcd. for C$_{30}$H$_{28}$RhS$_2$: 555.0687, found 555.0729 (90) [M–PMe$_3$ + H], 284.0738 (100) [C$_{20}$H$_{12}$S]. **IR (KBr):** ν_{max}/cm^{-1} 3044w (ν_{Ar-H}), 2906 m (ν_{C-H}), 1493 m, 1281 m, 949s, 815s, 747s, 672 m, 546w.

	1	2a	2b
Table 4. Crystallographic data for complexes 1, 2a and 2b			
Empirical formula	C$_{13}$H$_{24}$Br$_2$PRh	C$_{23}$H$_{30}$PRhS$_2$	C$_{25}$H$_{32}$PRhS$_2$
M	474	504	530
Crystal system	Tetragonal	Orthorhombic	Monoclinic
Space group	P4$_3$2$_1$2	P2$_1$2$_1$2$_1$	P2$_1$/n
a [Å]	11.994(9)	8.2099(19)	8.265(7)
b [Å]	11.994(9)	13.649(3)	8.706(7)
c [Å]	23.274(19)	19.602(4)	32.87(3)
α [°]	90.0	90.0	90.0
β [°]	90.0	90.0	95.663(7)
γ [°]	90.0	90.0	90.0
V [Å^3]	3348(4)	2196.5(8)	2354(3)
Z	8	4	4
ρ_{calcd} (g. cm^{-3})	1.881	1.525	1.497
μ [cm^{-1}]	58.817	10.448	9.791
Measured refln.	30194	29827	22775
Unique refln.	3088	4016	5378
R [I>2σ(I)]	0.0186	0.0291	0.0378
wR	0.0344	0.0689	0.0981

Table 5. Crystallographic data for complexes 2c and 2d

	2c	2d
Empirical formula	$C_{25}H_{32}PRhS_2$	$C_{33}H_{36}PRhS_2$
M	530	630
Crystal system	Orthorhombic	Monoclinic
Space group	$P2_12_12_1$	$P2_1/c$
a [Å]	9.5411(15)	10.4996(7)
b [Å]	15.202(3)	22.0450(15)
c [Å]	16.6474(19)	13.3815(9)
α [°]	90.0	90.0
β [°]	90.0	107.216(4)
γ [°]	90.0	90.0
V [Å³]	2414.6(7)	2958.6(4)
Z	4	4
ρ_{calcd} (g. cm⁻³)	1.459	1.416
μ [cm⁻¹]	9.545	7.918
Measured refln.	25505	25373
Unique refln.	5526	5437
R [I>2σ(I)]	0.0189	0.0277
wR	0.0454	0.0725

4.3. Crystal structure analysis

Tables 4 and 5 list the details of data collections and refinements. Data for **1**, **2b**, **2c** and **2d** were collected using a Rigaku SCX-Mini (Mo–Kα, graphite monochromator) at −100°C; for **2b** using a Rigaku Saturn724 at −148°C. Intensities were corrected for Lorentz polarisation, and adsorption. Structures were solved by direct methods and refined by full-matrix least-squares against F^2 (SHELXL) (Sheldrick, 2008). Hydrogen atoms were assigned riding isotropic displacements parameters and constrained to idealised geometries. Non-hydrogen atoms were refined anisotropically.

Funding
The authors received no direct funding for this research.

Author details
Phillip S. Nejman[1]
E-mail: pn46@st-andrews.ac.uk
Alexandra M.Z. Slawin[1]
E-mail: amzs@st-andrews.ac.uk
Petr Kilian[1]
E-mail: pk7@st-andrews.ac.uk
J.Derek Woollins[1]
E-mail: jdw3@st-andrews.ac.uk
ORCID ID: http://orcid.org/0000-0002-1498-9652
[1] EaStChem School of Chemistry, University of St Andrews, St Andrews, Fife KY16 9ST, UK.

References
Albers, A., Demeshko, S., Dechert, S., Saouma, C. T., Mayer, J. M., & Meyer, F. (2014). Fast proton-coupled electron transfer observed for a high-fidelity structural and functional [2Fe–2S] Rieske model. *Journal of the American Chemical Society, 136*, 3946–3954. http://dx.doi.org/10.1021/ja412449v

Armarego, W. L. F. (1960). 83. The synthesis of two dinaphthothiophens. *Journal of the Chemical Society (Resumed), 1960* 433–436.

Armarego, W. L. F., & Chai, C. L. L. (2009). *Purification of Laboratory Chemicals* (6th ed.). Oxford: Butterworth-Heinemann.

Ashe, A. J., Kampf, J. W., & Savla, P. M. (1994). The reaction of sulfur with dilithio compounds. The syntheses and structures of phenanthro [1, 10-cd]-1,2-dithiole and phenanthro[4,5-cde] [1,2]dithiin. *Heteroatom Chemistry, 5*, 113–119. http://dx.doi.org/10.1002/(ISSN)1098-1071

Aucott, S. M., Kilian, P., Milton, H. L., Robertson, S. D., Slawin, A. M. Z., & Woollins, J. D. (2005). Bis(Cyclopentadienyl) Titanium Complexes of Naphthalene−1,8-Dithiolates, Biphenyl 2,2'-Dithiolates, and Related Ligands. *Inorganic Chemistry, 44*, 2710–2718. http://dx.doi.org/10.1021/ic048483l

Aucott, S. M., Kilian, P., Robertson, S. D., Slawin, A. M. Z., & Woollins, J. D. (2006). Platinum complexes of dibenzo[1,2] dithiin, dibenzo[1,2]dithiin oxides and related polyaromatic hydrocarbon ligands. *Chemistry - A European Journal, 12*, 895–902. http://dx.doi.org/10.1002/(ISSN)1521-3765

Ballmann, J., Dechert, S., Demeshko, S., & Meyer, F. (2009). Tuning electronic properties of biomimetic [2Fe-2S] clusters by ligand variations. *European Journal of Inorganic Chemistry, 3219*–3225. http://dx.doi.org/10.1002/ejic.v2009:22

Benson, C. G. M., Schofield, C. M., Randall, R. A. M., Wakefield, L., Knight, F. R., Slawin, A. M. Z., & Woollins, J. D. (2013).

Platinum Complexes of 5,6-Dihydroacenaphtho[5,6-cd]-1,2-dichalcogenoles. *European Journal of Inorganic Chemistry, 427–437.* http://dx.doi.org/10.1002/ejic.201200967

Bond, A. M., & Martin, R. L. (1984). Electrochemistry and redox behaviour of transition metal dithiocarbamates *Coordination chemistry reviews, 54, 23–98.* http://dx.doi.org/10.1016/0010-8545(84)85017-1

Bosman, W. P., & van der Linden, H. G. M. (1977). Tris[triphenylp hosphinehexachloronaphthalene-1,8-dithiolatonickel(II)], [Ni$_3$(PPh$_3$)$_3$(S$_2$C$_{10}$Cl$_6$)$_3$], a Ni$_3$S$_6$ metal-sulphur cluster compound; preparation and crystal and molecular structure. *Journal of the Chemical Society, Chemical Communications, 714–715.* http://dx.doi.org/10.1039/C39770000714

Bruno, I. J., Cole, J. C., Edgington, P. R., Kessler, M., Macrae, C. F., McCabe, P., Pearson, J., & Taylor, R. (2002). New software for searching the Cambridge Structural Database and visualizing crystal structures. *Acta Crystallographica Section B Structural Science, 58, 389–397.* http://dx.doi.org/10.1107/S0108768102003324

Burns, R. P., & McAuliffe, C. A. (1979). 1,2-Dithiolene complexes of transition metals. *Advances in Inorganic Chemistry and Radiochemistry, 22, 303–348.*

Burns, R. P., McCullough, F. P., & McAuliffe, C. A. (1980). 1,1-Dithiolato complexes of the transition elements. *Advances in Inorganic Chemistry and Radiochemistry, 23, 211–280.*

Charreteur, K., Kdider, M., Capon, J.-F., Gloaguen, F., Pétillon, F. Y., Schollhammer, P., & Talarmin, J. (2010). Effect of electron-withdrawing dithiolate bridge on the electron-transfer steps in diiron molecules related to [2Fe](H) subsite of the [FeFe]-hydrogenases. *Inorganic Chemistry, 49, 2496–2501.* http://dx.doi.org/10.1021/ic902401k

Conry, R., & Tipton, A. A. (2001). A mononuclear molybdenum(V) mono-oxo biphenyl-2,2'-dithiolate complex in which the metal resides within a cleft formed by the ligands and that exhibits N-H … S hydrogen bonding in the solid state. *JBIC Journal of Biological Inorganic Chemistry, 6, 359–366.* http://dx.doi.org/10.1007/s007750100207

Cossu, S., Delogu, G., Fabbri, D., & Maglioli, P. (1991). A rapid preparation of 2,2'-dimercaptobiphenyl. *Organic Preparations and Procedures International, 23, 455–457.* http://dx.doi.org/10.1080/00304949109458237

Eisenberg, R. (2007). Structural systematics of 1,1- and 1,2-dithiolato chelates. In *Progress in inorganic chemistry* (pp. 295–369). John Wiley & Sons.

Figliola, C., Male, L., Horton, P. N., Pitak, M. B., Coles, S. J., Horswell, S. L., & Grainger, R. S. (2014). [FeFe]-Hydrogenase synthetic mimics based on peri-substituted dichalcogenides. *Organometallics, 33, 4449–4460.* http://dx.doi.org/10.1021/om500683p

Figliola, C., Male, L., Horswell, S. L., & Grainger, R. S. (2015). *N*-Derivatives of *peri*-Substituted Dichalcogenide [FeFe]-Hydrogenase Mimics: Towards Photocatalytic Dyads for Hydrogen Production. *European Journal of Inorganic Chemistry, 2015, 3146–3156.* http://dx.doi.org/10.1002/ejic.v2015.19

Jones, W. D., & Feher, F. J. (1984). Preparation and conformational dynamics of (C5Me5)Rh(PR13)RX. Hindered rotation about rhodium-phosphorus and rhodium-carbon bonds. *Inorganic Chemistry, 23, 2376–2388.* http://dx.doi.org/10.1021/ic00184a005

Kempe, R., Sieler, J., Hintzsche, E., & Schroth, W. (1993). Crystal structure of dinaphto[l,2-c][2',1'-e]-1,2- dithiin, C20H12S2. *Zeitschrift für Kristallographie, 207, 314–315.*

Macrae, C. F., Bruno, I. J., Chisholm, J. A., Edgington, P. R., McCabe, P., Pidcock, E., … Wood, P. A. (2008). *Mercury CSD*

2.0 – new features for the visualization and investigation of crystal structures. *Journal of Applied Crystallography, 41, 466–470.* http://dx.doi.org/10.1107/S0021889807067908

McNaughton, R. L., Tipton, A. A., Rubie, N. D., Conry, R. R., & Kirk, M. L. (2000). Electronic structure studies of oxomolybdenum tetrathiolate complexes: Origin of reduction potential differences and relationship to cysteine–molybdenum bonding in sulfite oxidase. *Inorganic Chemistry, 39, 5697–5706.* http://dx.doi.org/10.1021/ic0003729

Nejman, P. S., Morton-Fernandez, B., Black, N., Cordes, D. B., Slawin, A. M. Z., Kilian, P., & Woollins, J. D. (2015). The preparation and characterisation of rhodium(III) and Iridium(III) half sandwich complexes with napthalene-1,8-dithiolate, acenaphthene-5,6-dithiolate and biphenyl-2,2'-dithiolate. *Journal of Organometallic Chemistry, 776, 7–16.* http://dx.doi.org/10.1016/j.jorganchem.2014.10.023

Nejman, P. S., Morton-Fernandez, B., Moulding, D. J., Athukorala Arachchige, K. S., Cordes, D. B., Slawin, A. M. Z., … Woollins, J. D. (2015). Structural diversity of bimetallic rhodium and iridium half sandwich dithiolato complexes. *Dalton Transactions, 44, 16758–16766.* http://dx.doi.org/10.1039/C5DT02542G

Ojima, I., Vu, A. T., & Bonafoux, D. (2002). Product Class 5: Organometallic complexes of rhodium. *Science of Synthesis, 1, 531–616.*

Phillips, J. R., Poat, J. C., Slawin, A. M. Z., Williams, D. J., Wood, P. T., & Woollins, J. D. (1995). Polymeric and bimetallic complexes of diisopropyl monothiophosphate. *Journal of the Chemical Society, Dalton Transactions, 2369–2375.* http://dx.doi.org/10.1039/dt9950002369

Samuel, A. P. S., Co, D. T., Stern, C. L., & Wasielewski, M. R. (2010). Ultrafast photodriven intramolecular electron transfer from a zinc porphyrin to a readily reduced diiron hydrogenase model complex. *Journal of the American Chemical Society, 132, 8813–8815.* http://dx.doi.org/10.1021/ja100016v

Sheldrick, G. (2008). Acta Crystallographica Section A Foundations of Crystallography. *International Union of Crystallography, International Union of Crystallography, 64, 112–122.* http://dx.doi.org/10.1107/S0108767307043930

Stafford, P. R., Rauchfuss, T. B., Verma, A. K., & Wilson, S. R. (1996). Titanocene complexes of ring-opened dibenzothiophene and related dimercaptobiaryl ligands. *Journal of Organometallic Chemistry, 526, 203–214.* http://dx.doi.org/10.1016/S0022-328X(96)06581-3

Teo, B. K., & Snyder-Robinson, P. A. (1978). Metal tetrathiolenes. 3. Structural study of the first bimetallic tetrathiolene complex, tetrakis(triphenylphosphine)diplatinum tetrathionapthalene, (Ph3P)4Pt2(C10H4S4) revealing an unexpected molecular distortion. *Inorganic Chemistry, 17, 3489–3497.* http://dx.doi.org/10.1021/ic50190a034

Teo, B. K., & Snyder-Robinson, P. A. (1979a). Metal-tetrathiolene chemistry: the first disulphur bridge in the novel di-iridium cluster (Ph$_3$P)$_2$(CO)$_2$Br$_2$Ir$_2$(C$_{10}$H$_4$S$_4$). X-Ray crystal structure. *Journal of the Chemical Society, Chemical Communications, 255–256.* http://dx.doi.org/10.1039/C39790000255

Teo, B. K., & Snyder-Robinson, P. A. (1979b). Metal tetrathiolenes. 4. Synthesis, characterization, and electrochemistry of new discrete diplatinum-tetrathiolene complexes. *Inorganic Chemistry, 18, 1490–1495.* http://dx.doi.org/10.1021/ic50196a017

Teo, B. K., & Snyder-Robinson, P. A. (1981). Molecular systematics in metal tetrathiolenes. A novel 2-, 4-, 6-, 8-, and 12-electron-donating ligand system upon coordination to one, two, two, three and four metals, respectively. *Inorganic Chemistry, 20, 4235–4239.* http://dx.doi.org/10.1021/ic50226a041

Teo, B. K., & Snyder-Robinson, P. A. (1984). Metal tetrathiolenes. 8. Molecular structures of two isostructural two-electron systems: (Ph3P)2(CO)XIr(C10Cl4S4) (X–Cl,H). The first member of a novel series of metal tetrathiolene complexes. *Inorganic Chemistry, 23*, 32–39. http://dx.doi.org/10.1021/ic00169a009

Teo, B. K., Wudl, F., Hauser, J. J., & Kruger, A. (1977). Reactions of tetrathionaphthalene with transition metal carbonyls. Synthesis and characterization of two new organometallic semiconductors (C10H4S4Ni)x and [C10H4S4Co2(CO)2]x and a tetrairon cluster C10H4S4Fe4(CO)12. *Journal of the American Chemical Society, 99*, 4862–4863. http://dx.doi.org/10.1021/ja00456a073

Teo, B. K., Wudl, F., Marshall, J. H., & Kruger, A. (1977). Synthesis, characterization, and electrochemistry of a new platinum "tetrathiolene" cluster, tetrakis(triphenylphosphine)tetrathionaphthalenediplatinum(II,II): A novel system with five reversible oxidation states. *Journal of the American Chemical Society, 99*, 2349–2350. http://dx.doi.org/10.1021/ja00449a060

Teo, B. K., Bakirtzis, V., & Snyder-Robinson, P. A. (1983). Syntheses and structures of a novel series of tetrathiolene clusters: (.eta.5-C5H5)2Fe2(C10Cl4S4)(C10Cl4S3), (.eta.5-C5H5)2Co2(C10Cl4S4), and (.eta.5-C5H5)2Ni2(C10Cl4S4). *Journal of the American Chemical Society, 105*, 6330–6332. http://dx.doi.org/10.1021/ja00358a034

Thomas, I. R., Bruno, I. J., Cole, J. C., Macrae, C. F., Pidcock, E., & Wood, P. A. (2010). WebCSD: the online portal to the Cambridge Structural Database. *Journal of Applied Crystallography, 43*, 362–366. http://dx.doi.org/10.1107/S0021889810000452

Topf, C., Monkowius, U., & Knör, G. (2012). Design, synthesis and characterization of a modular bridging ligand platform for bio-inspired hydrogen production. *Inorganic Chemistry Communications, 21*, 147–150. http://dx.doi.org/10.1016/j.inoche.2012.04.034

Tuna, F., Smith, C. A., Bodensteiner, M., Ungur, L., Chibotaru, L. F., McInnes, E. J. L., Winpenny, R. E. P., Collison, D., & Layfield, R. A. (2012). A high anisotropy barrier in a sulfur-bridged organodysprosium single-molecule magnet. *Angewandte Chemie International Edition, 51*, 6976–6980. http://dx.doi.org/10.1002/anie.201202497

Wang, G.-L., Lin, Y.-J., Blacque, O., Berke, H., & Jin, G.-X. (2008). Helical Supramolecular Assemblies of {2,4,6-[Cp∗Rh(E2-1,2-C2B10H10)(NC5H4CH2S)]3(triazine)} (E = S, Se) Shaped by Cp∗–Toluene–Cp∗ π-Stacking Forces and BH–Pyridine Hydrogen Bonding. *Inorganic Chemistry, 47*, 2940–2942. http://dx.doi.org/10.1021/ic800105h

Woollins, J. D. (1996). Metallacycles with group 15/16 ligands. In S. C. Mitchell (Ed.), *Biological interactions of sulfur compounds* (pp. 1–19). London: Taylor & Francis.

Wright, R. J., Lim, C., & Tilley, T. D. (2009). Diiron proton reduction catalysts possessing electron-rich and electron-poor naphthalene-1,8-dithiolate ligands. *Chemistry - A European Journal, 15*, 8518–8525. http://dx.doi.org/10.1002/chem.v15:34

Xiao, X.-Q., & Jin, G.-X. (2008). Half-sandwich rhodium complexes containing both N-heterocyclic carbene and ortho-carborane-1,2-dithiolate ligands. *Journal of Organometallic Chemistry, 693*, 316–320. http://dx.doi.org/10.1016/j.jorganchem.2007.11.005

Yao, Z.-J., Xu, B., Huo, X.-K., & Jin, G.-X. (2013). Homonuclear half-sandwich iridium and rhodium complexes containing dichalcogenolato-functionalized o-carboranyl ligands. *Journal of Organometallic Chemistry, 747*, 85–89. http://dx.doi.org/10.1016/j.jorganchem.2013.03.002

Yui, K., Aso, Y., Otsubo, T., & Ogura, F. (1988). Syntheses and Properties of 2,2′-Binaphtho[1,8-de]-1,3-dithiinylidene and Its Selenium Analog, 2-(1,3-Dithiol-2-ylidene) naphtho[1,8-de]-1,3-dithiin, and 2-(4H-Thiopyran-4-ylidene)naphtho[1,8-de]-1,3-dithiin. *Bulletin of the Chemical Society of Japan, 61*, 953–959. http://dx.doi.org/10.1246/bcsj.61.953

Zhou, C.-L., Wang, Z.-M., Wang, B.-W., & Gao, S. (2011). Two Mn6 single-molecule magnets with sulfur-contained capping ligand. *Polyhedron, 30*, 3279–3283. http://dx.doi.org/10.1016/j.poly.2011.05.046

11

Synthesis, crystal structure, photoluminescent and antimicrobial properties of a thiocyanato-bridged copper(II) coordination polymer

Feudjio Tsague Chimaine[1], Divine Mbom Yufanyi[2], Amah Colette Benedicta Yuoh[1], Donatus Bekindaka Eni[1] and Moise Ondoh Agwara[1]*

*Corresponding author: Moise Ondoh Agwara, Department of Inorganic Chemistry, University of Yaounde I, P.O. Box 812 Yaounde, Yaounde, Cameroon
E-mail: agwara29@yahoo.com
Reviewing editor: J. Derek Woollins, University of St. Andrews, UK

Abstract: A Cu(II) 1D polymer with mixed ligands (SCN⁻ and pyridine) has been synthesized and characterized by elemental analyses, IR, UV–visible and TG-DTA analytical techniques. The crystal structure was determined by single-crystal X-ray diffraction analyses. The complex crystallizes in the triclinic crystal system with space group $P\bar{1}$ with one formula unit. Each Cu(II) atom is six coordinate with two N atoms of two pyridine molecules, two N atoms and two S-atoms from bridging SCN anions giving a distorted octahedral geometry with a CuS_2N_4 chromophore. The spectroscopic, photoluminescent and the antimicrobial activities of the synthesized complex were investigated.

Subjects: Applied & Industrial Chemistry; Inorganic Chemistry; Materials Chemistry

Keywords: crystal structure; copper(II); photoluminescence; antimicrobial properties; thiocyanate

1. Introduction

Research interest in coordination polymers has increased recently not only as a result of their fascinating topologies and intriguing frameworks but also due to their potential applications in gas storage, separation, biosensing, luminescence, magnetism, conductivity, nanoparticles, antimicrobial

ABOUT THE AUTHOR

Moise Ondoh Agwara is Associate Professor of Chemistry at the Department of Inorganic Chemistry, University of Yaounde I in Cameroon. He obtained a PhD in chemistry from the University of Ibadan, Nigeria in 1986. Research activities within his research group are focused on the development of the chemistry of transition metal complexes with heterocyclic N-, O- and N,O-donor ligands and some co-ligands. Such interest derives from the fascinating structural chemistry of the complexes obtained, their interesting physico-chemical properties and their diverse applications such as antimicrobials, in photoluminescence and as precursors for the development of nanostructured functional materials.

PUBLIC INTEREST STATEMENT

Interest in coordination polymers has increased because of their fascinating structures and potential applications in gas storage, separation, biosensing, luminescence, magnetism, conductivity, nanoparticles, antimicrobial activity and catalysis. Self-assembly of these coordination polymers through a mixed ligand strategy has progressively become an effective approach, which is expected to generate frameworks with more diverse structures and properties. The focus of this paper is the synthesis and determination of the structure and properties (thermal and photoluminescent) of a Cu(II)-pyridine thiocyanate coordination polymer. The complex exhibited photoluminescent properties in the solid state at room temperature because of charge transfer from ligand to the metal. The results of the preliminary antimicrobial screening against four pathogenic bacteria and four fungi species indicate that the complex is most active against *Salmonella typhi*.

activity and catalysis (Batten, Harris, Murray, & Smith, 2002; Eddaoudi et al., 2001; Etaiw & El-bendary, 2013; Henninger, Jeremias, Kummer, & Janiak, 2012; Janiak, 2003; Rowsell & Yaghi, 2004; Zaworotko, 2001). The topology and dimensionality of these frameworks is dependent on the rational choice of the metal ion and the ligands. Weak intermolecular interactions such as hydrogen bonding and π–π stacking interactions may also play an important role on the overall arrangement and influence the properties (Baca et al., 2008; Batten et al., 2002; Prins, Reinhoudt, & Timmerman, 2001; Whitesides & Boncheva, 2002). Furthermore, synthesis conditions (synthetic methods, reaction temperature, metal/ligand ratio, pH value and the types of solvents) can also greatly influence the crystal structure and its dimensionality (Etaiw & El-bendary, 2013).

The design and synthesis of one-dimensional coordination polymers is important since they can be used as examples for developing theoretical models of the exchange interaction in extended lattices (Demir, Yilmaz, Sariboga, Buyukgungor, & Mrozinski, 2010). Choice of organic ligands and metal ions is of great importance in the construction of these polymeric structures. Transition metals show rich diversity of oxidation states, coordination numbers and geometries, and their complexes and solid-state compounds possess an array of interesting redox, magnetic, optical, electrical and catalytic properties (Chattopadhyay et al., 2012; Chattopadhyay, Drew, Diaz, & Ghosh, 2007; Demir et al., 2010; Hong, 2008; Hu, Li, Wang, Du, & Guo, 2009).

Among several ligands employed, SCN⁻ is a highly versatile ambidentate ligand with two different donor atoms, which can coordinate through terminal modes or bridging modes: end-to-end (μ-1,3-NCS) and end-on (μ-1,1-NCS, μ-1,1-SCN) fashions via the nitrogen and sulphur atoms to generate coordination networks as well as interlink the one- or two-dimensional molecules into frameworks via non-covalent interactions (Li, Liang, & Tian, 2011; Shen & Feng, 2002). The rational design and construction of thiocyanato-brigded coordination polymers has been explored due to their fascinating topologies and remarkable properties (Bai, Shang, Dang, Sun, & Gao, 2009; Banerjee, Drew, & Ghosh, 2003; Chattopadhyay et al., 2007, 2012; Das et al., 2012; Hong, 2008; Li et al., 2011; Shi et al., 2005; You & Zhu, 2005; Yue et al., 2008). The thiocyanate group plays a key role in stabilizing a variety of transition metal centres and determining the structure of polymeric transition-metal complexes (Das et al., 2012). Recently, a large number of thiocyanato-bridged coordination polymers with intriguing topologies and fascinating properties have been reported (Bai et al., 2009; Banerjee et al., 2003; Chattopadhyay et al., 2007, 2012; Das et al., 2012; Li et al., 2011; Neumann, Jess, & Näther, 2014; Shen & Feng, 2002; Shi et al., 2005; You & Zhu, 2005; Yue et al., 2008). Among these are the thiocyanato-bridged copper(II) coordination polymers with N-donor auxiliary ligands (Das et al., 2012; Hong, 2008; Shen & Feng, 2002; Shi et al., 2005).

However, controlled synthesis of coordination polymers with preferred structures is still a challenge. Self-assembly of these coordination polymers through a mixed ligand strategy has progressively become an effective approach, which is expected to generate frameworks with more diverse structural motifs (Du, Li, Liu, & Fang, 2013; Shirdel, Marandi, Jalilzadeh, Huber, & Pfitzner, 2015; Yang & Sun, 2013). Auxiliary ligands such as N-donor heterocyclic ligands play a significant role in many biological systems, being a component of several vitamins and drugs (Dhaveethu & Ramachandramoothy, 2013). Nitrogen-containing heterocycles have been found to possess diverse pharmacological activities (Forood, Flatt, Chassaing, & Katritzky, 2002). Among this group of heterocycles are pyridine and its derivatives. Taking advantage of the coordination ability and properties of copper, pyridine and SCN, herein we report the synthesis, crystal structure and the luminescent properties of a thiocyanate-bridged copper(II) coordination polymer with pyridine. In addition, the *in vitro* antimicrobial activity of the complex against selected micro-organisms is reported.

2. Experimental

All chemicals and solvents were obtained from commercial sources and were used as received.

2.1. Synthesis of the complex

A 40 mL dry methanolic solution of $CuCl_2 \cdot 2H_2O$ (5.1144 g; 30 mmol) was poured into a three-neck round bottom flask under nitrogen atmosphere. To this solution was added drop wise, while stirring, a dry methanolic solution of a mixture of NH_4SCN (4.5672 g; 60 mmol) and pyridine (9.617 mL; 120 mmol). Upon addition of the mixed ligand solution into the dark green solution in the round bottom flask, a light green precipitate was formed. The mixture was refluxed for 3 h under nitrogen atmosphere at 30°C. The light green precipitate was filtered, washed with dry methanol, air dried and weighed. The powder was recrystallized by the diffusion method. It was dissolved in DMSO in a small vial and placed in a bigger vial containing DMF (in which it is insoluble). Shiny blue crystals, in good yield (92%), were afforded within 48 h.

2.2. Characterization techniques

The melting point was recorded using a Stuart SMP10 system. Conductivity measurement was carried out in distilled water using a HANNA multimeter H19811-5; pH/°C/EC/TDS meter at room temperature. Elemental analysis (C, H, N, S, Cu) was carried out on a Perkin-Elmer automated model 2400 series II CHNS/M analyser. The infrared spectrum was recorded using a Thermo Scientific Nicolet iS5 instrument directly on a small sample of the complex in the range of 400–4,000 cm^{-1}. The UV–vis spectrum of a DMSO solution of the complex was recorded using a Varian, Cary 50 UV–vis spectrophotometer at room temperature. Photoluminescence studies were carried out using a Perkin-Elmer, LS55 Luminescence Spectrometer, while thermal studies were carried out using the TGA/DSC 1 (STAR System) instrument. The TGA analyses were conducted between 30 and 600°C under nitrogen atmosphere at a flow rate of 10 mL min^{-1} and a temperature ramp of 10°C min^{-1}.

2.3. Single-crystal X-ray structure determination

Intensity data for the compound were collected using a Bruker AXS Kappa APEX II single crystal CCD diffractometer, equipped with graphite-monochromated CuKα radiation ($\lambda = 1.5418$ Å) at room temperature. The selected crystal for the diffraction experiment had a dimension of $0.21 \times 0.12 \times 0.06$ mm^3. The structure was solved by direct methods and refined by full-matrix least squares (Sheldrick, 1997) on F^2. The non-hydrogen atoms were refined anisotropically. H atoms were included in calculated positions with C–H lengths of 0.95(CH), 0.99(CH_2) and 0.98(CH_3) Å; $U_{iso}(H)$ values were fixed at $1.2U_{eq}(C)$ except for CH_3 where it was $1.5U_{eq}(C)$. They were assigned isotopic thermal parameters and allowed to ride on their parent carbon atoms. All calculations were carried out using the SHELXTL package (Bruker, 2001).

2.4. Antimicrobial tests

The antimicrobial tests were carried out in the laboratory of Phytobiochemical and Medicinal Plant Study, University of Yaounde I. The tests were done on eight pathogenic micro-organisms, 4 yeasts, *Candida albicans ATCC P37039*, *C. albicans 194B*, *Candida glabrata 44B* and *Cryptococcus neoformans* and 4 bacterial strains, Gram-positive *Staphylococcus aureus CIP 7625* and Gram-negatives, *Pseudomonas aeruginosa CIP 76110*, *Salmonella typhi* and *Escherichia coli ATCC25922* obtained from Centre Pasteur, Yaounde, Cameroon. Reference antibacterial drug chloramphenicol and antifungal drug nystatin were evaluated for their antibacterial and antifungal activities and their results were compared with those of the free ligands and the complex. The disc diffusion method, using Muller Hinton Agar, from the protocol described by the National Committee for Clinical Laboratory Standard was used for preliminary screening. Mueller-Hinton agar was prepared from a commercially available dehydrated base according to the manufacturer's instructions. Several colonies of each micro-organism were collected and suspended in saline (0.9% NaCl). Then, the turbidity of the test suspension was standardized to match that of a 0.5 McFarland standard (corresponds to approximately 1.5×10^8 CFU/mL for bacteria or 1×10^6 to 5×10^6 cells/mL for yeast). Each compound or reference was accurately weighed and dissolved in the appropriate diluents (DMSO at 10%, methanol at 10%, or distilled water) to yield the required concentration (2 mg mL^{-1} for compound or 1 mg mL for reference drug), using sterile glassware.

Whatman filter paper number 1 was used to prepare discs approximately 6 mm in diameter, which were packed up with aluminium paper and sterilized by autoclaving. Then, 25 µL of stock solutions of compound or positive control was delivered to each disc, leading to 50 µg of compound or 25 µg of reference drug. The dried surface of a Muller-Hinton agar plate was inoculated by flooding over the entire sterile agar surface with 500 µL of inoculum suspensions. The lid was left ajar for 3 to 5 min to allow for any excess surface moisture to be absorbed before applying the drug-impregnated discs. Discs containing the compounds or antimicrobial agents were applied within 15 min of inoculating the MHA plate. Six discs per Petri dish were plated. The plates were inverted and placed in an incubator set to 35°C. After 18 h (for bacteria) and 24 h (for yeasts) of incubation, each plate was examined. The diameters of the zones of complete inhibition (as judged by the unaided eye) were measured, including the diameter of the disc. Zones were measured to the nearest whole millimetre, using sliding callipers, which were held on the back of the inverted Petri plate. All experiments were carried out in duplicate. The compound was considered active against a microbe if the inhibition zone was 6 mm and above.

3. Results and discussion

3.1. Synthesis of the complex
The title complex was green in colour and air stable, with a sharp melting point (186°C) indicating its purity. The molar conductivity value of 60 Ωcm^{-2} mol^{-1} for the complex in water indicates that it is a nonelectrolyte. The complex was obtained in good yield (92%). The physicochemical properties of the title complex are summarized in Table 1.

3.2. X-ray crystal structure
The ORTEP representation of the asymmetric unit of $[Cu(py)_2(SCN)_2]_n$ is shown in Figure 1, the unit cell structure in Figure 2, while the 1-D polymeric chain structure together with the atomic numbering

Table 1. Physical data of the complexes

Complex	Nature	Colour	Yield (%)	Melting point (°C)	Molar conductivity ($\Omega^{-1}cm^2$ mol^{-1})	Elemental analyses: %found (%calc.)			
						%C	%H	%N	%Cu
$[Cu(py)_2(SCN)_2]_n$	Crystals	Green	92	186	60	40.88	3.36	16.27	19.18
						(42.65)	(2.98)	(16.58)	(18.81)

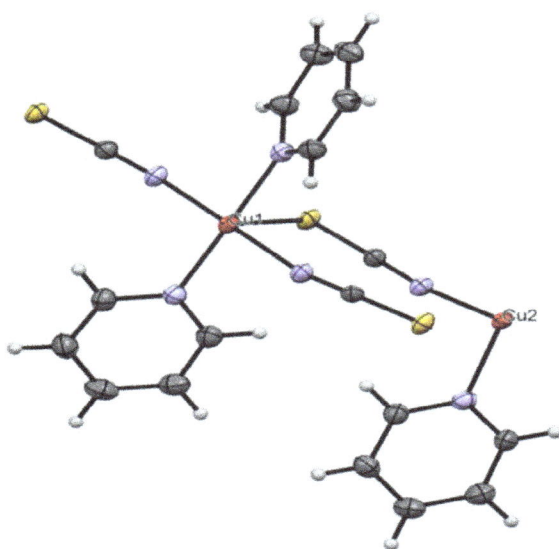

Figure 1. Asymmetric unit of the complex.

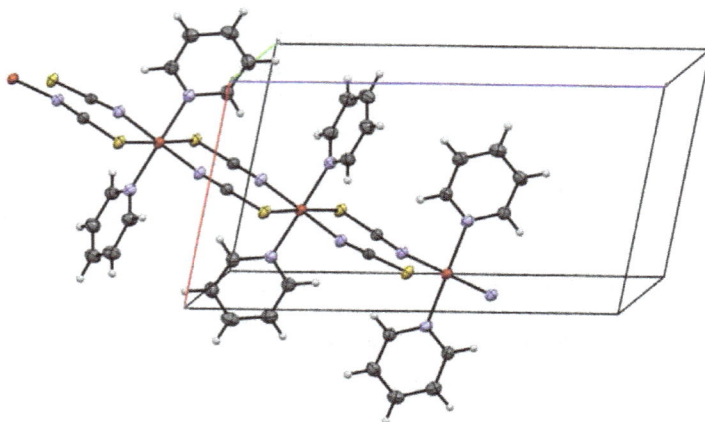

Figure 2. Unit cell diagram of the complex.

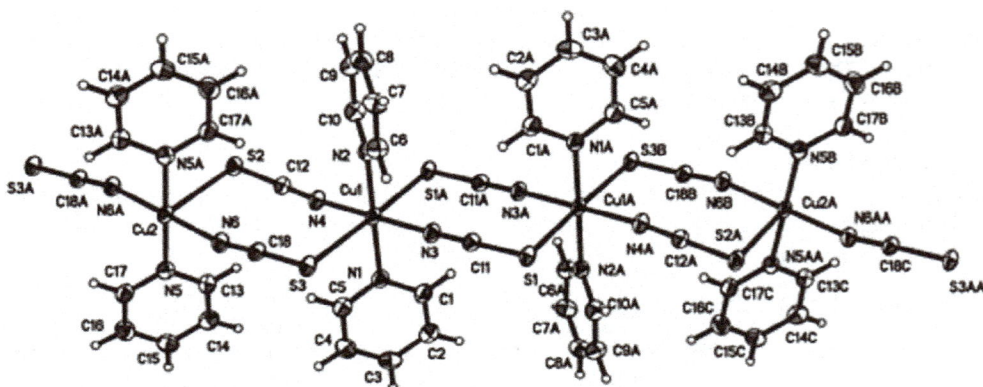

Figure 3. 1D polymeric structure of the complex with atom numbering scheme.

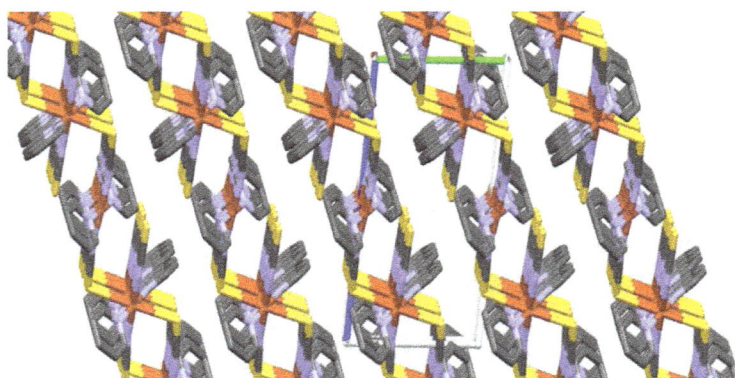

Figure 4. Packing diagram of complex showing 1-D chains parallel to the bc plane.

scheme of the complex is shown in Figure 3. The crystal packing diagram of $[Cu(py)_2(SCN)_2]_n$ is shown in Figure 4. The crystal data and structure refinement are presented in Table 2, while the selected bond lengths and bond angles are shown in Table 3.

Table 2. Crystal data and structure refinement for [Cu(Py)$_2$(SCN)$_2$]$_n$

Crystal data	
Chemical formula	$C_{36}H_{30}Cu_3N_{12}S_6$
M_r	1,013.70
Crystal system, space group	Triclinic, $P\bar{1}$
Temperature (K)	100 (2)
a, b, c (Å)	8.5381 (6), 8.6690 (8), 15.5456 (9)
α, β, γ (°)	93.367 (6), 96.385 (5), 114.746 (8)
V (Å^3)	1,031.40 (13)
Z	1
Radiation type	Cu $K\alpha$
μ (mm^{-1})	5.00
Crystal size (mm)	$0.21 \times 0.12 \times 0.06$
Crystal colour	Block blue
Data collection	
Diffractometer	Bruker *APEX*-II CCD diffractometer
Absorption correction	Multi-scan
T_{min}, T_{max}	0.781, 1.000
No. of measured, independent and observed [$I > 2\sigma(I)$] reflections	7,998, 3,899, 2,981
R_{int}	0.038
$(\sin \theta/\lambda)_{max}$ (Å^{-1})	0.626
Refinement	
$R[F^2 > 2\sigma(F^2)]$, $wR(F^2)$, S	0.048, 0.154, 1.07
No. of reflections	3,899
No. of parameters	259
No. of restraints	0
H-atom treatment	H-atom parameters constrained
$\Delta\rho_{max}, \Delta\rho_{min}$ (e Å^{-3})	1.07, −0.98

Notes: Computer programs: *CrysAlis PRO*, Agilent Technologies, Version 1.171.36.20 (release 27-06-2012 CrysAlis171 .NET).

The title compound is a one-dimensional thiocyanato-bridged polymeric structure. The complex crystallizes in the triclinic crystal system with space group $P\bar{1}$ and its asymmetric unit (Figure 1) consists of two crystallographically independent copper(II) atoms, of which one (Cu1) is located on a general position whereas the second (Cu2) is located on a crystallographic inversion centre. The structure is polymorphic to [Cu(Py)$_2$(SCN)$_2$]$_n$ (Chen, Bai, & Qu, 2005) and iso-structural with [Ni(NCS)$_2$(pyridine)$_2$]$_n$ (Neumann et al., 2014) found in the literature. There is one molecule in the triclinic unit cell as opposed to three in the previous report (Chen et al., 2005).

The crystal structure shows that Cu(II) is coordinated by four thiocyanate anions (μ-1,3) and two pyridine ligands adopting a slightly distorted octahedral coordination environment (CuS$_2$N$_4$). The Cu-N(pyridine) axial bonds are of length (Cu(1)-N(1) 2.045(3) Å, Cu(1)-N(2) 2.051(3) Å, Cu(2)-N(6)#1 1.943(3) Å, Cu(2)-N(5)#1 2.038(3) Å), while the equatorial Cu-N(thiocyanate) bonds (Cu(1)-N(3) 1.944(3) Å, Cu(1)-N(4) 1.951(3) Å). These bond lengths are similar to those found in the literature (Chen et al., 2005; Wohlert, Wriedt, Jess, & Nather, 2011). The Cu-S bonds (Cu1-S3 2.984(1) Å, Cu1-S1 2.923(1) Å, Cu2-S2 3.009 Å) are also close to reported values (Chen et al., 2005). Adjacent Cu centres are bridged by two SCN$^-$ ions resulting in a 1D polymeric chain structure extending along the crystallographic c-axis (Figure 4). The Cu1-Cu1 and Cu2-Cu2 distances within the chains are 8.669 Å while that of Cu1-Cu2 is 5.616 Å. In the equatorial plane, the SCN$^-$ ion is almost linear as evidenced by the bond angles N(3)-C(11)-S(1) 179.0(3)° and N(4)-C(12)-S(2) 179.4(4)°. The pyridine molecules are

Table 3. Selected bond lengths (Å) and angles (°) for [Cu(Py)₂(SCN)₂]ₙ			
Cu(1)-N(3)	1.944(3)	N(3)-Cu(1)-N(4)	179.74(11)
Cu(1)-N(4)	1.951(3)	N(3)-Cu(1)-N(1)	88.72(12)
Cu(1)-N(1)	2.045(3)	N(4)-Cu(1)-N(1)	91.16(12)
Cu(1)-N(2)	2.051(3)	N(3)-Cu(1)-N(2)	89.66(12)
N(1)-C(5)	1.327(5)	N(4)-Cu(1)-N(2)	90.45(11)
N(1)-C(1)	1.347(4)	N(1)-Cu(1)-N(2)	178.06(11)
N(2)-C(10)	1.337(5)	N(3)-C(11)-S(1)	179.0(3)
N(2)-C(6)	1.344(4)	N(4)-C(12)-S(2)	179.4(4)
N(3)-C(11)	1.164(5)	N(6)-Cu(2)-N(6)#1	179.999(1)
N(4)-C(12)	1.164(5)	N(6)-Cu(2)-N(5)#1	89.64(11)
Cu(2)-N(6)	1.943(3)	N(6)#1-Cu(2)-N(5)#1	90.36(11)
Cu(2)-N(6)#1	1.943(3)	N(6)-Cu(2)-N(5)	90.36(11)
Cu(2)-N(5)#1	2.038(3)	N(6)#1-Cu(2)-N(5)	89.64(11)
Cu(2)-N(5)	2.038(3)	N(5)#1-Cu(2)-N(5)	180.0
N(5)-C(17)	1.337(4)	C(17)-N(5)-C(13)	118.3(3)
N(5)-C(13)	1.346(4)	C(17)-N(5)-Cu(2)	121.7(2)
N(6)-C(18)	1.167(5)		

almost linearly arranged on the axial plane as confirmed by the bond angle N(1)-Cu(1)-N(2) (178.06°). These results are similar to those reported in literature (Małecki, Machura, Świtlicka, Gron, & Bałanda, 2011). The SCN⁻ and pyridine ligands are in different planes at right angles to each other as evidenced by the bond angles N(3)-Cu(1)-N(2) 89.66(12) and N(6)-Cu(2)-N(5) (90.36°). Selected crystal data of a polymorph of the complex are compared with crystal data of the title complex in Table 3. The title complex differs structurally from that in the literature (Chen et al., 2005) in terms of molecular weight, unit cell parameters and the number of atoms in a unit cell. The methods of syntheses of these complexes also differ.

Adjacent 1-D chains are further connected to form a 2-D supramolecular layer parallel to the bc plane by alternating S1...S3, S2...S2 (3.567(1) Å) and S3...S1 (3.541(1) Å) interactions as shown in Figure 5. This is similar to literature reports (Lu, Liu, Zhang, Wang, & Niu, 2010). The two-dimensional layers are further connected by off-set π–π stacking of pyridine rings, C–H...S and C–H...C interactions (Table 4) to form a three-dimensional supramolecular structure as shown in Figure 6 (Gerlach et al., 2015; Laachir, Bentiss, Guesmi, Saadi, & El Ammari, 2016; Trivedi, Pandey, & Rath, 2009). These

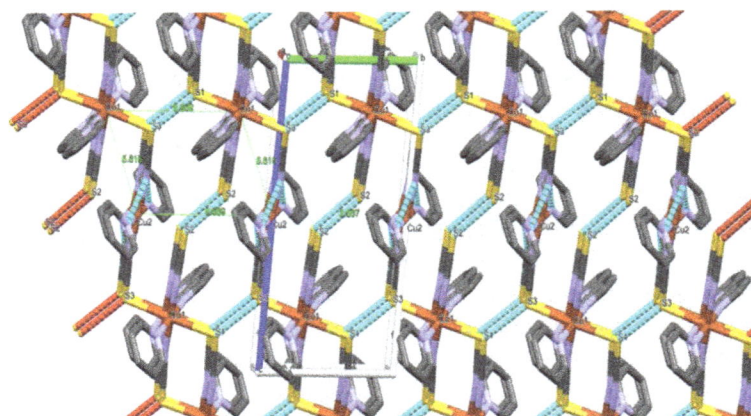

Figure 5. Part of the crystal structure showing the formation, through S...S contacts, of the two dimensional supramolecular sheet extending in the bc plane.

Table 4. Comparative crystal data of the complexes

Parameter	$[Cu_3(py)_6(SCN)_6]_n$ [this work]	$[Cu(py)_2(SCN)_2]_n$ [33]
Empirical formula	$C_{36}H_{30}Cu_3N_{12}S_6$	$C_{12}H_{10}CuN_4S_2$
Formula weight	1,013.70	337.90
Colour	Block blue	Block blue
Crystal system	Triclinic	Triclinic
Space group	$P\bar{1}$	$P\bar{1}$
Unit cell dimensions		
a	8.5381(6) Å	8.528(2) Å
b	8.6690(8) Å	9.128(1) Å
c	15.5456(9) Å	15.371(1) Å
α	93.367(6)°	91.737(1)°
β	96.385(5)°	97.043(1)°
γ	114.746(8)°	115.639(1)°
Unit cell volume, V	1,031.40(13) Å³	1,065.9(3) Å³
Calc. density	1.632 mg m⁻³	1.579 mg m⁻³
Z	1	3

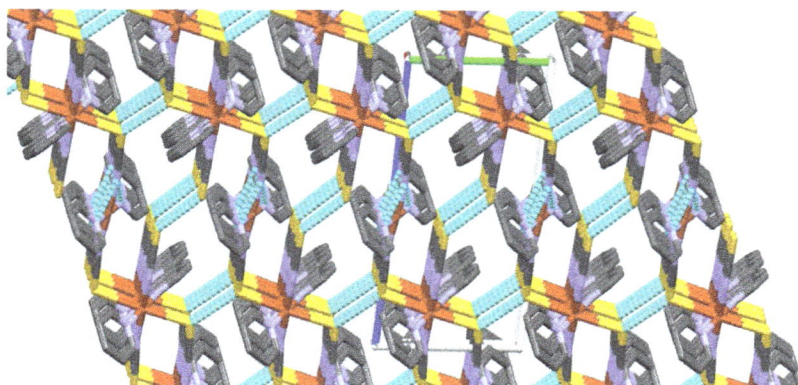

Figure 6. View of the 3D supramolecular layers of the complex down the a-axis via π–π and C–H…S interactions.

interactions, though weak compared to the metal–nitrogen and metal–sulphur coordination bonds, are crucial in the self-assembly of the 3D supramolecular structure.

3.3. IR spectroscopy

In the spectrum of the pyridine ligand as well as that of the complex, the absorption bands at 1,442 cm⁻¹ are assigned to the aryl C–H stretching vibrations. The $\nu_{C=N}$ stretching modes of the pyridine ring shifted from 1,595 to 1,604 cm⁻¹ in the spectrum of the complex, indicating its participation in bonding. The $\nu_{C\equiv N}$ asymmetric stretching vibrations of the thiocyanate have shifted from 2,063 to 2,087 cm⁻¹ in the spectrum of the complex, indicating it has taken part in bonding (Kabesova & Gazo, 1980). The ν_{SC} vibration frequency of the isothiocyanato ligand appears at 746 cm⁻¹ on the SCN⁻ spectrum and shifts to 753 cm⁻¹ on the spectrum of the complex, indicating NCS-M coordination in the complex (Kabesova & Gazo, 1980). These bands indicate the coordination of SCN⁻ in a bridging mode (Shen & Feng, 2002), which is confirmed by the crystal structure of the complex. The strong, well-resolved and sharp absorption bands found in the region of 1,495–1,000 cm⁻¹ in the spectrum of the complex are assigned to the coordinated pyridine ring (Das et al., 2012). The ν_{Cu-NPy} stretching mode is present at about 549 cm⁻¹.

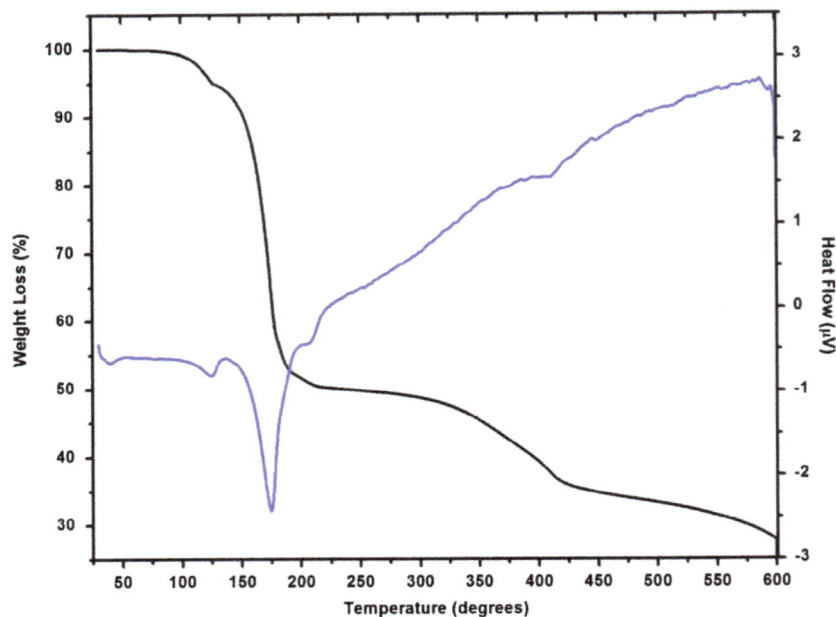

Figure 7. TG/DTA plots of the complex.

3.4. UV–Vis spectroscopy

The electronic absorption spectrum of the complex shows a single broad band centred at 15,625 cm^{-1} (640 nm). This d-d transition band in the Cu(II) ion has been assigned to $^2E_g \rightarrow ^2T_{2g}$ transition (Kurdziel, Głowiak, Materazzi, & Jezierska, 2003; Rapheal, Manoj, & Kurup, 2007; Reddy, Nethaji, & Chakravarty, 2002). The observed band is consistent with an octahedral geometry for Cu(II) complexes as confirmed by the single X-ray crystal structure. This value is smaller than that of $Cu(en)_2(SCN)_2$ (19,047 cm^{-1}; 525 nm) and $Cu(en)_2[Cd(SCN)_3]_2$ (18,248 cm^{-1}; 548 nm) (Shen & Feng, 2002) with analogous CuN_4S_2 chromophores. This shift in band position indicates some distortion from the perfect octahedral symmetry of Cu(II) (Bai et al., 2008).

3.5. Thermal analysis

In order to establish the thermal stability of the title complex, TG/DTA analyses were carried out in the temperature range of 30–600°C. The thermal decomposition thermogram (Figure 7) shows that the complex decomposes in several steps resulting in different phases of $[Cu_3(Py)_6(SCN)_6]_n$ as temperature was increased. The first weight loss of 5.62% from 80 to 120°C is probably due to the loss of adsorbed water molecules from the atmosphere. The second degradation step in the range of 130–220°C with mass loss of 43.92% is attributed to the loss of six pyridine molecules (calculated 46.76%). The sharp exothermic DTA peak at 170°C indicates that this is the major decomposition temperature. The third degradation step in the range of 260–460°C with weight loss of 14.92% is due to the loss of three SCN^- anions (calculated 17.16%). In the last decomposition step from 510 to 590°C with mass loss of 8.48% is attributed to the loss of one SCN^- and CN^- (calculated 8.29%). A stable mass is reached at 600°C. The residual mass 27.96% (calculated 28.56%) is probably due to CuS. The measured mass loss for each stage is in good agreement with the calculated values.

3.6. Photoluminescence studies

The fluorescence emission spectra of the ligand and the complex are shown in Figure 8. The results show that the ligand pyridine and the complex exhibit only one emission peak each at 23,923 cm^{-1} (418 nm) and 24,509 cm^{-1} (408 nm), respectively, when excited at 33,333 cm^{-1} (300 nm). For the ligand, the peak at 23,923 cm^{-1} is attributed to n→π* transition. The red shift of 10 nm in the spectrum of the complex indicates charge transfer from ligand to the metal (PyN→Cu) (Etaiw & Abdou, 2016; Rapheal et al., 2007).

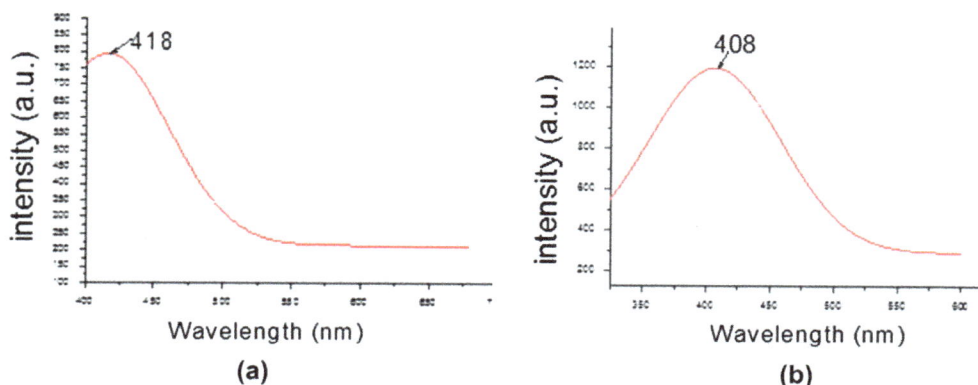

Figure 8. Photoluminescent emission spectra of (a) pyridine and (b) CuPySCN complex.

Table 5. Diameters of inhibition zones of compounds against microorganisms

Compounds	A	B	C	D	E	F	G	H
$CuCl_2.2H_2O$	0.0	7.0	6.0	7.0	6.0	9.0	10.0	9.0
SCN^-	6.0	8.0	6.0	11.5	9.0	6.0	6.0	6.0
Pyridine	6.0	7.5	3.0	3.0	6.5	6.0	6.0	6.0
$[Cu_3(py)_6(SCN)_6]_n$	6.0	9.0	9.0	7.5	6.0	10.5	10.5	9.0
Chloramphenicol	7.5	14.5	6.0	10.5	12.0	11.5	10.0	14.0
Nystatin	6.0	7.0	6.0	9.5	6.0	6.0	6.0	10.5

Notes: A = *C. albicans ATCC P37039*; B = *C. albicans 194B*; C = *C. glabrata 44B*; D = *C. neoformans*; E = *E. coli*; F = *P. aeruginosa*; G = *S. typhi*; H = *S. aureus*.

3.7. Antimicrobial studies

The effects of the starting materials, the resulting complex, the reference antibiotic (chloramphenicol) and antifungal (nystatin) were evaluated against some selected microbial pathogens (four bacteria and four fungi strains). The susceptibility of the bacteria and fungi strains towards these compounds was judged by measuring the size of the growth inhibition diameter. The diameter of the zone of inhibition (mm) was used to compare the antimicrobial activity of the test compound with that of the reference antibiotic and antifungal. Results of the preliminary screening are presented in Table 5.

The results indicate that SCN^- exhibits a high activity against the pathogens, especially the fungi species. The metal complex shows higher activity compared to that of the free ligand as well as SCN. It was found to be active against all the pathogens with high inhibition zones. The complex is most active against the fungi *C. albicans 194B*, *C. glabrata 44B* and the bacteria species *P. aeruginosa* and *S. typhi*. The complex is also more active than the reference drug nystatin towards the fungi species. The most sensitive bacteria species was *S. typhi*. This indicates that reaction of metal ions with the ligand plays an important role in enhancing its antimicrobial activity. This increase in activity could be due to the reduction of the polarity of the metal ion by partial sharing of the positive charge with the ligand's donor atoms so that there is electron delocalization within the metal complex. This may increase the hydrophobic and lipophilic character of the metal complex, enabling it to permeate the lipid layer of the organism killing them more effectively (Tabong et al., 2016; Yuoh et al., 2015).

4. Conclusion

A Cu(II) complex with SCN^- and pyridine $[Cu_3(py)_6(SCN)_6]_n$ has been synthesized and characterized. The structure is polymorphic to $[Cu(Py)_2(SCN)_2]_n$ (Chen et al., 2005) and iso-structural with $[Ni(NCS)_2(pyridine)_2]_n$ (Neumann et al., 2014). The crystal structure consists of two crystallographically independent copper(II) atoms, of which one (Cu1) is located on a general position whereas the second (Cu2) is located on a crystallographic inversion centre. Each Cu (II) atom adopts a slightly distorted octahedral coordination environment (CuS_2N_4) in which it is covalently bonded to two pyridine N-atoms in the axial position, two S-atoms of SCN^- and two N-atoms of SCN^- in the equatorial position. Adjacent Cu centres are bridged by two SCN^- (μ-1,3) ions resulting in a 1D polymeric chain structure. Adjacent 1-D chains are further connected to form a 2-D supramolecular layer parallel to the bc plane by alternating S1...S3, S2...S2 (3.567(1) Å) and S3...S1 (3.541(1) Å), interactions. The two-dimensional layers are further connected by off-set π–π stacking of pyridine rings, C–H...S and C–H...C interactions to form a three-dimensional supramolecular structure. The complex exhibited photoluminescent properties in the solid state at room temperature because of charge transfer from ligand to the metal. The results of the preliminary antimicrobial screening against four pathogenic bacteria and four fungi species indicate that the complex is most active against S. typhi.

Funding

The authors received no direct funding for this research.

Author details

Feudjio Tsague Chimaine[1]
E-mail: f.chimie@yahoo.fr
Divine Mbom Yufanyi[2]
E-mail: dyufanyi@yahoo.com
ORCID ID: http://orcid.org/0000-0001-8889-611X
Amah Colette Benedicta Yuoh[1]
E-mail: colette_amah@yahoo.fr
Donatus Bekindaka Eni[1]
E-mail: donatus_eni@yahoo.com
Moise Ondoh Agwara[1]
E-mail: agwara29@yahoo.com
ORCID ID: http://orcid.org/0000-0001-9112-7637

[1] Department of Inorganic Chemistry, University of Yaounde I, P.O. Box 812 Yaounde, Yaounde, Cameroon.

[2] Department of Chemistry, The University of Bamenda, P.O. Box 39 Bambili, Bamenda, Cameroon.

References

Baca, S. G., Malaestean, I. L., Keene, T. D., Adams, H., Ward, M. D., Hauser, J., … Decurtins, S. (2008). One-dimensional manganese coordination polymers composed of polynuclear cluster blocks and polypyridyl linkers: Structures and properties. Inorganic Chemistry, 47, 11108–11119. doi:10.1021/ic8014145

Bai, Y., Shang, W.-L., Dang, D.-B., Gao, H., Niu, X.-F., & Guan, Y.-F. (2008). Synthesis, crystal structure and luminescent properties of a thiocyanato bridged two-dimensional heteronuclear polymeric complex of cadmium(II) and copper(II). Inorganic Chemistry Communications, 11, 1470–1473. doi:10.1016/j.inoche.2008.10.016

Bai, Y., Shang, W.-L., Dang, D.-B., Sun, J.-D., & Gao, H. (2009). Synthesis, crystal structure and luminescent properties of one coordination polymer of cadmium(II) with mixed thiocyanate and hexamethylenetetramine ligands. Spectrochimica Acta Part A: Molecular and Biomolecular Spectroscopy, 72, 407–411. doi:10.1016/j.saa.2008.10.033

Banerjee, S., Drew, M. G. B., & Ghosh, A. (2003). Construction of coordination polymers of cadmium(II) with mixed hexamethylenetetramine and terephthalate or thiocyanate ligands. Polyhedron, 22, 2933–2941. doi:10.1016/S0277-5387(03)00404-2

Batten, S. R., Harris, A. R., Murray, K. S., & Smith, J. P. (2002). Crystal engineering with mercuric chloride. Crystal Growth & Design, 2, 87–89. doi:10.1021/cg0155696

Bruker. (2001). SMART (Version 5.625), SADABS (Version 2.03a) and SHELXTL (Version 6.12). Madison, Wisconsin, USA.: Bruker AXS Inc.

Chattopadhyay, S., Bhar, K., Choubey, S., Khan, S., Mitra, P., & Ghosh, B. K. (2012). A new luminous end-to-end thiocyanato bridged heptacoordinated coordination polymer of lead(II) containing a tetradentate Schiff base. Inorganic Chemistry Communications, 16, 21–24. doi:10.1016/j.inoche.2011.11.019

Chattopadhyay, S., Drew, M. G. B., Diaz, C., & Ghosh, A. (2007). The first metamagnetic thiocyanato-bridged one-dimensional nickel(ii) complex. Dalton Transactions, 24, 2492–2494. doi:http://dx.doi.org/10.1039/B702814H http://dx.doi.org/10.1039/b702814h

Chen, G., Bai, Z.-P., & Qu, S.-J. (2005). catena -Poly[[dipyridylcopper(II)]-di-μ-thiocyanato]. Acta Crystallographica Section E Structure Reports Online, 61, m2718–m2719. doi:10.1107/S1600536805036998

Das, K., Datta, A., Sinha, C., Huang, J.-H., Garribba, E., Hsiao, C.-S., & Hsu, C.-L. (2012). End-to-end thiocyanato-bridged helical chain polymer and dichlorido-bridged copper(II) complexes with a hydrazone ligand: Synthesis, characterisation by electron paramagnetic resonance and variable-temperature magnetic studies, and inhibitory effects on H. ChemistryOpen, 1, 80–89. doi:10.1002/open.201100011.

Demir, S., Yilmaz, V. T., Sariboga, B., Buyukgungor, O., & Mrozinski, J. (2010). Metal(II) nicotinamide complexes containing succinato, succinate and succinic acid: Synthesis, crystal structures, magnetic, thermal, antimicrobial and fluorescent properties. Journal of Inorganic and Organometallic Polymers and Materials, 20, 220–228. doi:10.1007/s10904-010-9340-2

Dhaveethu, K., Ramachandramoothy, T., & Thirunavukkarasu, k. (2013). Microwave-assisted synthesis of mixed ligand complexes of Zn(II), Cd(II) and Hg(II) derived from 4-aminopyridine and nitrite ion: Spectral, thermal and bilogical investigations. Journal of the Korean Chemical Society, 57, 341–351. doi:10.5012/jkcs.2013.57.3.341

Du, M., Li, C. P., Liu, C. S., & Fang, S. M. (2013). Design and construction of coordination polymers with mixed-ligand synthetic strategy. Coordination Chemistry Reviews, 257, 1282–1305. http://dx.doi.org/10.1016/j.ccr.2012.10.002

Eddaoudi, M., Moler, D. B., Li, H., Chen, B., Reineke, T. M., O'Keeffe, M., & Yaghi, O. M. (2001). Modular chemistry: Secondary building units as a basis for the design of highly porous and robust metal–organic carboxylate frameworks. Accounts of Chemical Research, 34, 319–330. doi:10.1021/ar000034b

Etaiw, S. E.-D. H., & El-bendary, M. M. (2013). The Influence of copper-copper interaction on the structure and applications of a metal-organic framework based on cyanide and 3-chloropyridine. Journal of Inorganic and Organometallic Polymers and Materials, 23, 510–518. doi:10.1007/s10904-012-9808-3

Etaiw, S. E. H., & Abdou, S. N. (2016). Double stranded helical organo-lead 3D-supramolecular coordination polymer containing copper cyanide and phenanthroline ligand as antimicrobial agent. *Journal of Inorganic and Organometallic Polymers and Materials, 26*, 117–126. doi:10.1007/s10904-015-0301-7

Forood, B., Flatt, B. T., Chassaing, C., & Katritzky, A. K. (2002). 2-aminopyridine derivatives and combinatorial libraries thereof. *United States Patent US 6458789 B1, Lion Bioscience AG.*

Gerlach, D. L., Nieto, I., Herbst-Gervasoni, C. J., Ferrence, G. M., Zeller, M., & Papish, E. T. (2015). Crystal structures of bis- and hexakis[(6,6'-dihydroxybipyridine)copper(II)] nitrate coordination complexes. *Acta Crystallographica Section E Crystallographic Communications, 71*, 1447–1453. doi:10.1107/S205698901502037X

Henninger, S. K., Jeremias, F., Kummer, H., & Janiak, C. (2012). MOFs for use in adsorption heat pump processes. *European Journal of Inorganic Chemistry, 2012*, 2625–2634. doi:10.1002/ejic.201101056.

Hong, Z. (2008). Synthesis, crystal structures, and antimicrobial activity of two thiocyanato-bridged dinuclear copper(II) complexes derived from 2,4-dibromo-6-[(2-diethylaminoethylimino)methyl]phenol and 4-nitro-2-[(2-ethylaminoethylimino)methyl]phenol. *Transition Metal Chemistry, 33*, 797–802. doi:10.1007/s11243-008-9113-8

Hu, X., Li, Y., Wang, Y., Du, W., & Guo, J. (2009). Synthesis, spectroscopic properties, and structures of copper(II) and manganese(II) complexes of pyridine-2,6-dicarboxylate and 1,10-phenanthroline. *Journal of Coordination Chemistry, 62*, 3438–3445. doi:10.1080/00958970903093686

Janiak, C. (2003). Engineering coordination polymers towards applications. *Dalton Transactions, 14*, 2781–2804. doi:10.1039/B305705B

Kabesova, M., & Gazo, J. (1980). Structure and classification of thiocyanates and the mutual influence of their ligands. *Chemicke Zvesti, 34*, 800–841.

Kurdziel, K., Głowiak, T., Materazzi, S., & Jezierska, J. (2003). Crystal structure and physico-chemical properties of cobalt(II) and manganese(II) complexes with imidazole-4-acetate anion. *Polyhedron, 22*, 3123–3128. http://dx.doi.org/10.1016/j.poly.2003.07.004

Laachir, A., Bentiss, F., Guesmi, S., Saadi, M., & El Ammari, L. (2016). Crystal structure of bis[2,5-bis(pyridin-2-yl)-1,3,4-thiadiazole-κ² N² , N³]bis(thiocyanato-κ S)copper(II). *Acta Crystallographica Section E Crystallographic Communications, 72*, 1176–1178. doi:10.1107/S2056989016011713

Li, L., Liang, J., & Tian, G. (2011). Synthesis and structure of a thiocyanato-bridged one-dimensional cdII coordination polymer. *Journal of Chemical Crystallography, 41*, 44–47. doi:10.1007/s10870-010-9834-3

Lu, J., Liu, H.-T., Zhang, X.-X., Wang, D.-Q., & Niu, M.-J. (2010). Important roles of weak interactions: Syntheses and supramolecular structures of four CoII/NiII-thiocyanato compounds. *Zeitschrift für anorganische und allgemeine Chemie, 636*, 641–647. doi:10.1002/zaac.200900286.

Małecki, J. G., Machura, B., Świtlicka A., Gron, T., & Bałanda, M., (2011). Thiocyanate manganese(II) complexes with pyridine and its derivatives ligands. *Polyhedron, 30*, 766–753.

Neumann, T., Jess, I., & Näther, C. (2014). catena-poly[[bis(pyridine-κN)nickel(II)]-di-μ-thiocyanato-κ(2) N:S;κ(2) S:N]. *Acta Crystallographica Section E: Structure Reports Online, 70*, m196–m196. doi:10.1107/S1600536814009611

Prins, L. J., Reinhoudt, D. N., & Timmerman, P. (2001). Noncovalent synthesis using hydrogen bonding. *Angewandte Chemie International Edition 40*, 2382–2426.

doi:10.1002/1521-3773(20010702)40:13<2382:AID-ANIE2382>3.0.CO;2-G

Rapheal, P. F., Manoj, E., & Kurup, M. R. P. (2007). Copper(II) complexes ofN(4)-substituted thiosemicarbazones derived from pyridine-2-carbaldehyde: Crystal structure of a binuclear complex. *Polyhedron, 26*, 818–828. doi:10.1016/j.poly.2006.09.091

Reddy, P. A. N., Nethaji, M., & Chakravarty, A. R. (2002). Synthesis, crystal structures and properties of ternary copper(II) complexes having 2,2'-bipyridine and α-amino acid salicylaldiminates as models for the type-2 sites in copper oxidases. *Inorganica Chimica Acta, 337*, 450–458. http://dx.doi.org/10.1016/S0020-1693(02)01108-8

Rowsell, J. L. C., & Yaghi, O. M. (2004). Metal–organic frameworks: A new class of porous materials. *Microporous and Mesoporous Materials, 73*, 3–14. doi:10.1016/j.micromeso.2004.03.034.

Sheldrick, G. M. (1997). *SHELXS-97 and SHELXL-97.* Germany: University of Gottingen.

Shen, L., & Feng, X. (2002). Synthesis and crystal structure of a novel polymeric thiocyanato-bridged heteronuclear complex of copper(II) and cadmium(II). *Structural Chemistry, 13*, 437–441. http://dx.doi.org/10.1023/A:1020509403583

Shi, J.-M., Xu, W., Zhao, B., Cheng, P., Liao, D.-Z., & Chen, X.-Y. (2005). A 2D thiocyanato-bridged copper(II)-manganese(II) bimetallic coordination polymer with ferromagnetic interactions. *European Journal of Inorganic Chemistry, 2005*, 55–58. doi:10.1002/ejic.200400335

Shirdel, H., Marandi, F., Jalilzadeh, A., Huber, S., & Pfitzner, A. (2015). effects of direction of bridging of thiocyanato on the dimension of coordination polymers: Synthesis, characterization and single-crystal X-ray structure determination of [Cd(4,4'-dm-2,2'-bpy)(NCS)₂]ₙ and [Cd(4,4'-dmo-2,2'-bpy)(NCS)₂]ₙ coordination polymers. *Chinese Journal of Structural Chemistry, 3*, 1135–1144.

Tabong, C. D., Yufanyi, D. M., Paboudam, A. G., Nono, K. N., Eni, D. B., & Agwara, M. O. (2016). Synthesis, crystal structure, and antimicrobial properties of [diaquabis(hexamethylen etetramine)diisothiocyanato-kN]nickel(II) Complex. *Advances in Chemistry, 2016, Article ID 5049718*, 8 pages. doi:http://dx.doi.org/10.1155/2016/5049718

Trivedi, M., Pandey, D. S., & Rath, N. P. (2009). catena -Poly[[(pyridine-κ N)copper(II)]-μ 3 -pyridine-2,6-dicarboxylato-κ³ O² : O²' , N , O⁶ : O⁶]. *Acta Crystallographica Section E Structure Reports Online, 65*, m303–m304. doi:10.1107/S1600536809005212.

Whitesides, G. M., & Boncheva, M. (2002). Beyond molecules: Self-assembly of mesoscopic and macroscopic components. *Proceedings of the National Academy of Sciences, 99*, 4769–4774. doi:10.1073/pnas.082065899.

Wohlert, S., Wriedt, M., Jess, I., & Nather, C. (2011). Chloridotetr apyridinecopper(II) dicyanamidate pyridine disolvate. *Acta Crystallographica Section E Structure Reports Online, 67*, m695. doi:10.1107/S1600536811016187

Yang, G. B., & Sun, Z. H. (2013). Tuning the structural topologies of two luminescent metal-organic frameworks through altering auxiliary ligand. *Inorganic Chemistry Communications, 29*, 94–96. http://dx.doi.org/10.1016/j.inoche.2012.12.022

You, Z.-L., & Zhu, H.-L. (2005). A novel thiocyanate-bridged dinuclear cadmium(II) complex: di-[mu]-thiocyanato-bis((methanol){4-nitro-2-[2-(dimethylamino) ethyliminomethyl]phenolato}cadmium(II)). *Acta Crystallographica Section C, 61*, m397–m399. doi:10.1107/S0108270105021906

12

Another application of newly prepared Brønsted-acidic ionic liquids as highly efficient reusable catalysts for neat synthesis of amidoalkyl naphthols

Maryam Dehghan[1], Abolghasem Davoodnia[1]*, Mohammad R. Bozorgmehr[1] and Niloofar Tavakoli-Hoseini[1]

*Corresponding author: Abolghasem Davoodnia, Department of Chemistry, Mashhad Branch, Islamic Azad University, Mashhad, Iran
E-mails: adavoodnia@mshdiau.ac.ir, adavoodnia@yahoo.com
Reviewing editor: George Weaver, University of Loughborough, UK

Abstract: In this work, two newly prepared Brønsted-acidic ionic liquids, [MPyrrSO$_3$H] Cl (**IL$_1$**) and [MMorSO$_3$H]Cl (**IL$_2$**), were efficiently used as catalysts for the synthesis of amidoalkyl naphthols through the one-pot, three-component reaction of β-naphthol, aryl aldehydes, and acetamide under neat conditions. High activity of the catalysts, excellent yields, short reaction times, simple procedure with an easy work-up, and the absence of any volatile and hazardous organic solvents are some advantages of the present methodology. Moreover, the catalysts are simply prepared and can be recovered conveniently and reused such that considerable catalytic activity can still be achieved after the fifth run.

Subjects: Organic Chemistry; Catalysis; Environmental

Keywords: Brønsted-acidic ionic liquids; amidoalkyl naphthols; solvent-free conditions

1. Introduction

A major challenge in modern chemistry is the design of highly efficient chemical reaction sequences that provide maximum structural complexity with a minimum number of synthetic steps in short reaction times (Dömling, 2006; Schreiber, 2000). Multicomponent reactions (MCRs) have gained considerable attention as a powerful method in organic synthesis and medicinal chemistry because they involve simultaneous reaction of more than two starting materials to yield a single product through one-pot reaction (Gore & Rajput, 2013; Slobbe, Ruijter, & Orru, 2012; Tavakoli-Hoseini &

ABOUT THE AUTHOR

Abolghasem Davoodnia was born in 1971, Mashhad, Iran. He studied chemistry at Tehran University, Tehran, Iran, where he received BSc in 1994. He received his MSc degree in organic chemistry in 1997 from Ferdowsi University of Mashhad, Mashhad, Iran, under the supervision of prof Majid M. Heravi and completed his PhD in organic chemistry in 2002 under the supervision of prof Mehdi Bakavoli at the same university. Currently, he is working as a professor at the Chemistry Department, Mashhad Branch, Islamic Azad University, Mashhad, Iran. He has published over 140 peer-reviewed articles in ISI journals. His current research interest is on heterocyclic chemistry, catalysis and new synthetic methodologies.

PUBLIC INTEREST STATEMENT

Application of new ionic Liquids in organic transformations is of great interest in recent years. Therefore, in this paper, two newly prepared Brønsted-acidic ionic liquids were efficiently used as catalysts for the synthesis of amidoalkyl naphthols through the one-pot, three-component reaction of β-naphthol, aryl aldehydes, and acetamide under neat conditions. Some advantages of this procedure are high yields, short reaction times, easy work-up, absence of volatile and hazardous solvents, and reusability of catalysts for a number of times without appreciable loss of activity.

Davoodnia, 2011). High atom economy, good selectivity, time and energy saving, low cost, minimum waste production, and short reaction time make MCRs suitable for the synthesis of complex molecules with potential biological activity (Chebanov & Desenko, 2012; Manjappa, Peng, Jhang, & Yang, 2016; Zang, Zhang, Zang, & Cheng, 2010). On the other hand, the nature of the catalyst plays a crucial role in the determination of the product and selectivity (Khan, Khan, & Bannuru, 2010; Mirzaei & Davoodnia, 2012; Shaterian & Mohammadnia, 2012). Therefore, development of inexpensive, mild, and reusable catalysts for MCRs such as the synthesis of amidoalkyl naphthols remains of interest to the synthetic organic chemists. It has been reported that amidoalkyl naphthols can convert to important biologically active aminoalkyl naphthol derivatives by amide hydrolysis. Later compounds have been evaluated for the hypotensive and bradycardiac effects (Dingermann, Steinhilber, & Folkers, 2004; Shen, Tsai, & Chen, 1999). Amidoalkyl naphthols are generally synthesized via the three-component reaction of β-naphthol, an aldehyde, and an amide in the presence of various catalysts, such as $Sb(OAc)_3$ (Hakimi, 2016), zirconocene dichloride (Cp_2ZrCl_2) (Khanapure, Jagadale, Salunkhe, & Rashinkar, 2016), $ZrOCl_2 \cdot 8H_2O$ (Sheik Mansoor, Aswin, Logaiya, & Sudhan, 2016), nano Al_2O_3 (Kiasat, Hemat-Alian, & Saghanezhad, 2016), carbon-based solid acid (Davoodnia, Mahjoobin, & Tavakoli-Hoseini, 2014), H_3BO_3 (Shahrisa, Esmati, & Nazari, 2012), iodine (Nagawade & Shinde, 2007), nano-sulfated zirconia (Zali & Shokrolahi, 2012), $K_5CoW_{12}O_{40} \cdot 3H_2O$ (Nagarapu, Baseeruddin, Apuri, & Kantevari, 2007), copper p-toluenesulfonate (Wang & Liang, 2011), $Al(H_2PO_4)_3$ (Shaterian, Amirzadeh, Khorami, & Ghashang, 2008), $Yb(OTf)_3$ in [bmim][BF_4] (Kumar, Rao, Ahmad, & Khungar, 2009), and nano silica phosphoric acid (Bamoniri, Mirjalili, & Nazemian, 2014). Although each of these individual methods has its own merits, many suffer from limitations such as long reaction times, unsatisfactory yields, and the use of relatively expensive catalysts. Thus, the exploration of novel methodologies using new efficient and reusable catalysts is still ongoing.

In recent years, ionic Liquids (ILs) have attracted rising interest as eco-friendly solvents, catalysts and reagents in organic transformations due to their advantageous properties, such as non-flammability, negligible vapor pressure, high thermal and chemical stability, and ability to dissolve a wide range of materials (Chowdhury, Mohan, & Scott, 2007; Olivier-Bourbigou, Magna, & Morvan, 2010; Pârvulescu & Hardacre, 2007). ILs are miscible with materials having very wide range of polarities and are simultaneously able to dissolve a wide range of organic, inorganic and organometallic substances. These features offer numerous opportunities for the improvement of organic reactions using ILs as solvents and catalysts. Moreover, their ionic character enhances the reaction rates to a great extent in many reactions. Among them, Brønsted acidic ILs, especially the SO_3H-functionalized ones, have designed as environmentally friendly catalysts to replace the traditional mineral liquid acids like sulfuric acid and hydrochloric acid in chemical processes (Greaves & Drummond, 2008; Qiu et al., 2016; Shirole, Kadnor, Tambe, & Shelke, 2017; Vafaeezadeh & Alinezhad, 2016; Zolfigol, Khazaei, Moosavi-Zare, & Zare, 2010).

Considering the unique properties of Brønsted-acidic ILs, recently, we have synthesized two sulfonic acid functionalized ILs, including 1-methyl-1-sulfonic acid pyrrolidinium chloride [MPyrrSO$_3$H]Cl (**IL$_1$**) and 4-methyl-4-sulfonic acid morpholinium chloride [MMorSO$_3$H]Cl (**IL$_2$**) (Figure 1), and successfully applied them as highly efficient catalysts in the synthesis of 1,8-dioxooctahydroxanthenes (Dehghan, Davoodnia, Bozorgmehr, & Bamoharram, 2016). These findings encouraged us to explore other applications of these ILs in the synthesis of organic compounds. Therefore, in line with our interest on the development of convenient methods using reusable catalysts (Davoodnia, 2011; Davoodnia, Allameh, Fazli, & Tavakoli-Hoseini, 2011; Davoodnia, Basafa, & Tavakoli-Hoseini, 2016; Davoodnia, Khojastehnezhad, Bakavoli, & Tavakoli-Hoseini, 2011; Emrani, Davoodnia, & Tavakoli-Hoseini, 2011; Khashi, Davoodnia, & Prasada Rao Lingam, 2015; Moghaddas, Davoodnia, Heravi, & Tavakoli-Hoseini, 2012; Nakhaei & Davoodnia, 2014; Taghavi-Khorasani & Davoodnia, 2015), herein, we report the results of our investigation on the application of **IL$_1$** and **IL$_2$** as catalysts in the synthesis of amidoalkyl naphthols through the one-pot, three-component reaction of β-naphthol, aryl aldehydes, and acetamide under neat conditions (Scheme 1).

[MPyrrSO$_3$H]Cl (**IL$_1$**)   [MMorSO$_3$H]Cl (**IL$_2$**)

Figure 1. Structures of IL$_1$ and IL$_2$.

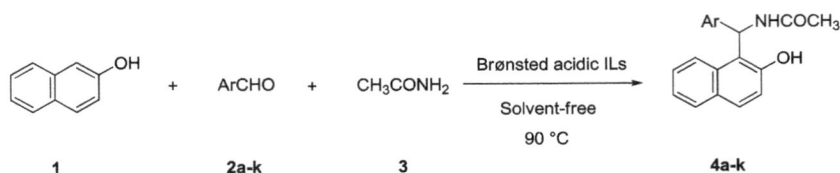

Scheme 1. Synthesis of amidoalkyl naphthols catalyzed by Brønsted acidic ILs.

2. Results and discussion

As a preliminary, we directed our studies toward examination of the effect of various parameters like catalyst composition, effect of solvent, and influence of temperature on the reaction of β-naphthol (1) (1.0 mmol), 4-chlorobenzaldehyde (**2d**) (1.0 mmol), and acetamide (3) (1.0 mmol) for the synthesis of compound **4d** as the model reaction in the absence or presence of IL$_1$ and IL$_2$ as catalysts. A summary of the optimization experiments is provided in Table 1. First, to illustrate the need for catalyst in the reaction, the model reaction was studied in the absence of catalyst under solvent-free condition. The yield of the product was trace at 90°C after 60 min (Table 1, entry 1). Next, the reaction was performed in the presence of IL$_1$ or IL$_2$ in different solvents as well as under solvent-free conditions. Among the solvents tested, those being EtOH, MeOH, CH$_2$Cl$_2$, and MeCN, the reaction proceeded most readily to give the highest yield of the product **4d** under solvent-free conditions. It was observed that the yield of the final product **4d** increased with increasing amount of catalyst in the reaction mixture. The best result was obtained with 10 mol% of the catalyst under solvent-free conditions, which gave the desired product in 95 and 98% yields after 3 and 2 min at 90°C, respectively, for IL$_1$ and IL$_2$ (Table 1, entry 12). Further increase in temperature and IL$_1$ or IL$_2$ amount were found to have an inhibitory effect on formation of the product (Table 1, entries 13, 16, 17).

With the optimized conditions in hand, β-naphthol was reacted with acetamide and a wide variety of aromatic aldehydes using IL$_1$ or IL$_2$ (Table 2). As it can be seen, the reaction is effective with a variety of aromatic aldehydes with electron-donating or withdrawing substituents. Although the kind of aromatic aldehyde had no significant effect on the reaction, in most cases, but not all, aromatic aldehydes substituted with electron-withdrawing group or none reacted slightly faster than those with electron-donating groups and gave the higher yields of the products. Furthermore, both catalysts were highly efficient, and gave the desired amidoalkyl naphthols in high yields and short reaction times. However, as depicted, IL$_2$ proved to be the better catalyst than IL$_1$ in terms of yield and reaction time.

We also investigated recycling of the catalysts under solvent-free conditions using the model reaction. After completion of the reaction, the reaction mixture was cooled to room temperature, and warm distilled water was added. The product was collected by filtration, and washed repeatedly with warm distilled water. The combined filtrate was evaporated to dryness under reduced pressure. The residual ionic liquid was repeatedly washed with diethyl ether, dried under vacuum at 60°C, and used for the subsequent catalytic runs. The recovered catalyst worked well for up to five catalytic runs without any significant loss of its activity (95/98, 95/96, 93/95, 92/93, and 91/93% yields for IL$_1$/IL$_2$ catalysts in first to fifth use, respectively).

Table 1. Screening of reaction condition for synthesis of compound 4d catalyzed by IL_1 or IL_2[a]

Entry	Catalyst (mol%)	Solvent	T (°C)	Time (min) IL_1/IL_2	Isolated yield (%) IL_1/IL_2
1	–	–	90	60/60	Trace/Trace
2	5	–	70	6/5	60/68
3	5	–	80	6/4	66/73
4	5	–	90	5/3	73/76
5	5	–	110	5/3	70/72
6	7	–	70	5/5	67/72
7	7	–	80	5/4	77/80
8	7	–	90	4/3	85/89
9	7	–	110	4/2	82/85
10	10	–	70	5/4	75/79
11	10	–	80	4/3	86/89
12	10	–	90	3/2	95/98
13	10	–	110	4/3	90/93
14	15	–	70	6/5	70/74
15	15	–	80	6/4	78/85
16	15	–	90	5/3	88/92
17	15	–	110	6/3	85/88
18	10	EtOH	Reflux	35/25	53/70
19	10	MeOH	Reflux	45/35	57/72
20	10	CH_2Cl_2	Reflux	40/30	54/60
21	10	MeCN	Reflux	30/20	65/72

[a]Reaction conditions: β-naphthol (1) (1.0 mmol), 4-chlorobenzaldehyde (2d) (1.0 mmol), and acetamide (3) (1.0 mmol).

Table 2. IL_1 or IL_2 catalyzed synthesis of amidoalkyl naphthols (4a-k)[a]

Entry	Ar	Product	Time (min) IL_1/IL_2	Isolated yield (%) IL_1/IL_2	m.p. (°C)	lit. m.p. (°C)
1	C_6H_5	4a	4/2	94/97	242–244	240–242 (Kiasat et al., 2016)
2	$4\text{-}O_2NC_6H_4$	4b	4/2	89/91	246–248	242–244 (Kiasat et al., 2016)
3	$3\text{-}O_2NC_6H_4$	4c	6/3	86/90	239–241	241–243 (Kiasat et al., 2016)
4	$4\text{-}ClC_6H_4$	4d	3/2	95/98	228–230	225–227 (Kiasat et al., 2016)
5	$2\text{-}ClC_6H_4$	4e	4/3	86/91	206–208	204–205 (Shahrisa et al., 2012)
6	$3\text{-}BrC_6H_4$	4f	4/2	90/91	227–229	229–230 (Shahrisa et al., 2012)
7	$4\text{-}BrC_6H_4$	4g	5/2	92/94	227–228	230–232 (Davoodnia et al., 2014)
8	$4\text{-}FC_6H_4$	4h	4/3	93/95	224–226	226–228 (Davoodnia et al., 2014)
9	$4\text{-}MeC_6H_4$	4i	6/4	85/87	218–220	217–220 (Wang & Liang, 2011)
10	$4\text{-}MeOC_6H_4$	4j	5/3	80/83	182–184	180–182 (Wang & Liang, 2011)
11	3-Pyridyl	4k	6/5	86/89	190–192	192–194 (Bamoniri et al., 2014)

[a]Reaction conditions: β-naphthol (1) (1.0 mmol), an aromatic aldehyde (2a-k) (1.0 mmol), acetamide (3) (1.0 mmol), IL_1 or IL_2 (0.1 mmol, 10 mol%), 90°C, solvent-free.

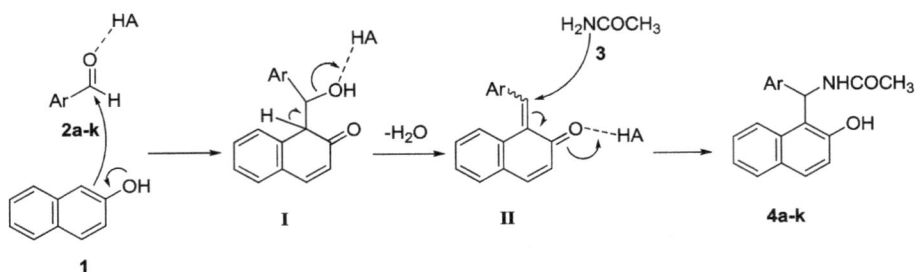

Scheme 2. Plausible mechanism for the formation of amidoalkyl naphthols in the presence of IL$_1$ or IL$_2$ ≡ HA.

In accordance with the literature (Kiasat et al., 2016; Shahrisa et al., 2012), the suggested mechanism is described in Scheme 2. We believe that these ILs can act as Brønsted acids and therefore promotes the reactions by increasing the electrophilic character of the electrophiles in the reaction. At first, *ortho*-quinone methide (*o*-QM) intermediate [II] is readily formed *in situ* by Knoevenagel condensation of β-naphthol (1) and aromatic aldehydes (2a-k) via the intermediate [I]. Subsequent Michael addition of acetamide (3) to the *o*-QM intermediate [II] afforded the final products 4a-k.

3. Conclusion

In conclusion, we showed that two newly synthesized Brønsted-acidic ILs, IL$_1$ and IL$_2$, efficiently catalyze the synthesis of amidoalkyl naphthols by increasing the electrophilic character of the electrophiles in the reaction β-naphthol, aryl aldehydes, and acetamide under solvent-free reactions. The kind of aldehyde had no significant effect on the reaction rates and products' yields. However, in general, electron-poor aldehydes reacted slightly faster than electron-rich ones and gave the higher yields of the products. Also, IL$_2$ proved to be the better catalyst than IL$_1$ in terms of yield and reaction time. Some advantages of this procedure are high yields, short reaction times, easy work-up, absence of volatile and hazardous solvents, and reusability of catalysts for a number of times without appreciable loss of activity.

4. Experimental

The IL$_1$ and IL$_2$ were synthesized according to the our previous report (Dehghan et al., 2016). All chemicals were available commercially and used without additional purification. Melting points were recorded on a Stuart SMP3 melting point apparatus. The ^{1}H NMR spectra were recorded with a Bruker 300 FT spectrometer.

4.1. General procedure for the synthesis of amidoalkyl naphthols (4a-k) catalyzed by IL$_1$ or IL$_2$

A mixture of β-naphthol (1) (1.0 mmol), an aromatic aldehyde (2a-k) (1.0 mmol), acetamide (3) (1.0 mmol), and IL$_1$ or IL$_2$ (0.1 mmol, 10 mol %) was heated in an oil bath at 90°C for 2–6 min. After completion of the reaction, monitored by TLC, the mixture was cooled to room temperature and warm distilled water was added. This resulted in the precipitation of the product, which was collected by filtration. The crude product was washed repeatedly with warm distilled water and then cold ethanol, and subsequently recrystallized from ethanol to give the pure products 4a-k in high yields. The products were characterized according to comparison of their melting points with those of authentic samples and for some of them by their ^{1}H NMR spectral data.

4.2. Selected ^{1}H NMR data

N-((2-hydroxynaphthalen-1-yl)(phenyl)methyl)acetamide (4a): ^{1}H NMR (300 MHz, DMSO-d$_6$): δ 2.01 (s, 3H, CH$_3$), 7.10–7.45 (m, 9H, arom-H and CH$_{sp}^3$), 7.76–7.87 (m, 3H, arom-H and NH), 8.52 (d, 1H, J = 8.1 Hz, arom-H), 10.08 (s br, 1H, OH).

N-((2-hydroxynaphthalen-1-yl)(4-nitrophenyl)methyl)acetamide (4b): ^{1}H NMR (300 MHz, DMSO-d$_6$): δ 2.04 (s, 3H, CH$_3$), 7.18–7.33 (m, 3H, arom-H and CH$_{sp}^3$), 7.38–7.46 (m, 3H, arom-H), 7.78–7.88 (m, 3H, arom-H and NH), 8.16 (d, 2H, J = 9.0 Hz, arom-H), 8.62 (d, 1H, J = 7.8 Hz, arom-H), 10.01 (br, 1H, OH).

N-((4-Chlorophenyl)(2-hydroxynaphthalen-1-yl)methyl)acetamide **(4d)**: ^{1}H NMR (300 MHz, DMSO-d$_6$): δ 2.03 (s, 3H, CH$_3$), 7.13–7.44 (m, 8H, arom-H and CH$_{sp^3}$), 7.82 (t, 2H, J = 8.7 Hz, arom-H), 7.87 (br, 1H, NH), 8.54 (d, 1H, J = 8.1 Hz, arom-H), 10.11 (s, 1H, OH).

N-((3-Bromophenyl)(2-hydroxynaphthalen-1-yl)methyl)acetamide **(4f)**: ^{1}H NMR (300 MHz, DMSO-d$_6$): δ 2.00 (s, 3H, CH$_3$), 7.12 (d, 2H, J = 6.9 Hz, arom-H), 7.18–7.45 (m, 6H, arom-H and CH$_{sp^3}$), 7.77–7.90 (m, 3H, arom-H and NH), 8.51 (d, 1H, J = 8.4 Hz, arom-H), 9.69 (br, 1H, OH).

N-((4-Bromophenyl)(2-hydroxynaphthalen-1-yl)methyl)acetamide **(4g)**: ^{1}H NMR (300 MHz, DMSO-d$_6$): δ 2.01 (s, 3H, CH$_3$), 7.08–7.32 (m, 5H, arom-H and CH$_{sp^3}$), 7.40 (t, 1H, J = 8.1 Hz, arom-H), 7.46 (d, 2H, J = 8.4 Hz, arom-H), 7.76–7.88 (m, 3H, arom-H and NH), 8.51 (d, 1H, J = 8.1 Hz, arom-H), 10.04 (s br, 1H, OH).

N-((2-hydroxynaphthalen-1-yl)(4-methoxyphenyl)methyl)acetamide **(4j)**: ^{1}H NMR (300 MHz, DMSO-d$_6$): δ 2.00 (s, 3H, CH$_3$), 3.69 (s, 3H, OCH$_3$), 6.83 (d, 2H, J = 8.7 Hz, arom-H), 7.08–7.31 (m, 5H, arom-H and CH$_{sp^3}$), 7.38 (t, 1H, J = 7.2 Hz, arom-H), 7.75–7.94 (m, 3H, arom-H and NH), 8.49 (d, 1H, J = 8.4 Hz, arom-H), 10.05 (br, 1H, OH).

N-((2-hydroxynaphthalen-1-yl)(pyridin-3-yl)methyl)acetamide **(4k)**: ^{1}H NMR (300 MHz, DMSO-d$_6$): δ 2.02 (s, 3H, CH$_3$), 7.15–7.35 (m, 4H, arom-H and CH$_{sp^3}$), 7.42 (t, 1H, J = 7.5 Hz, arom-H), 7.55 (d, 1H, J = 8.1 Hz, arom-H), 7.78–7.95 (m, 3H, arom-H and NH), 8.37–8.44 (m, 2H, arom-H), 8.57 (d, 1H, J = 8.1 Hz, arom-H), 10.14 (s br, 1H, OH).

Funding
We gratefully acknowledge financial support from the Islamic Azad University, Mashhad Branch, Iran.

Author details
Maryam Dehghan[1]
E-mail: m.dehghan32@yahoo.com
Abolghasem Davoodnia[1]
E-mails: adavoodnia@mshdiau.ac.ir, adavoodnia@yahoo.com
ORCID ID: http://orcid.org/0000-0002-1425-8577
Mohammad R. Bozorgmehr[1]
E-mail: mr_bozorgmehr@yahoo.com
Niloofar Tavakoli-Hoseini[1]
E-mail: niloofartavakoli@ymail.com
[1] Department of Chemistry, Mashhad Branch, Islamic Azad University, Mashhad, Iran.

References
Bamoniri, A., Mirjalili, B. F., & Nazemian, S. (2014). Nano silica phosphoric acid: an efficient catalyst for the one-pot synthesis of amidoalkyl naphthols under solvent-free condition. *Journal of the Iranian Chemical Society, 11*, 653–658. doi:10.1007/s13738-013-0336-z

Chebanov, V. A., & Desenko, S. M. (2012). Multicomponent heterocyclization reactions with controlled selectivity (Review). *Chemistry of Heterocyclic Compounds, 48*, 566–583. doi:10.1007/s10593-012-1030-2

Chowdhury, S., Mohan, R. S., & Scott, J. L. (2007). Reactivity of ionic liquids. *Tetrahedron, 63*, 2363–2389. doi:10.1016/j.tet.2006.11.001

Davoodnia, A. A. (2011). Highly efficient and fast method for the synthesis of biscoumarins using tetrabutylammonium hexatungstate [TBA]$_2$[W$_6$O$_{19}$] as green and reusable heterogeneous catalyst. *Bulletin of the Korean Chemical Society, 32*, 4286–4290. doi:10.5012/bkcs.2011.32.12.4286

Davoodnia, A., Allameh, S., Fazli, S., & Tavakoli-Hoseini, N. (2011). One-pot synthesis of 2-amino-3-cyano-4-arylsubstituted tetrahydrobenzo[b]pyrans catalysed by silica gel-supported polyphosphoric acid (PPA-SiO$_2$) as an efficient and reusable catalyst. *Chemical Papers, 65*, 714–720. doi:10.2478/s11696-011-0064-8

Davoodnia, A., Basafa, S., & Tavakoli-Hoseini, N. (2016). Neat synthesis of octahydroxanthene-1,8-diones, catalyzed by silicotungstic acid as an efficient reusable inorganic catalyst. *Russian Journal of General Chemistry, 86*, 1132–1136. doi:10.1134/S107036321605025X

Davoodnia, A., Khojastehnezhad, A., Bakavoli, M., & Tavakoli-Hoseini, N. (2011). SO3H-functionalized ionic liquids: Green, efficient and reusable catalysts for the facile dehydration of aldoximes into nitriles. *Chinese Journal of Chemistry, 29*, 978–982. doi:10.1002/cjoc.201190199

Davoodnia, A., Mahjoobin, R., & Tavakoli-Hoseini, N. (2014). A facile, green, one-pot synthesis of amidoalkyl naphthols under solvent-free conditions catalyzed by a carbon-based solid acid. *Chinese Journal of Catalysis, 35*, 490–495. doi:10.1016/S1872-2067(14)60011-5

Dehghan, M., Davoodnia, A., Bozorgmehr, M. R., & Bamoharram, F. F. (2016). Synthesis, characterization and application of two novel sulfonic acid functionalized ionic liquids as efficient catalysts in the synthesis of 1,8-dioxo-octahydroxanthenes. *Heterocyclic Letters, 6*, 251–257.

Dingermann, T., Steinhilber, D., & Folkers, G. (2004). *In molecular biology in medicinal chemistry*. Weinheim: Wiley-VCH.

Dömling, A. (2006). Recent developments in isocyanide based multicomponent reactions in applied chemistry. *Chemical Preview, 106*, 17–89. doi:10.1021/cr0505728

Emrani, A., Davoodnia, A., & Tavakoli-Hoseini, N. (2011). Alumina supported ammonium dihydrogenphosphate (NH$_4$H$_2$PO$_4$/Al$_2$O$_3$): Preparation, characterization and its application as catalyst in the synthesis of 1,2,4,5-tetrasubstituted imidazoles. *Bulletin of the Korean Chemical Society, 32*, 2385–2390. doi:10.5012/bkcs.2011.32.7.2385

Gore, R. P., & Rajput, A. P. (2013). A review on recent progress in multicomponent reactions of pyrimidine synthesis. *Drug Invention Today, 5*, 148–152. doi:10.1016/j.dit.2013.05.010

Greaves, T. L., & Drummond, C. J. (2008). Protic Ionic Liquids: Properties and Applications. *Chemical Reviews, 108*, 206–237. doi:10.1021/cr068040u

Hakimi, F. (2016). Antimony(III) acetate an efficient catalyst for the synthesis of 1-amidoalkyl-2-naphthols. *Journal of Chemical Research, 40*, 489–491. doi:10.3184/174751916X14665064359398

Khan, A. T., Khan, M. M., & Bannuru, K. K. R. (2010). Iodine catalyzed one-pot five-component reactions for direct synthesis of densely functionalized piperidines. *Tetrahedron, 66*, 7762–7772. doi:10.1016/j.tet.2010.07.075

Khanapure, S., Jagadale, M., Salunkhe, R., & Rashinkar, G. (2016). Zirconocene dichloride catalyzed multicomponent synthesis of 1-amidoalkyl-2-naphthols at ambient temperature. *Research on Chemical Intermediates, 42*, 2075–2085. doi:10.1007/s11164-015-2136-9

Khashi, M., Davoodnia, A., & Prasada Rao Lingam, V. S. (2015). DMAP catalyzed synthesis of some new pyrrolo[3,2-e][1,2,4]triazolo[1,5-c]pyrimidines. *Research on Chemical Intermediates, 41*, 5731–5742. doi:10.1007/s11164-014-1697-3

Kiasat, A. R., Hemat-Alian, L., & Saghanezhad, S. J. (2016). Nano Al2O3: An efficient and recyclable nanocatalyst for the one-pot preparation of 1-amidoalkyl-2-naphthols under solvent-free conditions. *Research on Chemical Intermediates, 42*, 915–922. doi:10.1007/s11164-015-2062-x

Kumar, A., Rao, M. S., Ahmad, I., & Khungar, B. (2009). A simple and facile synthesis of amidoalkyl naphthols catalyzed by Yb(OTf)3 in ionic liquids. *Canadian Journal of Chemistry, 87*, 714–719. doi:10.1139/V09-049

Manjappa, K. B., Peng, Y. T., Jhang, W. F., & Yang, D. Y. (2016). Microwave-promoted, catalyst-free, multi-component reaction of proline, aldehyde, 1,3-diketone: One pot synthesis of pyrrolizidines and pyrrolizinones. *Tetrahedron, 72*, 853–861. doi:10.1016/j.tet.2015.12.056

Mirzaei, H., & Davoodnia, A. (2012). Microwave assisted sol-gel synthesis of MgO nanoparticles and their catalytic activity in the synthesis of hantzsch 1,4-dihydropyridines. *Chinese Journal of Catalysis, 33*, 1502–1507. doi:10.1016/S1872-2067(11)60431-2

Moghaddas, M., Davoodnia, A., Heravi, M. M., & Tavakoli-Hoseini, N. (2012). Sulfonated carbon catalyzed biginelli reaction for one-pot synthesis of 3,4-dihydropyrimidin-2(1H)-ones and -thiones. *Chinese Journal of Catalysis, 33*, 706–710. doi:10.1016/S1872-2067(11)60377-X

Nagarapu, L., Baseeruddin, M., Apuri, S., & Kantevari, S. (2007). Potassium dodecatungstocobaltate trihydrate ($K_5CoW_{12}O_{40} \cdot 3H_2O$): A mild and efficient reusable catalyst for the synthesis of amidoalkyl naphthols in solution and under solvent-free conditions. *Catalysis Communications, 8*, 1729–1734. doi:10.1016/j.catcom.2007.02.008

Nagawade, R. R., & Shinde, D. B. (2007). Synthesis of amidoalkyl naphthols by an iodine-catalyzed multicomponent reaction of β-naphthol. *Mendeleev Communications, 17*, 299–300. doi:10.1016/j.mencom.2007.09.018

Nakhaei, A., & Davoodnia, A. (2014). Application of a Keplerate type giant nanoporous isopolyoxomolybdate as a reusable catalyst for the synthesis of 1,2,4,5-tetrasubstituted imidazoles. *Chinese Journal of Catalysis, 35*, 1761–1767. doi:10.1016/S1872-2067(14)60174-1

Olivier-Bourbigou, H., Magna, L., & Morvan, D. (2010). Ionic liquids and catalysis: Recent progress from knowledge to applications. *Applied Catalysis A: General, 373*, 1–56. doi:10.1016/j.apcata.2009.10.008

Pârvulescu, V. I., & Hardacre, C. (2007). Catalysis in ionic liquids. *Chemical Reviews, 107*, 2615–2665. doi:10.1021/cr050948h

Qiu, T., Guo, X., Yang, J., Zhou, L., Li, L., Wang, H., & Niu, Y. (2016). The synthesis of biodiesel from coconut oil using novel Brønsted acidic ionic liquid as green catalyst. *Chemical Engineering Journal, 296*, 71–78. doi:10.1016/j.cej.2016.03.096

Schreiber, S. L. (2000). Target-oriented and diversity-oriented organic synthesis in drug discovery. *Science, 287*, 1964–1969. doi:10.1126/science.287.5460.1964

Shahrisa, A., Esmati, S., & Nazari, M. G. (2012). Boric acid as a mild and efficient catalyst for one-pot synthesis of 1-amidoalkyl-2-naphthols under solvent-free conditions. *Journal of Chemical Sciences, 124*, 927–931. doi:10.1007/s12039-012-0285-6

Shaterian, H. R., Amirzadeh, A., Khorami, F., & Ghashang, M. (2008). Environmentally friendly preparation of amidoalkyl naphthols. *Synthetic Communications, 38*, 2983–2994. doi:10.1080/00397910802006396

Shaterian, H. R., & Mohammadnia, M. (2012). Mild basic ionic liquids catalyzed new four-component synthesis of 1H-pyrazolo[1,2-b]phthalazine-5,10-diones. *Journal of Molecular Liquids, 173*, 55–61. doi:10.1016/j.molliq.2012.06.007

Sheik Mansoor, S., Aswin, K., Logaiya, K., & Sudhan, S. P. N. (2016). $ZrOCl_2 \cdot 8H_2O$: An efficient and recyclable catalyst for the three-component synthesis of amidoalkyl naphthols under solvent-free conditions. *Journal of Saudi Chemical Society, 20*, 138–150. doi:10.1016/j.jscs.2012.06.003

Shen, A. Y., Tsai, C. T., & Chen, C. L. (1999). Synthesis and cardiovascular evaluation of N-substituted 1-aminomethyl-2-naphthols. *European Journal of Medicinal Chemistry, 34*, 877–882. doi:10.1016/S0223-5234(99)00204-4

Shirole, G. D., Kadnor, V. A., Tambe, A. S., & Shelke, S. N. (2017). Brønsted-acidic ionic liquid: green protocol for synthesis of novel tetrasubstituted imidazole derivatives under microwave irradiation via multicomponent strategy. *Research on Chemical Intermediates, 43*, 1089–1098. doi:10.1007/s11164-016-2684-7

Slobbe, P., Ruijter, E., & Orru, R. V. A. (2012). Recent applications of multicomponent reactions in medicinal chemistry. *MedChemComm, 3*, 1189–1218. doi:10.1039/c2md20089a

Taghavi-Khorasani, F., & Davoodnia, A. (2015). A fast and green method for synthesis of tetrahydrobenzo[a]xanthene-11-ones using $Ce(SO_4)_2 \cdot 4H_2O$ as a novel, reusable, heterogeneous catalyst. *Research on Chemical Intermediates, 41*, 2415–2425. doi:10.1007/s11164-013-1356-0

Tavakoli-Hoseini, N., & Davoodnia, A. (2011). Carbon-based solid acid as an efficient and reusable catalyst for one-pot synthesis of tetrasubstituted imidazoles under solvent-free conditions. *Chinese Journal of Chemistry, 29*, 203–206. doi:10.1002/cjoc.201190053

Vafaeezadeh, M., & Alinezhad, H. (2016). Brønsted acidic ionic liquids: Green catalysts for essential organic reactions. *Journal of Molecular Liquids, 218*, 95–105. doi:10.1016/j.molliq.2016.02.017

Wang, M., & Liang, Y. (2011). Solvent-free, one-pot synthesis of amidoalkyl naphthols by a copper p-toluenesulfonate catalyzed multicomponent reaction. *Monatshefte für Chemie - Chemical Monthly, 142*, 153–157. doi:10.1007/s00706-010-0429-7

Zali, A., & Shokrolahi, A. (2012). Nano-sulfated zirconia as an efficient, recyclable and environmentally benign catalyst for one-pot three component synthesis of amidoalkyl naphthols. *Chinese Chemical Letters, 23*, 269–272. doi:10.1016/j.cclet.2011.12.002

13

P_2O_5 supported on SiO_2 as an efficient and reusable catalyst for rapid one-pot synthesis of carbamatoalkyl naphthols under solvent-free conditions

Atefeh Ghasemi[1], Abolghasem Davoodnia[1]*, Mehdi Pordel[1] and Niloofar Tavakoli-Hoseini[1]

*Corresponding author: Abolghasem Davoodnia, Department of Chemistry, Mashhad Branch, Islamic Azad University, Mashhad, Iran
E-mails: adavoodnia@mshdiau.ac.ir, adavoodnia@yahoo.com
Reviewing editor: George Weaver, University of Loughborough, UK

Abstract: Under mild conditions and without any additional organic solvent, synthesis of carbamatoalkyl naphthols by the one-pot three-component reaction of β-naphthol with a wide range of aromatic aldehydes and methyl carbamate could be carried out in the presence of P_2O_5 supported on SiO_2 (P_2O_5/SiO_2). The results showed that the catalyst has high activity and the desired products were obtained in high yields in short reaction times. Other beneficial features of this protocol include inexpensive and easily obtained catalyst, simple work-up, and the recyclability and reusability of the catalyst for up to five consecutive runs.

Subjects: Organic Chemistry; Environmental Chemistry; Inorganic Chemistry

Keywords: P_2O_5/SiO_2; carbamatoalkyl naphthols; solvent-free conditions

1. Introduction

Multi-component reactions (MCRs) have attracted much interest and are highly regarded in medicinal chemistry and discovery and synthesis of natural products because they are one-pot processes that bring together three or more components and show high atom economy and high selectivity (Brauch, van Berkel, & Westermann, 2013; Dömling, 2006; Slobbe, Ruijter, & Orru, 2012; Thompson, 2000; Touré & Hall, 2009). They consist of two or more steps which are carried out without isolation of any intermediate. They also provide a rapid and efficient approach to organic synthesis (Davoodnia, Tavakoli-Nishaburi, & Tavakoli-Hoseini, 2011; Gholipour, Davoodnia, & Nakhaei-Moghaddam, 2015; Meerakrishna, Periyaraja, & Shanmugam, 2016). Still, great efforts are being made to develop new

ABOUT THE AUTHOR

Abolghasem Davoodnia was born in 1971, Mashhad, Iran. He studied chemistry at Tehran University, Tehran, Iran, where he received BSc in 1994. He received his MSc degree in organic chemistry in 1997 from Ferdowsi University of Mashhad, Mashhad, Iran, under the supervision of Professor Majid M. Heravi and completed his PhD in organic chemistry in 2002 under the supervision of Prof. Mehdi Bakavoli at the same university. Currently, he is working as a professor at the Chemistry Department, Mashhad Branch, Islamic Azad University, Mashhad, Iran. He has published over 140 peer-reviewed articles in ISI journals. His current research interest is on heterocyclic chemistry, catalysis and new synthetic methodologies.

PUBLIC INTEREST STATEMENT

Synthesis of carbamatoalkyl naphthols which can be converted to important biologically active 1-aminomethyl-2-naphthol derivatives by carbamate hydrolysis is of great interest. Therefore, in this paper, a simple and efficient method for the synthesis of these compounds by the one-pot three-component reaction of β-naphthol, aromatic aldehydes, and methyl carbamate using P_2O_5/SiO_2 as catalyst has been reported. The method was fast and the desired products were obtained within a few minutes in high yields under solvent-free conditions. Other advantages of this protocol include inexpensive and easily obtained catalyst, simple work-up, and the recyclability and reusability of the catalyst.

MCRs and improve known ones such as the synthesis of carbamatoalkyl naphthols. These compounds can be converted to important biologically active 1-aminomethyl-2-naphthol derivatives by carbamate hydrolysis. The hypotensive and bradycardiac effects of later compounds have been evaluated (Dingermann, Steinhilber, & Folkers, 2004; Shen, Tsai, & Chen, 1999). A literature survey revealed that a number of methods were known about the synthesis of carbamatoalkyl naphthols via the one-pot three-component reaction of β-naphthol, an aldehyde, and a carbamate in the presence of a variety of catalysts such as cerium ammonium nitrate (CAN) (Wang, Liu, Song, & Zhao, 2013), Mg(HSO$_4$)$_2$ (Ghashang, 2014), Zwitterionic salt (Kundu, Majee, & Hajra, 2010), Tween® 20 (Yang, Jiang, Dong, & Fang, 2013), ionic liquids (Shaterian & Hosseinian, 2014; Zare, Yousofi, & Moosavi-Zare, 2012), PPA-SiO$_2$ (Shaterian, Hosseinian, & Ghashang, 2009), CuCl$_2$·2H$_2$O (Song, Liu, Sun, & Cui, 2014), aluminum methanesulfonate (Al(MS)$_3$·4H$_2$O) (Song, Sun, Liu, & Cui, 2013), SnCl$_4$·5H$_2$O (Wang, Wang, Zhao, & Wan, 2013), and Mg(OOCCF$_3$)$_2$ (Mohammad Shafiee, Moloudi, & Ghashang, 2011). Though each of these methods has its own advantage, the discovery of new and efficient catalysts with high catalytic activity, short reaction times, recyclability, and simple reaction work-up for the preparation of carbamatoalkyl naphthols is of great interest.

Phosphorus pentoxide supported on silica gel (P$_2$O$_5$/SiO$_2$) has received considerable attention as an efficient, heterogeneous, eco-friendly, highly reactive, stable, easy to handle, and non-toxic catalyst for various organic transformations including acetalization of carbonyl compounds (Mirjalili, Zolfigol, Bamoniri, Amrollahi, & Hazar, 2004), sulfonylation and nitration of aromatic compounds (Hajipour & Ruoho, 2005; Hajipour et al., 2005), N-acylation of sulfonamides (Massah et al., 2009), the Ritter and Schmidt Reactions (Eshghi & Hassankhani, 2006; Tamaddon, Khoobi, & Keshavarz, 2007), Fries and Beckmann rearrangement (Eshghi & Gordi, 2003; Eshghi, Rafie, Gordi, & Bohloli, 2003), the cross-aldol condensation (Hasaninejad, Zare, Balooty, Mehregan, & Shekouhy, 2010), and also the preparation of bis(indolyl)methanes (Hasaninejad, Zare, Sharghi, Niknam, & Shekouhy, 2007), highly substituted imidazoles (Shaterian, Ranjbar, & Azizi, 2011), 4,4'-epoxydicoumarins (Wu & Wang, 2011), Schiff bases (Naeimi, Sharghi, Salimi, & Rabiei, 2008), 1-substituted 1H-1,2,3,4-tetrazoles (Habibi, Nasrollahzadeh, Mehrabi, & Mostafaee, 2013), and β-enaminones (Mohammadizadeh, Hasaninejad, Bahramzadeh, & Sardari Khanjarlou, 2009), affording the corresponding products in excellent yields and high selectivity. Other applications of P$_2$O$_5$/SiO$_2$ in organic synthesis have been reviewed by Eshghi and Hassankhani (2012).

On the other hand, in recent years, considerable interest has been devoted to finding new methodologies for the synthesis of organic compounds in solvent-free condition (Davoodnia, Basafa, & Tavakoli-Hoseini, 2016; Kumar et al., 2016). The toxicity and volatile nature of many organic solvents have posed a serious threat to the environment. Thus, design of solvent-free catalytic reaction has received tremendous attention in recent times in the area of green synthesis (Bettanin, Botteselle, Godoi, & Braga, 2014).

Considering the above facts and also in extension of our previous studies on the development of new environmental friendly methodologies in the synthesis of organic compounds using reusable catalysts (Davoodnia, Allameh, Fazli, & Tavakoli-Hoseini, 2011; Davoodnia, Khashi, & Tavakoli-Hoseini, 2013; Davoodnia, Zare-Bidaki, & Behmadi, 2012; Dehghan, Davoodnia, Bozorgmehr, & Bamoharram, 2016; Khashi, Davoodnia, & Prasada Rao Lingam, 2015; Moghaddas, Davoodnia, Heravi, & Tavakoli-Hoseini, 2012; Nakhaei & Davoodnia, 2014; Nakhaei, Yadegarian, & Davoodnia, 2016; Taghavi-Khorasani & Davoodnia, 2015), we report here the first application of P$_2$O$_5$/SiO$_2$ as an efficient, low cost and reusable catalyst for the efficient solvent-free synthesis of carbamatoalkyl naphthols by the one-pot three-component reaction of β-naphthol (1) with aromatic aldehydes (2a-i) and methyl carbamate (3) (Scheme 1).

2. Results and discussion

To begin our study P$_2$O$_5$/SiO$_2$ was prepared according to the method reported by Eshghi (Eshghi, Rafei, & Karimi, 2001). Grinding of the mixture of P$_2$O$_5$ and SiO$_2$ in dry conditions for 30 min gave the P$_2$O$_5$/SiO$_2$ reagent as white powder. As shown in Scheme 2, the hydroxyl groups in silica gel can be

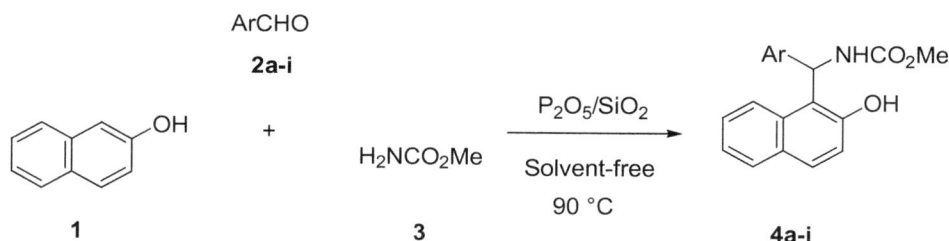

Scheme 1. P₂O₅/SiO₂ catalyzed synthesis of carbamatoalkyl naphthols.

phosphorylated to give the relatively stable P_2O_5/SiO_2. This reagent that can act as an acid catalyst has less sensitivity to moisture than P_2O_5 (Eshghi & Hassankhani, 2012).

Different reaction parameters were optimized for the synthesis of compound **4c** by the one-pot three-component reaction of β-naphthol (**1**) (1 mmol), 4-chlorobenzaldehyde (**2c**) (1 mmol), and methyl carbamate (**3**) (1.1 mmol) as a model reaction in the absence and presence of P_2O_5/SiO_2 as catalyst. The results are summarized in Table 1. Only trace amounts of the product **4c** was formed in the absence of the catalyst in refluxing H_2O or EtOH and also under solvent-free conditions (Entries 1–3) indicating that the catalyst is necessary for the reaction. Several reactions were scrutinized using various solvents, such as H_2O, MeOH, EtOH, CH_3CN, CH_2Cl_2, and also under solvent-free conditions in the presence of P_2O_5/SiO_2 as catalyst. As shown in Table 1, the trial reaction gives the best yield in the presence of 0.05 g of P_2O_5/SiO_2 under solvent-free conditions and proceeds smoothly at 90°C to afford the desired product **4c** in 2 min (entry 14). With tested solvents, the reaction gave low yields within comparable reaction times. Therefore, 0.05 g of the catalyst P_2O_5/SiO_2 under solvent-free condition at 90°C were found to be the optimized conditions. All subsequent reactions were carried out in these optimized conditions.

Under the optimized reaction conditions, we investigated the scope and the limitations of the reaction employing a variety of aromatic aldehydes. The results are summarized in Table 2. Almost all reactions worked well and the desired compounds were obtained in high yields within short reaction time. Under the same conditions however, low yields of the products were obtained using aliphatic aldehydes.

Because of importance of recyclability and reusability of catalysts in organic reactions, the recovery and catalytic activity of recycled P_2O_5/SiO_2 was explored. For this purpose, the synthesis of compound **4c** was again studied under optimized conditions. The P_2O_5/SiO_2 catalyst was readily recovered from the reaction mixture using the procedure outlined in the experimental section. The separated catalyst was washed with hot ethanol and then dried at 50°C under vacuum for 1 h before being reused in a similar reaction. The catalyst could be used at least four times with only a slight reduction in activity

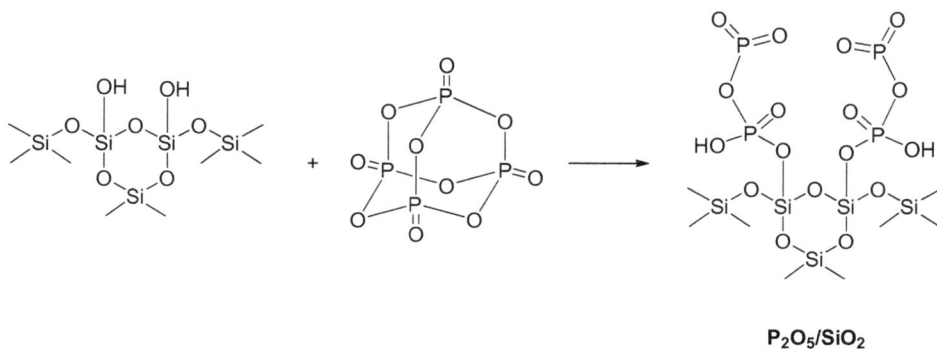

P₂O₅/SiO₂

Scheme 2. Preparation of P₂O₅/SiO₂.

Table 1. Optimization of reaction conditions for synthesis of compound 4c catalyzed by P$_2$O$_5$/SiO$_2$[a]

Entry	Catalyst (g)	Solvent	T (°C)	Time (min)	Isolated yield (%)
1	–	–	90	90	Trace
2	–	H$_2$O	Reflux	90	Trace
3	–	EtOH	Reflux	90	Trace
4	0.01	–	70	8	48
5	0.01	–	90	7	59
6	0.01	–	110	8	59
7	0.02	–	70	8	68
8	0.02	–	90	7	73
9	0.02	–	110	6	74
10	0.04	–	70	8	80
11	0.04	–	90	5	84
12	0.04	–	110	6	83
13	0.05	–	70	6	88
14	0.05	–	90	2	94
15	0.05	–	110	4	93
16	0.07	–	90	4	92
17	0.07	–	110	2	92
18	0.05	H$_2$O	Reflux	20	16
19	0.05	MeOH	Reflux	20	22
20	0.05	EtOH	Reflux	20	30
21	0.05	CH$_3$CN	Reflux	20	15
22	0.05	CH$_2$Cl$_2$	Reflux	20	20

[a]Reaction conditions: β-naphthol (1) (1 mmol), 4-chlorobenzaldehyde (2c) (1 mmol), and methyl carbamate (3) (1.1 mmol).

Table 2. Synthesis of carbamatoalkyl naphthols (4a-i) using P$_2$O$_5$/SiO$_2$[a]

Entry	R	Product	Time (min)	Isolated yield (%)	m.p. (°C)	
					Found	**Reported**
1	2-ClC$_6$H$_4$	4a	4	91	180–182	181–183 (Ghashang, 2014)
2	3-ClC$_6$H$_4$	4b	2	93	202–204	200–202 (Ghashang, 2014)
3	4-ClC$_6$H$_4$	4c	2	94	202–203	203–205 (Ghashang, 2014)
4	2,4-Cl$_2$C$_6$H$_3$	4d	2	92	191–193	189–191 (Ghashang, 2014)
5	4-BrC$_6$H$_4$	4e	3	89	196–199	197–199 (Ghashang, 2014)
6	4-FC$_6$H$_4$	4f	4	90	204–205	203–205 (Ghashang, 2014)
7	3-O$_2$NC$_6$H$_4$	4g	3	94	254–256	249–251 (Zare et al., 2012)
8	4-O$_2$NC$_6$H$_4$	4h	3	95	200–202	203–205 (Zare et al., 2012)
9	4-MeC$_6$H$_4$	4i	4	88	186–188	187–189 (Ghashang, 2014)

[a]Reaction conditions: β-naphthol (1) (1 mmol), an aromatic aldehyde (2a-i) (1 mmol), methyl carbamate (3) (1.1 mmol), P$_2$O$_5$/SiO$_2$ (0.05 g), 90°C, solvent-free.

(94, 93, 92, 92, and 91% yields for first to fifth use, respectively) which clearly demonstrates the practical reusability of this catalyst. This reusability demonstrates the high stability and turnover of P$_2$O$_5$/SiO$_2$ under the employed conditions. The stability of P$_2$O$_5$/SiO$_2$ has been also confirmed in several papers reviewed by Eshghi and Hassankhani (2012) and also two papers reported by Habibi et al. (2013) and Eshghi, Rahimizadeh, Ghadamyari, and Shiri (2012). While P$_2$O$_5$ is very sensitive to moisture, P$_2$O$_5$/SiO$_2$ is stable in various reaction mixtures containing water, amines and alcohols.

3. Conclusion

In this paper, a simple, efficient, and eco-friendly method for the synthesis of carbamatoalkyl naphthols by the one-pot three-component reaction of β-naphthol with a wide range of aromatic aldehydes and methyl carbamate using P_2O_5/SiO_2 as catalyst has been successfully developed. The method was fast and the desired products were obtained within a few minutes in high yields under solvent-free conditions at 90°C. The catalyst can be recycled after a simple work-up, and used at least five times without substantial reduction in its catalytic activity. The procedure is also advantageous in the sense that it is a solvent-free reaction and therefore operates under environmentally friendly conditions.

4. Experimental

All chemicals were available commercially and used without additional purification. Melting points were recorded on a Stuart SMP3 melting point apparatus. The IR spectra were obtained using a Tensor 27 Bruker spectrophotometer as KBr disks. The ^{1}H NMR spectra were recorded with a Bruker 300 FT spectrometer.

4.1. Preparation of P_2O_5/SiO_2

A mixture of P_2O_5 (3 g) and SiO_2 (4 g, 230–400 mesh) was ground vigorously in a mortar for 30 min to give P_2O_5/SiO_2 as a white powder (Eshghi et al., 2001).

4.2. General procedure for the synthesis of carbamatoalkyl naphthols (4a-i) catalyzed by P_2O_5/SiO_2

A mixture of β-naphthol (1) (1 mmol), an aromatic aldehyde (2a-i) (1 mmol), methyl carbamate (3) (1.1 mmol), and P_2O_5/SiO_2 (0.05 g) was heated in the oil bath at 90°C for 2–4 min and monitored by TLC. On completion of the transformation, the reaction mixture was cooled to room temperature and hot ethanol was added. The catalyst was collected by filtration, and the filtrate was cooled to room temperature. The crude product was collected and recrystallized from ethanol to give compounds 4a-i in high yields.

4.3. Selected spectral data

Methyl (3-chlorophenyl)(2-hydroxynaphthalen-1-yl)methylcarbamate (4b): IR (KBr disc): υ 3417 (NH), 3293 (OH), 1690 (C=O) cm^{-1}; ^{1}H NMR (300 MHz, DMSO-d$_6$): δ 3.59 (s, 3H, OCH$_3$), 6.87 (d, 1H, J = 5.7 Hz, CH), 7.16 (d, 1H, J = 6.9 Hz, arom-H), 7.20–7.33 (m, 6H, arom-H and NH), 7.42 (t, 1H, J = 7.2 Hz, arom-H), 7.81 (t, 2H, J = 8.4 Hz, arom-H), 7.92 (d, 1H, J = 8.4 Hz, arom-H), 10.24 (br, 1H, OH).

Methyl (4-chlorophenyl)(2-hydroxynaphthalen-1-yl)methylcarbamate (4c): IR (KBr disc): υ 3421 (NH), 3212 (OH), 1686 (C=O) cm^{-1}; ^{1}H NMR (300 MHz, DMSO-d$_6$): δ 3.58 (s, 3H, OCH$_3$), 6.83 (broadened doublet, 1H, CH), 7.18–7.36 (m, 6H, arom-H), 7.41 (t, 1H, J = 7.0 Hz, arom-H), 7.71 (br, 1H, NH), 7.76–7.85 (m, 2H, arom-H), 7.90 (d, 1H, J = 9.0 Hz, arom-H), 10.20 (br, 1H, OH).

Methyl (2,4-dichlorophenyl)(2-hydroxynaphthalen-1-yl)methylcarbamate (4d): IR (KBr disc): υ3402 (NH), 3259 (OH), 1678 (C=O) cm^{-1}; ^{1}H NMR (300 MHz, DMSO-d$_6$): δ 3.55 (s, 3H, OCH$_3$), 6.82 (broadened doublet, 1H, CH), 7.14 (d, 1H, J = 9.0 Hz, arom-H), 7.29 (t, 1H, J = 7.2 Hz, arom-H), 7.37–7.57 (m, 4H, arom-H), 7.77 (d, 1H, J = 8.7 Hz, arom-H), 7.82 (d, 1H, J = 7.5 Hz, arom-H), 7.95 (br, 1H, NH), 8.01 (d, 1H, J = 8.7 Hz, arom-H), 9.99 (br, 1H, OH).

Methyl (4-bromophenyl)(2-hydroxynaphthalen-1-yl)methylcarbamate (4e): IR (KBr disc): υ 3417 (NH), 3288 (OH), 1688 (C=O) cm^{-1}; ^{1}H NMR (300 MHz, DMSO-d$_6$): δ 3.58 (s, 3H, OCH$_3$), 6.82 (broadened doublet, 1H, CH), 7.18 (d, 2H, J = 8.4 Hz, arom-H), 7.23 (d, 1H, J = 9.0 Hz, arom-H), 7.45 (t, 1H, J = 8.4 Hz, arom-H), 7.37–7.50 (m, 3H, arom-H), 7.70–7.85 (m, 3H, arom-H and NH), 7.90 (d, 1H, J = 7.8 Hz, arom-H), 10.12 (br, 1H, OH).

Methyl (4-fluorophenyl)(2-hydroxynaphthalen-1-yl)methylcarbamate (4f): IR (KBr disc): υ 3422 (NH), 3224 (OH), 1685 (C=O) cm^{-1}; ^{1}H NMR (300 MHz, DMSO-d$_6$): δ 3.58 (s, 3H, OCH$_3$), 6.85 (d, 1H, J = 9.0

Hz, CH), 7.09 (t, 2H, J = 8.1 Hz, arom-H), 7.20–7.45 (m, 5H, arom-H), 7.70–7.85 (m, 3H, arom-H and NH), 7.92 (d, 1H, J = 8.1 Hz, arom-H), 10.23 (br, 1H, OH).

Methyl (2-hydroxynaphthalen-1-yl)(4-nitrophenyl)methylcarbamate (**4h**): IR (KBr disc): v 3422 (NH), 3268 (OH), 1683 (C=O), 1517 and 1345 (NO_2) cm^{-1}; ^{1}H NMR (300 MHz, DMSO-d$_6$): δ 3.61 (s, 3H, OCH$_3$), 6.96 (d, 1H, J = 7.5 Hz, CH), 7.23 (d, 1H, J = 9.0 Hz, arom-H), 7.30 (t, 1H, J = 7.5 Hz, arom-H), 7.36–7.52 (m, 3H, arom-H), 7.80–7.95 (m, 4H, arom-H and NH), 8.16 (d, 2H, J = 8.7 Hz, arom-H), 10.26 (br, 1H, OH).

Funding

We gratefully acknowledge financial support from the Islamic Azad University, Mashhad Branch, Iran.

Author details

Atefeh Ghasemi[1]
E-mail: ghasemi2314@yahoo.com
Abolghasem Davoodnia[1]
E-mails: adavoodnia@mshdiau.ac.ir, adavoodnia@yahoo.com
ORCID ID: http://orcid.org/0000-0002-1425-8577
Mehdi Pordel[1]
E-mail: mehdipordel58@yahoo.com
Niloofar Tavakoli-Hoseini[1]
E-mail: niloofartavakoli@ymail.com

[1] Department of Chemistry, Mashhad Branch, Islamic Azad University, Mashhad, Iran.

References

Bettanin, L., Botteselle, G. V., Godoi, M., & Braga, A. L. (2014). Green synthesis of 1,3-diynes from terminal acetylenes under solvent-free conditions. *Green Chemistry Letters and Reviews, 7*, 105–112. doi:10.1080/17518253.2014.895868

Brauch, S., van Berkel, S. S., & Westermann, B. (2013). Higher-order multicomponent reactions: Beyond four reactants. *Chemical Society Reviews, 42*, 4948–4962. doi:10.1039/c3cs35505e

Davoodnia, A., Allameh, S., Fazli, S., & Tavakoli-Hoseini, N. (2011). One-pot synthesis of 2-amino-3-cyano-4-arylsubstituted tetrahydrobenzo[b]pyrans catalysed by silica gel-supported polyphosphoric acid (PPA-SiO$_2$) as an efficient and reusable catalyst. *Chemical Papers, 65*, 714–720. doi:10.2478/s11696-011-0064-8

Davoodnia, A., Basafa, S., & Tavakoli-Hoseini, N. (2016). Neat synthesis of octahydroxanthene-1,8-diones, catalyzed by silicotungstic acid as an efficient reusable inorganic catalyst. *Russian Journal of General Chemistry, 86*, 1132–1136. doi:10.1134/S107036321605025X

Davoodnia, A., Khashi, M., & Tavakoli-Hoseini, N. (2013). Tetrabutylammonium hexatungstate [TBA]$_2$[W$_6$O$_{19}$]: Novel and reusable heterogeneous catalyst for rapid solvent-free synthesis of polyhydroquinolines via unsymmetrical Hantzsch reaction. *Chinese Journal of Catalysis, 34*, 1173–1178. doi:10.1016/S1872-2067(12)60547-6

Davoodnia, A., Tavakoli-Nishaburi, A., & Tavakoli-Hoseini, N. (2011). Carbon-based solid acid catalyzed one-pot mannich reaction: A facile synthesis of β-amino carbonyl compounds. *Bulletin of the Korean Chemical Society, 32*, 635–638. doi:10.5012/bkcs.2011.32.2.635

Davoodnia, A., Zare-Bidaki, A., & Behmadi, H. (2012). A rapid and green method for solvent-free synthesis of 1,8-dioxodecahydroacridines using tetrabutylammonium hexatungstate as a reusable heterogeneous catalyst. *Chinese Journal of Catalysis, 33*, 1797–1801. doi:10.1016/S1872-2067(11)60449-X

Dehghan, M., Davoodnia, A., Bozorgmehr, M. R., & Bamoharram, F. F. (2016). Synthesis, characterization and application of two novel sulfonic acid functionalized ionic liquids as efficient catalysts in the synthesis of 1,8-dioxo-octahydroxanthenes. *Heterocyclic Letters 6*, 251–257.

Dingermann, T., Steinhilber, D., & Folkers, G. (2004). *In molecular biology in medicinal chemistry*. Weinheim: Wiley-VCH.

Dömling, A. (2006). Recent developments in isocyanide based multicomponent reactions in applied chemistry. *Chemical Preview, 106*, 17–89. doi:10.1021/cr0505728

Eshghi, H., & Gordi, Z. (2003). An easy method for the generation of amides from ketones by a beckmann type rearrangement mediated by microwave. *Synthetic Communications, 33*, 2971–2978. doi:10.1081/SCC-120022469

Eshghi, H., & Hassankhani, A. (2006). P$_2$O$_5$/SiO$_2$-catalyzed one-pot synthesis of amides from ketones via schmidt reaction under microwave irradiation in dry media. *Synthetic Communications, 36*, 2211–2216. doi:10.1080/00397910600638978

Eshghi, H., & Hassankhani, A. (2012). Phosphorus pentoxide supported on silica gel and alumina (P$_2$O$_5$/SiO$_2$, P$_2$O$_5$/Al$_2$O$_3$) as useful catalysts in organic synthesis. *Journal of the Iranian Chemical Society, 9*, 467–482. doi:10.1007/s13738-011-0057-0

Eshghi, H., Rafei, M., & Karimi, M. H. (2001). Document P$_2$O$_5$/SiO$_2$ as an efficient reagent for esterification of phenols in dry media. *Synthetic Communications, 31*, 771–774. doi:10.1081/SCC-100103268

Eshghi, H., Rafie, M., Gordi, Z., & Bohloli, M. (2003). Improvement of selectivity in the Fries rearrangement and direct acylation reactions by means of P$_2$O$_5$/SiO$_2$ under microwave irradiation in solvent-free media. *Journal of Chemical Research*, 763–764. doi:10.3184/030823503772913674

Eshghi, H., Rahimizadeh, M., Ghadamyari, Z., & Shiri, A. (2012). P$_2$O$_5$/SiO$_2$ as an efficient and mild catalyst for trimethylsilylation of alcohols using hexamethyldisilazane. *Synthesis and Reactivity in Inorganic, Metal-Organic, and Nano-Metal Chemistry, 42*, 1435–1439. doi:10.1080/15533174.2012.682686

Ghashang, M. (2014). Preparation of methyl/benzyl (2-hydroxynaphthalen-1-yl)(aryl)methylcarbamate derivatives using magnesium hydrogen sulfate. *Research on Chemical Intermediates, 40*, 1357–1364. doi:10.1007/s11164-013-1044-0

Gholipour, S., Davoodnia, A., & Nakhaei-Moghaddam, M. (2015). Synthesis, characterization, and antibacterial evaluation of new alkyl 2-amino-4-aryl-4H-chromene-3-carboxylates. *Chemistry of Heterocyclic Compounds, 51*, 808–813. doi:10.1007/s10593-015-1779-1

Habibi, D., Nasrollahzadeh, M., Mehrabi, L., & Mostafaee, S. (2013). P$_2$O$_5$-SiO$_2$ as an efficient heterogeneous catalyst for the solvent-free synthesis of 1-substituted 1H-1,2,3,4-tetrazoles under conventional and ultrasound irradiation conditions. *Monatshefte für Chemie - Chemical Monthly, 144*, 725–728. doi:10.1007/s00706-012-0871-9

Hajipour, A. R., & Ruoho, A. E. (2005). Nitric acid in the presence of P2O5 supported on silica gel—a useful reagent for nitration of aromatic compounds under solvent-free conditions. *Tetrahedron Letters, 46*, 8307–8310. doi:10.1016/j.tetlet.2005.09.178

Hajipour, A. R., Zarei, A., Khazdooz, L., Pourmousavi, S. A., Mirjalili, B. F., & Ruoho, A. E. (2005). Direct sulfonylation of aromatic rings with aryl or alkyl sulfonic acid using supported P$_2$O$_5$/

Al$_2$O$_3$. *Phosphorus, Sulfur and Silicon and the Related Elements, 180*, 2029–2034. doi:10.1080/104265090902796

Hasaninejad, A., Zare, A., Balooty, L., Mehregan, H., & Shekouhy, M. (2010). Solvent-free, cross-aldol condensation reaction using silica-supported, phosphorus-containing reagents leading to α, α′-Bis(arylidene) cycloalkanones. *Synthetic Communications, 40*, 3488–3495. doi:10.1080/00397910903457282

Hasaninejad, A., Zare, A., Sharghi, H., Niknam, K., & Shekouhy, M. (2007). P$_2$O$_5$/SiO$_2$ as an efficient, mild, and heterogeneous catalytic system for the condensation of indoles with carbonyl compounds under solvent-free conditions. *Arkivoc, 14*, 39–50.

Khashi, M., Davoodnia, A., & Prasada Rao Lingam, V. S. (2015). DMAP catalyzed synthesis of some new pyrrolo[3,2-e][1,2,4]triazolo[1,5-c]pyrimidines. *Research on Chemical Intermediates, 41*, 5731–5742. doi:10.1007/s11164-014-1697-3

Kumar, S., Aggarwal, R., Kumar, V., Sadana, R., Patel, B., Kaushik, P., & Kaushik, D. (2016). Solvent-free synthesis of bacillamide analogues as novel cytotoxic and anti-inflammatory agents. *European Journal of Medicinal Chemistry, 123*, 718–726. doi:10.1016/j.ejmech.2016.07.033

Kundu, D., Majee, A., & Hajra, A. (2010). Zwitterionic-type molten salt: An efficient mild organocatalyst for synthesis of 2-amidoalkyl and 2-carbamatoalkyl naphthols. *Catalysis Communications, 11*, 1157–1159. doi:10.1016/j.catcom.2010.06.001

Massah, A. R., Dabagh, M., Shahidi, S., Javaherian Naghash, H., Momeni, A. R., & Aliyan, H. (2009). P$_2$O$_5$/SiO$_2$ as an efficient and recyclable catalyst for N-Acylation of sulfonamides under heterogeneous and solvent-free conditions. *Journal of the Iranian Chemical Society, 6*, 405–411. doi:10.1007/BF03245851

Meerakrishna, R. S., Periyaraja, S., & Shanmugam, P. (2016). Copper-Catalyzed Multicomponent Synthesis of Fluorescent 2′-Phenyl-1′H-spiro[fluorene-9,4′-naphtho[2,3-h]quinoline]-7′,12′-dione Derivatives. *European Journal of Organic Chemistry, 2016*, 4516–4525. doi:10.1002/ejoc.201600647

Mirjalili, B. F., Zolfigol, M. A., Bamoniri, A., Amrollahi, M. A., & Hazar, A. (2004). An efficient procedure for acetalization of carbonyl compounds with P$_2$O$_5$/SiO$_2$. *Phosphorus, Sulfur and Silicon and the Related Elements, 179*, 1397–1401. doi:10.1080/10426500490463583

Moghaddas, M., Davoodnia, A., Heravi, M. M., & Tavakoli-Hoseini, N. (2012). Sulfonated carbon catalyzed biginelli reaction for one-pot synthesis of 3,4-dihydropyrimidin-2(1H)-ones and -thiones. *Chinese Journal of Catalysis, 33*, 706–710. doi:10.1016/S1872-2067(11)60377-X

Mohammadizadeh, M. R., Hasaninejad, A., Bahramzadeh, M., & Sardari Khanjarlou, Z. (2009). P$_2$O$_5$/SiO$_2$ as a new, efficient, and reusable catalyst for preparation of β-enaminones under solvent-free conditions. *Synthetic Communications, 39*, 1152–1165. doi:10.1080/00397910802513052

Naeimi, H., Sharghi, H., Salimi, F., & Rabiei, Kh. (2008). Facile and efficient method for preparation of schiff bases catalyzed by P$_2$O$_5$/SiO$_2$ under free solvent conditions. *Heteroatom Chemistry, 19*, 43–47. doi:10.1002/hc.20383

Nakhaei, A., & Davoodnia, A. (2014). Application of a Keplerate type giant nanoporous isopolyoxomolybdate as a reusable catalyst for the synthesis of 1,2,4,5-tetrasubstituted imidazoles. *Chinese Journal of Catalysis, 35*, 1761–1767. doi:10.1016/S1872-2067(14)60174-1

Nakhaei, A., Yadegarian, S., & Davoodnia, A. (2016). Efficient and rapid hantzsch synthesis of 1,4-dihydropyridines using a nano isopolyoxomolybdate as a reusable catalyst under solvent-free condition. *Heterocyclic Letters 6*, 429–439.

Shafiee, M. R. M., Moloudi, R., & Ghashang, M. (2011). Document Preparation of methyl (2-hydroxynaphthalen-1-yl)(aryl)methyl/ benzylcarbamate derivatives using magnesium (II) 2,2,2-trifluoroacetate as an efficient catalyst. *Journal of Chemical Research*, 622–625. doi:10.3184/174751911X13182405888457

Shaterian, H. R., & Hosseinian, A. (2014). Efficient synthesis of 1-carbamatoalkyl-2-naphthols using Brønsted acidic ionic liquid as reusable catalyst. *Research on Chemical Intermediates, 40*, 3011–3019. doi:10.1007/s11164-013-1147-7

Shaterian, H. R., Hosseinian, A., & Ghashang, M. (2009). PPA-SiO$_2$ catalyzed multi-component synthesis of n-[α-(β-hydroxy-α-naphthyl)(benzyl)] o-alkyl carbamate derivatives. *Chinese Journal of Chemistry, 27*, 821–824. doi:10.1002/cjoc.200990137

Shaterian, H. R., Ranjbar, M., & Azizi, K. (2011). Efficient multi-component synthesis of highly substituted imidazoles utilizing P$_2$O$_5$/SiO$_2$ as a reusable catalyst. *Chinese Journal of Chemistry, 29*, 1635–1645. doi:10.1002/cjoc.201180293

Shen, A. Y., Tsai, C. T., & Chen, C. L. (1999). Synthesis and cardiovascular evaluation of N-substituted 1-aminomethyl-2-naphthols. *European Journal of Medicinal Chemistry, 34*, 877–882. doi:10.1016/S0223-5234(99)00204-4

Slobbe, P., Ruijter, E., & Orru, R. V. A. (2012). Recent applications of multicomponent reactions in medicinal chemistry. *MedChemComm, 3*, 1189–1218. doi:10.1039/c2md20089a

Song, Z. G., Sun, X. H., Liu, L. L., & Cui, Y. (2013). Efficient one-pot synthesis of 1-carbamatoalkyl-2-naphthols using aluminum methanesulfonate as a reusable catalyst. *Research on Chemical Intermediates, 39*, 2123–2131. doi:10.1007/s11164-012-0744-1

Song, Z., Liu, L., Sun, X., & Cui, Y. (2014). Copper chloride-catalyzed efficient three-component one-pot synthesis of carbamatoalkyl naphthols under solvent-free conditions. *Indian Journal of Chemistry - Section B Organic and Medicinal Chemistry, 53B*, 740–745.

Taghavi-Khorasani, F., & Davoodnia, A. (2015). A fast and green method for synthesis of tetrahydrobenzo[a]xanthene-11-ones using Ce(SO$_4$)$_2$·4H$_2$O as a novel, reusable, heterogeneous catalyst. *Research on Chemical Intermediates, 41*, 2415–2425. doi:10.1007/s11164-013-1356-0

Tamaddon, F., Khoobi, M., & Keshavarz, E. (2007). (P$_2$O$_5$/SiO$_2$): A useful heterogeneous alternative for the Ritter reaction. *Tetrahedron Letters, 48*, 3643–3646. doi:10.1016/j.tetlet.2007.03.134

Thompson, L. A. (2000). Recent applications of polymer-supported reagents and scavengers in combinatorial, parallel, or multistep synthesis. *Current Opinion in Chemical Biology, 4*, 324–337. doi:10.1016/S1367-5931(00)00096-X

Touré, B. B., & Hall, D. G. (2009). Natural product synthesis using multicomponent reaction strategies. *Chemical Reviews, 109*, 4439–4486. doi:10.1021/cr800296p

Wang, M., Liu, Y., Song, Z., & Zhao, S. (2013). A convenient three-component synthesis of carbamatoalkyl naphthols catalyzed by cerium ammonium nitrate. *Bulletin of the Chemical Society of Ethiopia, 27*, 421–426. doi:10.4314/bcse.v27i3.11

Wang, M., Wang, Q. L., Zhao, S., & Wan, X. (2013). Solvent-free one-pot synthesis of 1-carbamatoalkyl-2-naphthols by a tin tetrachloride catalyzed multicomponent reaction. *Monatshefte für Chemie - Chemical Monthly, 144*, 975–980. doi:10.1007/s00706-013-0927-5

Wu, L., & Wang, X. (2011). P$_2$O$_5$/SiO$_2$ as a new, efficient and reusable catalyst for preparation of 4,4′-epoxydicoumarins under solvent-free conditions.

14

Gallium (III) chloride-catalyzed synthesis of 3,4-dihydropyrimidinones for Biginelli reaction under solvent-free conditions

Haixin Yuan[1], Kehua Zhang[2], Jingjing Xia[2], Xianhai Hu[2] and Shizhen Yuan[2]*

*Corresponding author: Shizhen Yuan, Department of Chemistry, Anhui JianZhu University, Hefei, 230022, China
E-mail: yuanshzh3@hotmail.com
Reviewing editor: Chris Smith, University of Reading, UK

Abstract: We describe the syntheses of 3,4-dihydropyrimidinones or analogous thioketones by a one-pot cyclocondensation of acetoacetates, aldehydes and urea or thiourea using gallium (III) chloride as catalyst under solvent-free conditions. The improved Biginelli reaction not only features a simple procedure, high yields and easy purification of production, but also the recycled catalyst could be directly reused for many times while the yields of reaction would not decrease.

Subjects: Medicinal & Pharmaceutical Chemistry; Organic Chemistry; Applied & Industrial Chemistry; Materials Chemistry

Keywords: gallium (III) chloride; Biginelli reaction; dihydropyrimidinones; solvent-free

1. Introduction

Dihydropyrimidinone (DHPM) and its derivatives have attracted considerable interest in recent years due to their diverse pharmacological properties such as calcium channel blockers, antihypertensive

PROF. SHIZHEN YUAN'S GROUP'S KEY RESEARCH ACTIVITIES

Our research interests are focused on the development of new reactions for building carbon–carbon, carbon–hydrogen, and carbon–heteroatom bonds with particular emphasis on selectivity, cost, and environmental impact. The contents of research include the following areas:

1. Green organic metal reactions: Barbier–Grignard reaction and Pinancol couple reaction in aqueous media, Knoevenagel condensation under solvent-free conditions.

2. Green synthesis of heterocyclic compounds: novel synthetic methods of Pyrroles, Oxazoles, Quinolinones, Dihydropyrimidinones.

3. The asymmetric synthesis of chiral reagent: such as Preparation of L-Proline Chiral Bonded Silica Gels and Their Application in Beer.

4. Chemistry of natural products: extraction and separation of effective components and structure identification.

We hope that research in our lab could help improve some traditional synthetic methods, and in particular provide better solutions to current existing issues related to medicines, materials, and energy.

PUBLIC INTEREST STATEMENT

A traditional concept in chemistry primarily seeks to increase reaction yield. With the development of chemical industry, people have come to realize chemical hazards to health, community safety and the environment. Now researchers in chemistry are more concerned with benign chemical process, such as solvent-free reaction, recyclable catalyst, aqueous organic reaction and so on.

This article was based on cycling economy and concept of green chemistry, and studied the syntheses of Dihydropyrimidinones or analogous thioketones by using gallium (III) chloride to replace HCl as catalyst under solvent-free conditions. The improved Biginelli reaction not only features a simple procedure, high yields and easy purification of production, but also the recycled catalyst could be directly reused for many times.

Green environment is our goal and we are always trying to come up with new, green and safe ways to resolve the pollution problem of conventional chemical methods.

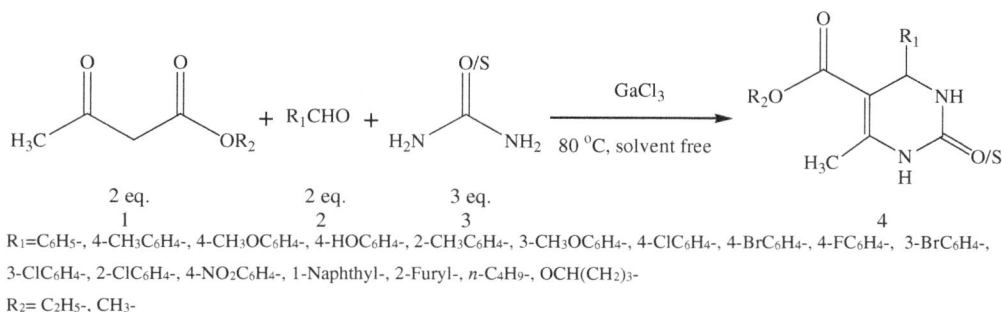

2 eq.   2 eq.   3 eq.
1   2   3   4

R$_1$=C$_6$H$_5$-, 4-CH$_3$C$_6$H$_4$-, 4-CH$_3$OC$_6$H$_4$-, 4-HOC$_6$H$_4$-, 2-CH$_3$C$_6$H$_4$-, 3-CH$_3$OC$_6$H$_4$-, 4-ClC$_6$H$_4$-, 4-BrC$_6$H$_4$-, 4-FC$_6$H$_4$-, 3-BrC$_6$H$_4$-, 3-ClC$_6$H$_4$-, 2-ClC$_6$H$_4$-, 4-NO$_2$C$_6$H$_4$-, 1-Naphthyl-, 2-Furyl-, n-C$_4$H$_9$-, OCH(CH$_2$)$_3$-

R$_2$= C$_2$H$_5$-, CH$_3$-

Scheme 1. Gallium (III) chloride-catalyzed one-pot synthesis of 3,4-dihydropyrimidinones for Biginelli reaction under solvent-free conditions.

agents, neuropeptide antagonists, and α-1a-antagonists (Atwal et al., 1990, 1991). In 1893, Biginelli first reported the synthesis of DHPMs by a simple one-pot condensation reaction of an aromatic aldehyde, β-ketoester, and urea in ethanol containing a catalytic amount of HCl (Biginelli, 1893), however, the yields were low (20–50%).

The acid catalyst is the key in the Biginelli reaction. If there is no acid catalyst, the condensation reaction of aromatic aldehydes can hardly react with urea in the first step. However, it is difficult to freely transmit a protonic acid (H$^+$) catalyst in the non-aqueous medium found in the classical Biginelli reaction, but Lewis acid catalysts can freely bond and separate in non-aqueous medium.

Recently, the Biginelli reaction for the synthesis of DHPMs has gained in popularity again and feature Lewis acid catalysts, such as BF$_3$·OEt$_2$/CuCl (Hu, Sidler, & Dolling, 1998), CuI (Kalita & Phukan, 2007), CuSO$_4$·5H$_2$O (Gohain, Prajapati, & Sandhu, 2004), InBr$_3$ (Fu et al., 2002), InCl$_3$ (Ranu, Hajra, & Jana, 2000), GaI$_3$ (Li, Mao, An, Zhao, & Zou, 2010), GaBr$_3$/GaCl$_3$ (Saini et al., 2007), LaCl$_3$·7H$_2$O (Lu, Bai, Wang, Yang, & Ma, 2000), LiBr (Maiti, Kundu, & Guin, 2003), CeCl$_3$·7H$_2$O (Bose, Fatima, & Mereyala, 2003), MgBr$_2$ (Salehi & Guo, 2004), CdCl$_2$ (Narsaiah, Basak, & Nagaiah, 2004), Al(HSO$_4$)$_3$ (Khodaei, Salehi, Zolfigol, & Sirouszadeh, 2004), FeCl$_3$·6H$_2$O/NiCl$_2$·6H$_2$O/HCl (Lu & Bai, 2002), SmI$_2$ (Han, Xu, Luo, & Shen, 2005), ZnI$_2$ (Jenner, 2004), ZnCl$_2$ (El Badaoui, Bazi, Tahir, Lazrek, & Sebti, 2005), BiCl$_3$ (Ramalinga, Vijayalakshmi, & Kaimal, 2001), Bi(NO$_3$)$_3$ (Slimi, Moussaoui, & ben Salem, 2016), which allow to the preparation of DHPM in high yields.

Recently, gallium (III) halides have emerged as a powerful Lewis catalyst for implementing the Biginelli reaction under solvent-free or microwave-assisted conditions (Maiti et al., 2003; Bose et al., 2003). After considering the gallium (III) halides, gallium (III) chloride, with lower melting point (78°C), is readily melted, and suitable for solvent-free Biginelli reactions.

Herein, we wish to report our study of using gallium (III) chloride for as catalyst for Biginelli reaction under solvent-free conditions (Scheme 1).

2. Results and discussion

Our initial study was started by reacting benzaldehyde (2 mmol), urea (3 mmol) with ethyl acetoacetate (2 mmol) under different conditions. The results are summarized in Table 1.

As shown in Table 1, dihydropyrimidinone **4a** was observed in low yield at 80°C with no catalyst (entry 1, Table 1). When indium (III) halides, gallium (III) bromide, and gallium (III) iodide were used as catalysts, the yields of dihydropyrimidinone **4a** were also lower at 80°C (entries 2–6, Table 1). Encouraging, the dihydropyrimidinone **4a** was obtained in high yields using gallium (III) chloride as catalyst under the same condition (entry 7, Table 1). Increasing the catalyst loading moderately helped to increase yields of 3,4-dihydropyrimidinone **4a** (entries 8–10, Table 1). The effective reaction time was 6 h (entries 11–12, Table 1) and when the reaction temperature was below 75°C, the yields declined sharply (entries 13–14, Table 1). Raising the reaction temperature hardly increased the yield of dihydropyrimidinone **4a** with respect to 80°C (entry 15, Table 1). Gratifyingly, the recycled

Table 1. The synthesis of dihydropyrimidinone (4a) by reacting benzaldehyde (2 mmol), urea (3 mmol) with ethyl acetoacetate (2 mmol) under solvent-free conditions

Entry	Catalyst	Catalyst Mpt. (°C)	Time (h)	Temp(°C)	Yield (4a)/%[a,b]
1	/		5	80	21
2	InCl₃ (0.05 mmol)	586	5	80	33
3	InBr₃ (0.05 mmol)	436	5	80	34
4	InI₃ (0.05 mmol)	355	5	80	37
5	GaBr₃ (0.05 mmol)	124	5	80	49
6	GaI₃ (0.05 mmol)	213	5	80	43
7	GaCl₃ (0.05 mmol)	78	5	80	82
8	GaCl₃ (0.1 mmol)	78	5	80	91
9	GaCl₃ (0.15 mmol)	78	5	80	94
10	GaCl₃ (0.2 mmol)	78	5	80	94
11	GaCl₃ (0.15 mmol)	78	6	80	96
12	GaCl₃ (0.15 mmol)	78	7	80	95
13	GaCl₃ (0.15 mmol)	78	6	75	73
14	GaCl₃ (0.15 mmol)	78	6	65	56
15	GaCl₃ (0.15 mmol)	78	6	90	96
16[c]	GaCl₃ (0.15 mmol, recycled)	78	6	80	95

[a]Isolated yields are reported.

[b]The pure product was identified by IR, ^{1}H, ^{13}C NMR and HRMS.

[c]Recycled catalysts were reused for 4 times.

catalyst could be directly reused and the yields for the four cycles were as high as for the first cycle (entry 16, Table 1).

After this successful survey of the reaction conditions, gallium (III) chloride was considered as an appropriate catalyst for the Biginelli reaction, and was used to synthesize a series dihydropyrimidinone **4a-4r** by reacting benzaldehydes (2 mmol), urea/thiourea (3 mmol) with ethyl/methyl acetoacetate (2 mmol) at 80°C under solvent-free conditions (Scheme 1) and these results are listed in Table 2. The results are listed in Table 2.

It was found that benzaldehyde and both electron withdraw groups or electron donating groups could implement Biginelli reaction with urea and ethyl (methyl) acetoacetate in good yields at 80°C (entries 1–13, Table 2). The position of the substitution group seldomly had an effect on reaction yields, although it was necessary to prolong the reaction time when *ortho*- and *meta*-substituted substrates was used for this reaction (entries 5, 6, 10, 11, and 12, Table 2), which may be caused by the increase in steric hindrance around the carbonyl group and the effect of electron withdrawing substituent, respectively. However, aromatic aldehydes substituted by electron-withdrawing group gave higher yields of DHPMs (4) than by electron-donating group at the same position (entries 7–13, Table 2). Moreover, naphthaldehyde, furfural also could afford the corresponding DHPMs (4) in high yields under the same conditions smoothly (entries 14–15, Table 2). Inspiringly, when an aliphatic aldehyde was used in the present reaction, the corresponding DHPM (4p) also was given the better yield (entry 16, Table 2). Finally, a reaction of aromatic aldehydes with ethyl acetoacetate and thiourea also gave high yields of the desired product (entries 17–18, Table 2).

Table 2. Gallium (III) chloride-catalyzed Biginelli reaction between aldehydes, urea or thiourea and acetoacetate

Entry	R_1	R_2	O/S	Time (h)	Yield (4)/(%) [a,b]	Ref.
1	C_6H_5	OC_2H_5	O	6	96 (4a)	Fu et al. (2002)
2	$4\text{-}CH_3C_6H_4$	OC_2H_5	O	7	93 (4b)	Fu et al. (2002)
3	$4\text{-}CH_3OC_6H_4$	OC_2H_5	O	6	83 (4c)	Fu et al. (2002)
4	$4\text{-}OHC_6H_4$	OC_2H_5	O	7	90 (4d)	Fu et al. (2002)
5	$3\text{-}CH_3OC_6H_4$	OC_2H_5	O	8	78 (4e)	Fu et al. (2002)
6	$2\text{-}CH_3C_6H_4$	OC_2H_5	O	9	89 (4f)	Fu et al. (2002)
7	$4\text{-}ClC_6H_4$	OC_2H_5	O	6	95 (4g)	Fu et al. (2002)
8	$4\text{-}BrC_6H_4$	OC_2H_5	O	6	97 (4h)	Fu et al. (2002)
9	$4\text{-}FC_6H_4$	OC_2H_5	O	6	89 (4i)	Ahmed, Khan, and Habibullah (2009)
10	$3\text{-}BrC_6H_4$	OC_2H_5	O	7	93 (4j)	Lu and Bai (2002)
11	$3\text{-}ClC_6H_4$	OC_2H_5	O	7	91 (4k)	Fu et al. (2002)
12	$2\text{-}ClC_6H_4$	OC_2H_5	O	8	90 (4l)	Fu et al. (2002)
13	$4\text{-}NO_2C_6H_4$	OC_2H_5	O	6	96 (4m)	Fu et al. (2002)
14	1-Naphthyl	OCH_3	O	9	85 (4n)	Cepanec, Litvić, Bartolinčić, and Lovrić (2005)
15	2-Furyl	OC_2H_5	O	6	92 (4o)	Fu et al. (2002)
16	$n\text{-}C_4H_9$	OC_2H_5	O	8	76 (4p)	Fu et al. (2002)
17	$4\text{-}OHC_6H_4$	OC_2H_5	S	7	91 (4q)	Ladole, Salunkhe, and Aswar (2016)
18	2-Furyl	OC_2H_5	S	7	90 (4r)	Ladole, Salunkhe, and Aswar (2016)

[a]Isolated yields are reported.

[b]The pure product was identified by IR, ^{1}H, ^{13}C NMR and HRMS.

3. Conclusions

In summary, we have developed efficient catalyst for the Biginelli reaction using gallium (III) chloride, the condensation of aromatic aldehydes, ethyl acetoacetate, and urea or thiourea generated the corresponding products in excellent yields under free-solvent conditions. The $GaCl_3$ can be recovered by filtration, and untreated $GaCl_3$ was reused directly for Biginelli reaction for at least four cycles. The procedure will find important applications in the synthesis of dihydropyrimidinones to cater the needs of academic and pharmaceutical industries.

4. Experimental section

All melting points were determined on a WT-melting point apparatus. IR (Perkin-Elmer, 2000 FTIR), ^{1}H NMR (DMSO-d_6, 400 MHz), ^{13}C NMR (DMSO-d_6, 100 MHz) were recorded on a Bruker AC-300 FT spectrometer and MS-GC (HP5890 (II)/HP 5972, EI) spectra were obtained at the Center of Analytical Configuration of Anhui Jianzhu University. Flash chromatographic sheet employed was purchased from Anhui Liangchen Silicon Material Co., Ltd. and all materials from Aldrich and used directly as received.

4.1. General procedure for the synthesis of 5-alkoxycarbonyl-6-methyl- 4-aryl-3,4-dihydropyrimidin-2(1H)-ones

The mixture of aldehyde (2 mmol), urea or thiourea (3 mmol), ethyl (methyl) acetoacetate (2 mmol), and gallium (III) chloride (0.15 mmol) was heated with stirring at 80°C for the stated tine, and the progress of the reaction was monitored by TLC. After completion of the reaction, the reaction mixture was cooled to reach room temperature, and then was poured into crushed ice and stirred for 10 min. The solid separated was filtered under suction, washed with ice-cold water. The pure products were obtained by recrystallized from hot ethanol or by flash chromatography on silica gel eluting with petroleum ether/EtOAc (4:1, V: V) and identified by IR, ^{1}H, ^{13}C NMR, and HRMS. All compounds obtained were consistent with authentic ones in the literatures (Ahmed et al., 2009; Cepanec et al., 2005; Fu et al., 2002; Ladole et al., 2016; Lu & Bai, 2002).

4.2. 4-(4-bromophenyl)-5-Ethoxycarbonyl-6-methyl-3,4-dihydropyrimidin-2(1H)-one(4h)

[9]Mp 212–214°C, lit. [9]212–214°C; IR (KBr): v 3,233, 1,727, 1,659 cm^{-1}; ^{1}H NMR (DMSO-d_6, 400 MHz) δ: 9.22 (s, 1H, NH), 7.51 (d, 1H, NH), 7.50–7.18 (m, 4H, C$_6H_4$), 5.11 (d, 1H, CH), 3.97 (q, J = 6.9 Hz, 2H, OCH_2CH$_3$), 2.23 (s, 3H, CH_3), 1.08 (t, J = 6.9 Hz, 3H, OCH$_2$CH_3); ^{13}C NMR (DMSO-d_6, 100 MHz): δ 165.7, 152.4, 149.2, 144.7, 131.8, 129.0, 120.8, 99.3, 59.7, 54.0, 18.3, 14.6; HRMS: calcd for $C_{14}H_{15}N_2O_3Br$: 338.9769, found: 338.9771.

4.3. 5-Ethoxycarbonyl-4-(2-furyl)-6-methyl-3,4-dihydropyrimidin-2(1H)-one(4o)

[9]Mp 211–213°C, lit. [9]210–212°C; IR (KBr): v 3,225, 1,711, 1,640 cm^{-1}; ^{1}H NMR (DMSO-d_6, 400 MHz): δ 9.21 (s, 1H, NH), 7.73 (s, 1H, NH), 7.53 (m, 1H, furyl), 6.33 (d, 1H, furyl), 6.07 (d, 1H, furyl), 5.20 (d, J = 3.6 Hz, 1H, CH), 4.00 (q, J = 7.2 Hz, 2H, OCH_2CH$_3$), 2.22 (s, 3H, CH_3), 1.13 (t, J = 7.0 Hz, 3H, OCH$_2$CH_3); ^{13}C NMR (DMSO-d_6, 100 MHz): δ 165.4, 156.4, 152.8, 149.7, 142.5, 110.7, 105.7, 97.2, 59.6, 48.2, 18.1, 14.6; MS: m/z 250 (M$^+$), 221, 177; Anal. (%): calcd for $C_{12}H_{14}O_4N_2$: C, 57.57; H, 5.64; N, 11.20. Found: C, 57.67; H, 5.67; N, 11.26; HRMS: calcd for $C_{12}H_{14}N_2O_4$: 350.7458, found: 350.7452.

4.4. 5-Ethoxycarbonyl-4-(2-furyl)-6-methyl-3,4-dihydropyrimidin-2(1H)-thione(4r)

[41]Mp 226–227°C, lit. [41]225–227°C; IR (KBr): v 3,244, 1,703, 1,652 cm^{-1}; ^{1}H NMR (DMSO-d_6, 400 MHz): δ 9.61 (s, 1H, NH), 7.55 (s, 1H, NH), 7.53 (m, 1H, furyl), 6.35 (d, 1H, furyl), 6.11 (d, 1H, furyl), 5.20 (d, J = 3.2 Hz, 1H, CH), 4.01 (q, J = 4.2 Hz, 2H, OCH_2CH$_3$), 2.22 (s, 3H, CH_3), 1.11 (t, J = 5.2 Hz, 3H, OCH$_2$CH_3); ^{13}C NMR (DMSO-d_6, 100 MHz): δ 175.3, 165.4, 155.2, 146.5, 143.2, 111.1, 106.8, 98.8, 60.2, 17.7, 14.6; MS: m/z 250 (M$^+$), 221, 177; Anal. (%): calcd for $C_{12}H_{14}O_4N_2$: C, 57.57; H, 5.64; N, 11.20. Found: C, 57.67; H, 5.67; N, 11.26; HRMS: calcd for $C_{12}H_{14}N_2O_3S$: 366.4321, found: 366.4317.

Funding

We appreciate financial support from the Natural Science Foundation of China [grant number 51573001], the Natural Science Foundation of Anhui Province [grant number 1508085MB36], and the Center of Analytical Configuration of Anhui Jianzhu University.

Author details

Haixin Yuan[1]
E-mail: yuanhaixin7@hotmail.com
Kehua Zhang[2]
E-mail: zhangkehua@ahjzu.edu.cn
Jingjing Xia[2]
E-mail: xiajj@ahjzu.edu.cn
Xianhai Hu[2]
E-mail: hxyh@ahjzu.edu.cn
Shizhen Yuan[2]
E-mail: yuanshzh3@hotmail.com
[1] School of Medical Economics & Management, Anhui University of Chinese Medicine, Hefei 230012, China.
[2] Department of Chemistry, Anhui JianZhu University, Hefei 230022, China.

References

Ahmed, B., Khan, R. A., & Habibullah, M. (2009). An improved synthesis of Biginelli-type compounds via phase-transfer catalysis. *Tetrahedron Letters, 50*, 2889–2892. https://doi.org/10.1016/j.tetlet.2009.03.177

Atwal, K. S., Rovnyak, G. C., Kimball, S. D., Floyd, D. M., Moreland, S., Swanson, B. N., ... Malley, M. F., (1990). Dihydropyrimidine calcium channel blockers. II. 3-Substituted-4-aryl-1,4-dihydro-6-methyl-5-pyrimidinecarboxylic acid esters as potent mimics of dihydropyridines. *Journal of Medicinal Chemistry, 33*, 2629–2635. https://doi.org/10.1021/jm00171a044

Atwal, K. S., Swanson, B. N., Unger, S. E., Floyd, D. M., Moreland, S., Hedberg, A., & O'Reilly, B. C. (1991). Dihydropyrimidine calcium channel blockers. 3. 3-Carbamoyl-4-aryl-1, 2, 3, 4-tetrahydro-6-methyl-5-pyrimidinecarboxylic acid esters

as orally effective antihypertensive agents. *Journal of Medicinal Chemistry, 34*, 806–811. https://doi.org/10.1021/jm00106a048

Biginelli, P. (1893). Aldehyde-urea derivatives of aceto-and oxaloacetic acids. *Gazzetta Chimica Italiana, 23*, 360–413.

Bose, D. S., Fatima, L., & Mereyala, H. B. (2003). Green chemistry approaches to the synthesis of 5-alkoxycarbonyl-4-aryl-3,4- dihydropyrimidin-2(1 H)-ones by a three-component coupling of one-pot condensation reaction: Comparison of ethanol, water, and solvent-free conditions. *The Journal of Organic Chemistry, 68*, 587–590. https://doi.org/10.1021/jo0205199

Cepanec, I., Litvić, M., Bartolinčić, A., & Lovrić, M. (2005). Ferric chloride/tetraethyl orthosilicate as an efficient system for synthesis of dihydropyrimidinones by Biginelli reaction. *Tetrahedron, 61*, 4275–4280. https://doi.org/10.1016/j.tet.2005.02.059

El Badaoui, H., Bazi, F., Tahir, R., Lazrek, H. B., & Sebti, S. (2005). Synthesis of 3, 4-dihydropyrimidin-2-ones catalysed by fluorapatite doped with metal halides. *Catalysis Communications, 6*, 455–458. https://doi.org/10.1016/j.catcom.2005.04.003

Fu, N. Y., Yuan, Y. F., Cao, Z., Wang, S. W., Wang, J. T., & Peppe, C. (2002). Indium(III) bromide-catalyzed preparation of dihydropyrimidinones: Improved protocol conditions for the Biginelli reaction. *Tetrahedron, 58*, 4801–4807. https://doi.org/10.1016/S0040-4020(02)00455-6

Gohain, M., Prajapati, D., & Sandhu, J. S. (2004). A novel Cu-catalysed three-component one-pot synthesis of dihydropyrimidin-2 (1H)-ones using microwaves under solvent-free conditions. *Synlett, 2004*, 235–238.

Han, X., Xu, F., Luo, Y., & Shen, Q. (2005). An efficient one-pot synthesis of dihydropyrimidinones by a samarium diiodide catalyzed Biginelli reaction under solvent-free conditions. *European Journal of Organic Chemistry, 2005*, 1500–1503. https://doi.org/10.1002/(ISSN)1099-0690

Hu, E. H., Sidler, D. R., & Dolling, U. H. (1998). Unprecedented catalytic three component one-pot condensation reaction: An efficient synthesis of 5-alkoxycarbonyl-4-aryl-3, 4-dihydropyrimidin-2 (1H)-ones. *The Journal of Organic Chemistry, 63*, 3454–3457.

Jenner, G. (2004). Effect of high pressure on Biginelli reactions. Steric hindrance and mechanistic considerations. *Tetrahedron Letters, 45*, 6195–6198. https://doi.org/10.1016/j.tetlet.2004.05.106

Kalita, H. R., & Phukan, P. (2007). CuI as reusable catalyst for the Biginelli reaction. *Catalysis Communications, 8*, 179–182. https://doi.org/10.1016/j.catcom.2006.06.004

Khodaei, M. M., Salehi, P., Zolfigol, M. A., & Sirouszadeh, S. (2004). Efficient synthesis of 3, 4-dihydropyrimidin-2 (1H)-ones by aluminum hydrogensulfate. *Polish journal of chemistry, 78*, 385–388.

Ladole, C. A., Salunkhe, N. G., & Aswar, A. S. (2016). A microwave assisted synthesis of 3,4-dihydropyrimidin-2(1H)-ones and thiones using NiFe$_2$O$_4$ ferrite as an effective and reusable catalys. *Journal of Indian Chemical Society, 93*, 337.

Li, D., Mao, H., An, L., Zhao, Z., & Zou, J. (2010). Gallium(III) iodide-catalyzed Biginelli-type reaction under solvent-free conditions: Efficient synthesis of dihydropyrimidine-2(1H)-one derivatives. *Chinese Journal of Chemistry, 28*, 2025–2032. https://doi.org/10.1002/cjoc.201090338

Lu, J., & Bai, Y. (2002). Catalysis of the Biginelli reaction by ferric and nickel chloride hexahydrates. One-pot synthesis of 3,4-dihydropyrimidin-2(1H)-ones. *Synthesis, 2002*, 466–470. https://doi.org/10.1055/s-2002-20956

Lu, J., Bai, Y., Wang, Z., Yang, B., & Ma, H. (2000). One-pot synthesis of 3,4-dihydropyrimidin-2(1H)-ones using lanthanum chloride as a catalyst. *Tetrahedron Letters, 41*, 9075–9078. https://doi.org/10.1016/S0040-4039(00)01645-2

Maiti, G., Kundu, P., & Guin, C. (2003). One-pot synthesis of dihydropyrimidinones catalysed by lithium bromide: An improved procedure for the Biginelli reaction. *Tetrahedron Letters, 44*, 2757–2758. https://doi.org/10.1016/S0040-4039(02)02859-9

Narsaiah, A. V., Basak, A. K., & Nagaiah, K. (2004). Cadmium chloride: An efficient catalyst for one-pot synthesis of 3, 4-dihydropyrimidin-2 (1H)-ones. *Synthesis, 2004*, 1253–1256.

Ramalinga, K., Vijayalakshmi, P., & Kaimal, T. N. B. (2001). Bismuth(III)-catalyzed synthesis of dihydropyrimidinones: Improved protocol conditions for the Biginelli reaction. *Synlett, 2001*, 863–865. https://doi.org/10.1055/s-2001-14587

Ranu, B. C., Hajra, A., & Jana, U. (2000). Indium(III) chloride-catalyzed one-pot synthesis of dihydropyrimidinones by a three-component coupling of 1,3-dicarbonyl compounds, aldehydes, and urea: An improved procedure for the Biginelli reaction. *The Journal of Organic Chemistry, 65*, 6270–6272. https://doi.org/10.1021/jo000711f

Saini, A., Kumar, S., & Sandhu, J. S. (2007). Aluminium(III) halides mediated synthesis of 5-unsustituted 3,4-dihydropyrimidin-2(1H)-ones via three component Biginelli-like reaction [J]. *Indian Journal of Chemistry B, 46*, 1690–1694.

Salehi, H., & Guo, Q. X. (2004). A facile and efficient one-pot synthesis of dihydropyrimidinones catalyzed by magnesium bromide under solvent-free conditions. *Synthetic Communications, 34*, 171–179. https://doi.org/10.1081/SCC-120027250

Slimi, H., Moussaoui, Y., & ben Salem, R. (2016). Synthesis of 3,4-dihydropyrimidin-2(1H)-ones/thiones via Biginelli reaction promoted by bismuth(III)nitrate or PPh3 without solvent. *Arabian Journal of Chemistry, 9*, S510–S514. https://doi.org/10.1016/j.arabjc.2011.06.010

15

Silver rubber-hydrogel nanocomposite as pH-sensitive prepared by gamma radiation

Mohamed Mohamady Ghobashy[1]*, A. Awad[1], Mohamed A. Elhady[1] and Ahmed M. Elbarbary[1]

*Corresponding author: Mohamed Mohamady Ghobashy, Radiation Research of Polymer Department, National Center for Radiation Research and Technology (NCRRT), Atomic Energy Authority, P.O. Box 29, Nasr City, Cairo, Egypt
E-mail: mohamed.ghobashy@eaea. org.eg

Reviewing editor: Seth C. Rasmussen, North Dakota State University, USA

Abstract: Silver rubber-hydrogel nanocomposite based on silver/styrene butadiene rubber/polyvinylpyrrolidone/methacrylic acid (SBR/PVP/MAA)/Ag was prepared by gamma radiation-induced crosslinking. During the radiation crosslinking of SBR/PVP/MAA solution containing silver nitrate $AgNO_3$ (0.01 mol), *in situ* reduction of Ag^+ ions was performed under the radiolysis of water. The properties of sliver rubber-hydrogel nanocomposite were investigated by FT-IR, XRD, TEM, SEM, DSC and TGA techniques. Transmission electron microscope (TEM) reveals that AgNPs have uniform distribution and spherical shape with mean diameter in the range of 8–10 nm. Differential scanning calorimetry (DSC) results of the nanocomposite showed one phase suggesting the miscibility between rubber and hydrogel phases. The swelling measurement of the synthesized silver rubber-hydrogel nanocomposite in different pHs at room temperature was performed. The results showed that it has pH-sensitivity.

Subjects: Chemistry; Material Science; Materials Science; Composites; Polymers & Plastics

Keywords: rubber; hydrogel; nanocomposite; AgNPs; gamma irradiation; pH-sensitive

1. Introduction

Recent advances in the design and synthesis of hydrogels for a wide range of applications have been made by introducing other compatible materials with them. Hydrogels have high degrees of swelling, environmental sensitivity and high permeability (Gupta & Shivakumar, 2012). Hydrogels can be defined as polymeric networks that can retain a significant amount of water within their structures, and swell without dissolving in it. Relatively high water content, hydrophilicity, expandability, selective permeability, soft rubbery consistency and low interfacial tension are among the advantages of hydrogels which enable them to resemble soft living tissues.

ABOUT THE AUTHORS

Specialist: 1-nanotechnology applications 2-specific interest, Ionic Polymer Metal Composite (IPMC), as electroactive polymer (EAP) and magnetic polymer. 3-application of radiation chemistry for synthesis of polymer and hydrogel with a wide range biomedical application. Another kind of polymeric matrials likes (Organogel, fabric polymer and grafted films for different applications like self cleaning and self healing hydrogel.

PUBLIC INTEREST STATEMENT

- Uses of radiation technology for prepare sliver rubber-hydrogel (styrene butadiene rubber/ poly vinyl pyrrolidone/poly methacrylic acid)/ silver nanocomposites.
- *In situ* reduction of Ag+ ions was performed under the radiolysis of water.
- The swelling behavior of (SBR/PVP/PMAA)/Ag nanocomposites as a function of pH at room temperature was performed.
- Ag nanoparticles were in the range of 8–10 nm.

A hydrogel is a crosslinked polymer that may be synthesized in a number of chemical ways. These include one-step procedures like polymerization and crosslinking of multifunctional monomers, as well as multiple step procedures involving synthesis of polymer molecules having reactive groups with suitable crosslinking agents. The polymer design and synthesize of polymer networks with molecular-scale control over structure such as crosslinking density and with tailored properties, such as biodegradation, mechanical strength, and chemical and biological response to stimuli (Burkert, Schmidt, Gohs, Dorschner, & Arndt, 2007).

Ionizing radiation has been used as an initiator to prepare the hydrogels. This includes gamma rays (Elbarbary & Ghobashy, 2017; Ghobashy & Khafaga, 2016; Karadað, Saraydin, & Guven, 2001) and electron beams (Ajji, Mirjalili, Alkhatab, & Dada, 2008). On the other meaning, Gamma irradiation is a promising technique to fabricate a wide scale of different materials especially polymeric materials (Ghobashy & Abdeen, 2016; Ghobashy & Ehab, 2016; Ghobashy & Elhady, 2017). The irradiation of an aqueous polymer in the presence of vinyl monomers solutions leads to the formation of radicals on the polymer chains and/or on the vinyl groups. The recombination of the radicals on different molecules results in the formation of covalent bonds or crosslinked structure.

Hybrid hydrogel with other materials exhibited high stiffness for different applications (Ahmad & Al-Malaika, 2014; Zheng, Sun, Yuan, Duan, & Tian, 2016) such as polypeptide–polyethylene glycol for extracellular matrix (ECM) mimics (Wang, Cai, Paul, Enejder, & Heilshorn, 2014), graphene oxide-hybrid hydrogels for bio-inspired smart actuator (Wang et al., 2015), chitin/hydroxylapatite hybrid hydrogels as scaffold nano-materials (Chang et al., 2013) and laponite/alginate hybrid hydrogels as pH-sensitive (Li et al., 2011) and amino clay hybrid hydrogel which act as novel supramolecular scaffolds for light-harvesting (Venkatarao, Datta, Eswaramoorthy, & George, 2011).

The response of hydrogels to environmental parameters is an active area for researchers such as pH (Ghobashy, 2017; Wang, Turhan, & Gunasekaran, 2004), temperature, electric field (Ali, El-Rehim, Hegazy, & Ghobashy, 2006), ionic strength (El-Hag Ali, El-Rehiem, Hegazy, & Ghobashy, 2007), etc. A pH-sensitive hydrogel is a gel structure that is sensitive to the changes in pH. Depending on the nature of functional groups, the hydrogel will swell or contract in response to a change in the local chemical environment. These pH-sensitive hydrogels have applications in controlling valves that are sensitive to a change in pH such as in systems that can release a compound (Peterson, 2014), controlled drug delivery systems (Kim & Peppas, 2003), and in membrane separations.

Metal nanoparticles such as gold, silver, iron, zinc and metal oxide nanoparticles have been of interest in bioapplications (Bhattacharya & Mukherjee, 2008). Silver nanoparticles (AgNPs) can be prepared by several methods such as chemical reduction (Lorestani, Shahnavaz, Mn, Alias, & Manan, 2015), electrochemical reduction (Hadipour-Goudarzi, Montazer, Latifi, & Aghaji, 2014), photochemical reduction (Kim & Lee, 2015), UV irradiation (Yang, Zhai, Wang, & Wei, 2014), ultrasonic method (He et al., 2014) and γ-irradiation (Akhavan, Sheikh, Khoylou, Naimian, & Ataeivarjovi, 2014; Malkar, Mukherjee, & Kapoor, 2014). Recently, radiolytic method was used in the preparation of metal nanoparticles in which, the hydrated electrons produced during γ-irradiation can reduce metal ions to metal particles of zero valences (Li, Park, & Choi, 2007). Radiolytic *in situ* formation of AgNPs is rarely described in previous literature with extended analysis (Jovanović et al., 2011), such as radiolytic synthesis of AgNPs within poly(N-vinyl-2-pyrrolidone) (PVP) hydrogel for biomedical application (Kacarevic-Popovic et al., 2010), rubber composite fibers containing AgNPs were prepared through combination of electrospinning and *in situ* chemical crosslinking (Hu, Wu, Zhang, Fong, & Tian, 2012).

Rubber is a soft material that has glass transition temperatures lower than room temperature, and due to the high flexibility of rubber. It could be used for enhancing the elasticity of much composite materials (Fong & Reneker, 1999; Sarkawi, Kaewsakul, Sahakaro, Dierkes, & Noordermeer, 2016; Yong, 2015; Yong & Mustafa, 2014). Styrene-butadiene rubber (SBR) describes families of synthetic rubbers derived from styrene and butadiene. These materials have good abrasion resistance and good aging stability when protected by additives (Ansarifar, Wang, Yoong, Osmani, & Pappu, 2010).

Poly(N-vinyl-2-pyrrolidone) (PVP) has important applications in various biotechnology areas such as tissue engineering, controlled drug release, separation of bio-macromolecules as well as biosensor (Ding, Kamulegeya, Chen, Chen, & Liu, 2007; Wan, Xu, Huang, Huang, & Yao, 2007). Poly(methacrylic acid) (PMAA) hydrogels can undergo a volume change in response to pH stimuli; due to this unique property, PMAA-related polymers have been developed to design pH-sensitive hydrogels that can modulate drug release by change in pH (Liu, Fan, Kang, & Sun, 2004; Liu, Liu, Chen, Sun, & Zhang, 2007; Mullarney, Seery, & Weiss, 2006).

The compatibility was expected between SBR and hydrogel matrices, whereas the latter would be reactive with the SBR matrix due to the polar groups. The pH sensitivity of rubber-hydrogel matrices was carried out to evaluate the compatibility between them. A high miscibility was observed between the MAA/PVP hydrogel and the matrix of SBR. This study is the first step in preparing a hydrogel-rubber with a pH-sensitive material.

In this study, silver rubber-hydrogel (SBR/PVP/MAA)/Ag nanocomposite was prepared by mixing SBR and copolymer hydrogel (polyvinylpyrrolidone -co-methacrylic acid) crosslinked by gamma irradiation in the presence of silver nitrate ($AgNO_3$) (0.01 mol). The changes in hydrogel properties due to SBR rubber matrices and/or AgNPs were investigated by FT-IR, XRD, TEM, SEM, DSC, and TGA. The swelling of rubber-hydrogel matrices embedded and non-embedded AgNPs in water or in different buffers at various pH's were investigated.

2. Experimental

2.1. Materials

Emulsion of styrene butadiene rubber (E-SBR) including styrene and butadiene monomers, water, emulsifier, initiator system, modifier, shortstop and a stabilizer system was used as it is. E-SBR is commercial grades (BUNATM SB 1502-Schkopau), the percentage of styrene is 23.5, the glass transition temperature (Tg) of SBR is –50°C and this material has been supplied from Trinseo, USA. Polyvinyl pyrrolidone (PVP) with molecular weight (100 kDa), methacrylic acid monomer (MAA) 99% and silver nitrate ($AgNO_3$) were provided by Sigma-Aldrich. Other chemicals and solvents were used without further purification.

2.2. Preparation of (SBR/PVP/MAA)/Ag nanocomposite

Firstly, SBR/PVP/MAA was prepared by dissolving 0.5 g of PVP in 10 ml solution of (30/70) (methanol/distilled water) followed by adding 0.5 ml of MAA. Secondly, 10 ml of commercial solution of SBR was mixed well with the above mixture solution and stirred for 30 min. The reactant mixtures were placed into test tubes (Scheme 1), deaerated by bubbling with nitrogen gas for 3–5 min, sealed, and then subjected to γ-radiation at dose of 20 kGy. The preparation of (SBR/PVP/MAA)/Ag nanocomposite was carried out by repeating the above steps by adding 1.7 mg $AgNO_3$ (0.01 mol) after the polymerization process has been carried out. The resulting rubber-hydrogel samples were cut into disks (10 mm in diameter). Then, the disk of samples were soaked in distilled water for 24 h to remove the unreacted molecules, the water has been renewed three times every 24 h and then the samples were dried in oven at 50°C for 24 h.

2.3. Gamma irradiation cell

The polymerization process of rubber-hydrogel was carried out in the cobalt-60 gamma cell at a dose rate of 2.05 kGy/h. The irradiation facility is located at the National Center for Radiation Research and Technology (NCRRT), Atomic Energy Authority of Egypt.

2.4. Determination of swelling ratio

A dried sample of a known weight was immersed in distilled water or in buffers of different pHs for 24 h, the equilibrium swelling ratio was determined by the amount of water absorbed by samples after 24 h ($W_s - W_d$) divided by the weight of dried samples W_d, the swelling ratio was calculated as follows:

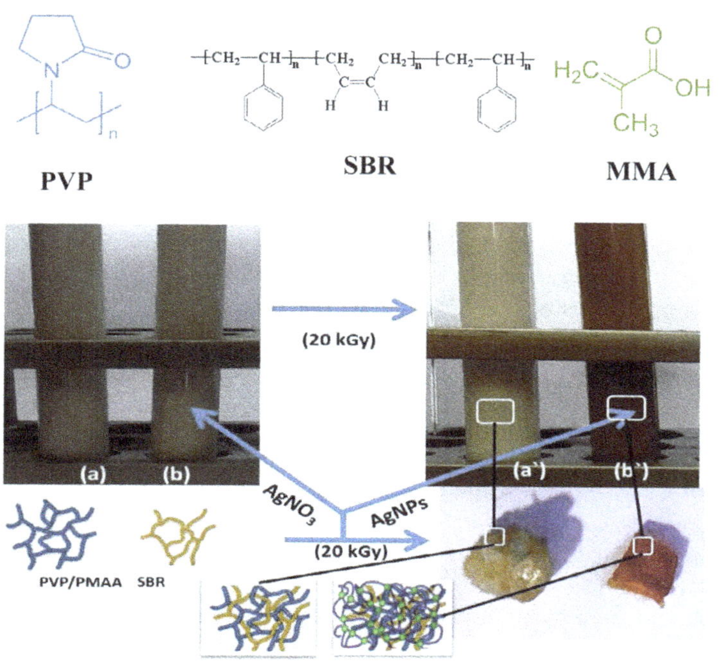

Scheme 1. The procedure of polymerization and *in situ* nanoparticle formation induced by gamma irradiation (a) and (b) are the solution of (SBR/PVP/MAA) and (SBR/PVP/MAA/AgNO$_3$), respectively, before exposure to γ-rays and (a`) and (b`) the hybrid rubber-hydrogel and rubber-hydrogel silver nanocomposite of (SBR/PVP/PMAA) and (SBR/PVP/PMMA)/Ag, respectively, after exposed to 20 kGy γ-rays.

$$\text{Swelling ratio (g/g)} = \frac{(W_s - W_d)}{W_d}$$

where, W_s and W_d are the weights of wet and dry rubber-hydrogel, respectively. Note: the measurement of swelling was repeated three times with error ±5%.

2.5. Preparation of buffer

To investigate the effect of pH on the equilibrium swelling ratio, the samples were immersed in buffer solution at pH ranged from 1 to 11 at room temperature. Swelling at various pHs individual solutions with acidic and basic pHs were prepared as follows: pH 1 and 2 prepared by 0.1 M of HCl (dilute 4.23 ml of 36.5% HCl to a final volume of 500 ml with distilled water), and 0.1 M KCl (7.46 g + H$_2$O to 1 liter). pH 1 was obtained by 97 ml of HCl mixed with 3 ml of KCl, and pH 2 was obtained by mixing 10.6 ml of HCl with 89.4 ml of KCl . pH 3, 4, 5 and 6 have been prepared by 0.1 M Citric acid: 19.21 g/l (M.W.: 192.1) and 0.1 M Sodium citrate dihydrate: 29.4 g/l (M.W.: 294.0). Citric acid and sodium citrate solutions were mixed in the proportions (46.5:3.5) ml for pH3,(31:19) ml for pH 4 (20.5:29.5) ml for pH 5 and for pH 6 (7:43) ml. Mixing and adjusting the final volume to 100 ml with deionized water have been carried out to get the required pH 7, 8, 9, 10 and 11 made up the following solutions; 0.1 M disodium hydrogen phosphate (14.2 g/l), 0.1 M HCl and 0.1 M NaOH (4.0 g/l), mixed in the following proportions for pH 7 (75.6:24.4:0) ml, for pH 8 (95.51: 4.49:0) ml, for pH 9 (95.50:4.50:0) ml for pH 10 (96.64:0:3.36) ml and for pH 11 (96.53:0:34.7) ml, respectively.

2.6. Fourier-transform infrared spectroscopy (ATR-FTIR)

The chemical structure of samples was identified using attenuated total reflectance Fourier-transform infrared ATR-FTIR spectroscopy Vertex 70 FTIR spectrometer equipped with HYPERION™ series microscope, Bruker Optik GmbH, Ettlingen, Germany, over the 4000–400 cm^{-1} range, at a resolution of 4 cm^{-1} has been utilized. A software OPUS 6.0 (BRUKER) was used for data processing, which was baseline corrected by rubber band method with CO$_2$ bands excluded.

2.7. X-ray diffraction (XRD)

X-ray diffraction patterns were obtained with a XRD-6000 series, Shimadzu apparatus using Ni-filter and Cu-Kα target.

2.8. Transmission electron microscopy

The particle size distribution of AgNPs was investigated using a transmission electron microscopy (TEM), JEOL JSM-100 CX, Japan, with an acceleration voltage of 80 kV. For TEM observations, the samples were prepared by making a suspension from the Ag nanocomposite in distilled water using ultrasonic water bath. The silver rubber-hydrogel sample was frozen on liquid nitrogen and quickly ground to a fine powder. This powder was suspended in acetone solution and centrifuged to separate the aggregated large size particle. Then a drop of the suspension was put into the carbon grid and left to dry at room temperature.

2.9. Scanning electron microscopy

The surface morphology of the swelling sample was investigated using Scanning Electron Microscopy (SEM) which is a powerful technique to provide information about miscibility of both the rubber and hydrogel phases. The surface topology was measured by SEM JEOL JSM-5400, Japan, with accelerating voltage of 30 kV. The surfaces of dried samples were sputter-coated with gold for 3 min for forming ~3–5 nm of thick gold on the sample surface. Sputtering time 5 min., current of 10–40 mA (discharge power 3–15 W), total pressure about 5 Pa, and the electrode distance of 15 mm. The power density was 0.13 W cm^{-2}.

2.10. Thermal analysis

Thermal analysis was performed using differential scanning calorimetry (DSC) and Thermo gravimetric analysis- Derivative thermo-gravimetric (TGA-DTG) from TA Instruments Waters—LLC, 159 Lukens Drive, New Castle, DE 19720. The melting temperature (T_m) measured by heating the samples from 25°C up to 300°C with heating rate 10°C/min. All measurements were conducted under a nitrogen atmosphere (20 ml/min). The cell was calibrated using an indium standard. The weight of the sample was 5–10 mg and heated up to 600°C, at a heating rate of 10°C/min under N_2 atmosphere.

3. Results and discussion

Scheme 1 demonstrates the procedure of polymerization and *in situ* nanoparticle formation induced by gamma irradiation. From Scheme 1, it is clear that the color after polymerization has been changed. The color of silver salt (AgNO$_3$) changes from its original color to a golden yellow color due to the reduction process. Before and after polymerization process, it was observed that (1) the miscibility of both rubber and hydrogel phases took place. (2) Good dispersion of AgNPs colloid thereby indicated that the AgNPs have no tendency to agglomerate because of effective capping by rubber hydrogel matrices.

3.1. Characterization of sliver rubber-hydrogel nanocomposites

3.1.1. Chemical structure by FT-IR studies

The ATR-FT-IR spectrum (Figure 1(b)) for (SBR/PVP/PMAA)/Ag showed that there is a chemical bonding between rubber hydrogel matrices and silver nanoparticles (Sobczak-Kupiec, Malina, Wzorek, & Zimowska, 2011). This causes shifts of the main characteristic peak as shown in Figure 1(a) and (b). The stretching peak of (C=O) groups at 1,655 cm^{-1} shifted to 1,675 cm^{-1} with a slight broadening. C-N has a peak at 1,030 cm^{-1} which shifted to 1,026 cm^{-1}, while no shift was observed in the peak attributed to C-H at 2,922 cm^{-1}. From the above-mentioned, N and O atoms in the molecule of function groups have affinity for Ag ions and metallic Ag coordination (Khanna, Gokhale, & Subbarao, 2004; Silvert, Herrera-Urbina, Duvauchelle, Vijayakrishnan, & Tekaia-Elhsissen, 1996). This results give spot about *in situ* chemical reduction of Ag$^+$ ions to AgNPs were performed by rubber hydrogel compositions like PVP. The PVP reduction mechanism of AgNPs formation has been described by two steps. The first step involves the formation of coordinative bonds between Ag$^+$ ions and the N and O atoms in polymer function groups, as a result of donation of lone pair electrons from O and N to empty

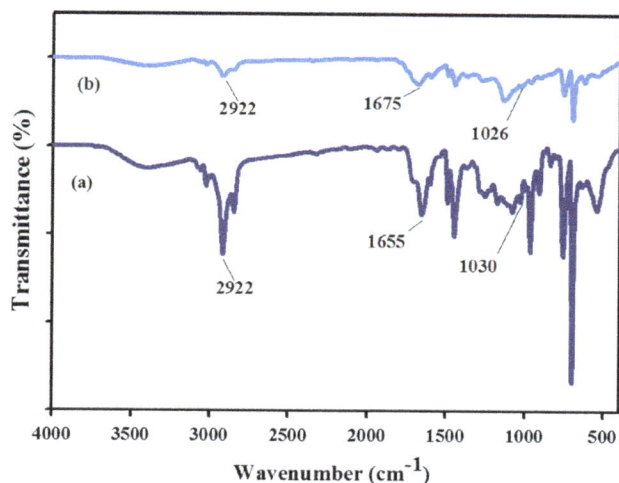

Figure 1. ATR-FTIR of (a) (SBR/PVP/PMAA) and (b) (SBR/PVP/PMAA)/Ag nanocomposite.

orbital of Ag^+. The second step, the complex is then thought to promote nucleation of Ag^+ atoms that are exposed to γ-irradiation are reduced to AgNPs (Chahal, Mahendia, Tomar, & Kumar, 2010; Malina, Sobczak-Kupiec, Wzorek, & Kowalski, 2012; Sadeghi, Sadjadi, & Pourahmad, 2008; Shin, Yang, Kim, & Lee, 2004; Zhang, Zhao, & Hu, 1996). The role of PVP in the irradiated solution considerably remains as one of the holding matrix beside COOH functional group of PMAA that may temporarily bind with the Ag^+ ions this made stabilization of AgNPs.

3.2. XRD studies

Figure 2 shows the XRD pattern of SBR/PVP/PMAA and (SBR/PVP/PMAA)/Ag nanocomposite. The XRD of SBR/PVP/PMAA shows a strong and broad diffraction peak located at $2\theta = 20°$. The XRD of irradiated Ag (SBR/PVP/PMAA) nanocomposite shows the diffraction peaks at $2\theta = 38.11°$, $44.28°$, $64.61°$ and $77.9°$ corresponding to (111), (200), (220), and (311), lattice planes of Ag is observed and compared with the standard powder diffraction card (silver file No. 04–0783) of Joint Committee on Powder Diffraction Standards (JCPDS). Silver file No. 04–0783.The XRD study confirms that the result-ant particles are (FCC) AgNPs corresponding to Face Center Cubic (FCC) (Sulaiman, Mohammad, Abdul-Wahed, & Ismail, 2013).

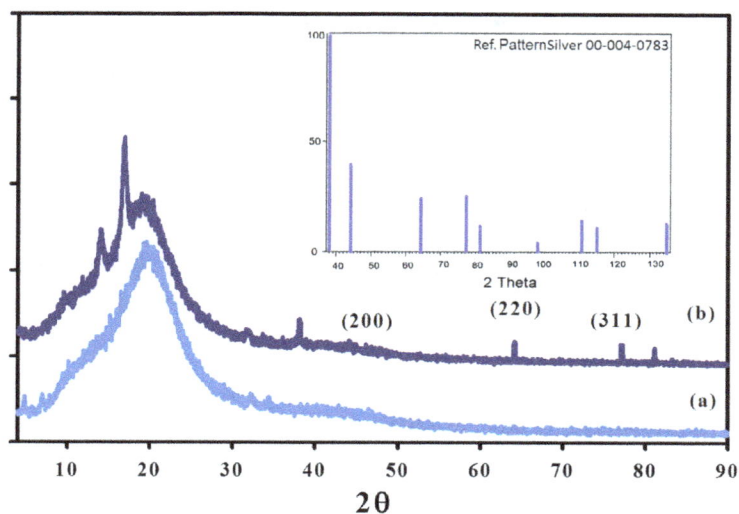

Figure 2. XRD pattern of (a) SBR/PVP/PMAA and (b) (SBR/PVP/PMAA)/Ag nanocomposite.

Note: The upper side curve of standard powder diffraction pattern of Ag.

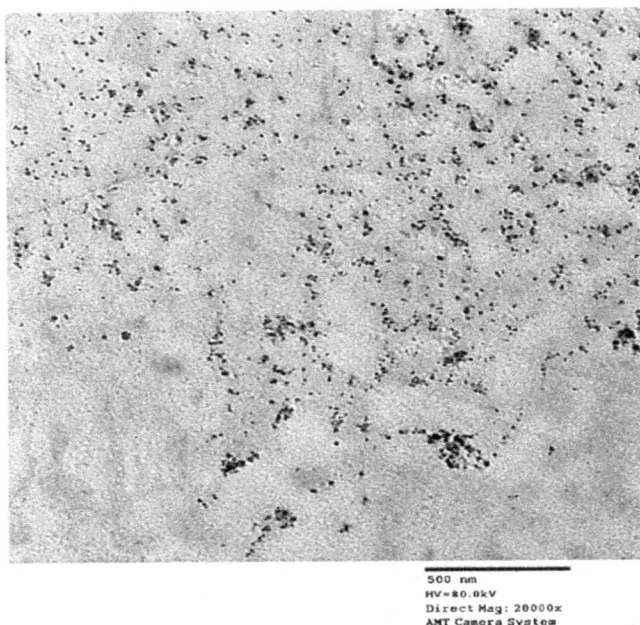

```
500 nm
HV=80.0kV
Direct Mag: 20000x
AMT Camera System
```

Figure 3. TEM images for (SBR/PVP/PMAA)/Ag nanocomposite.

3.3. Particle size measurements using TEM

A typical TEM image of AgNPs formed is displayed in Figure 3. The Ag nanoparticles were readily generated during gamma irradiation. The silver nanoparticles are well distributed in the rubber-hydrogel nanocomposite, where TEM image reveals that the uniform sphere structure shape of AgNPs was in the range of 8–10 nm.

3.4. Swelling characterization

Swelling is one of the most important parameters affecting the characteristic properties of the prepared rubber-hydrogel composite and, therefore, determines their applicability for practical use. The influence of the swelling time on the water uptake was investigated for SBR/PVP/PMAA without silver and silver rubber-hydrogel (SBR/PVP/PMAA)/Ag nanocomposite prepared as shown in Figure 4. The water uptake rapidly increases as the swelling time increases, reaching a certain limiting value after almost 4 h (25 and 24 (g/g), respectively). As a result, the diffusion and swelling properties are shown to be new features of rubber when hydrogel is introduced. However, in case of embedding AgNps, this leads to a decrease in the swelling efficiency due to silver particles filling the network gaps of rubber hydrogel matrices.

3.5. Surface morphology (SEM)

Figure 5(a) and (b) shows SEM analysis of freeze-swelled rubber-hydrogel and Ag nanocomposite. SEM for rubber-hydrogel (Figure 5(a)) verified a uniform nature of surface morphology indicating good compatibility between the rubber and hydrogel materials. The porous structure affected by elasticity of SBR chains. SEM for silver rubber-hydrogel nanocomposite (Figure 5(b)) revealed a uniform dispersion of AgNPs in rubber-hydrogel matrix and no agglomerated regions were noticed.

3.6. Thermal analysis DSC and TGA-DTG

The TGA-DTG thermogram of dried samples of rubber hydrogel with and without AgNPs exhibits major one degradation steps from room temperature to 420°C which may be attributed to the higher thermal stability of SBR (Varkey, Augustine, & Thomas, 2000; Wang, Wang, & Cheng, 2009). Figure 6(a) and (b) shows that the decomposition occurred gradually for all polymer compositions (SBR, PVP and PMAA) at the same time. The same behavior observed in case embedded AgNPs. The results accounts for a single degradation step due to the higher miscibility and crosslink density of rubber-hydrogel matrices.

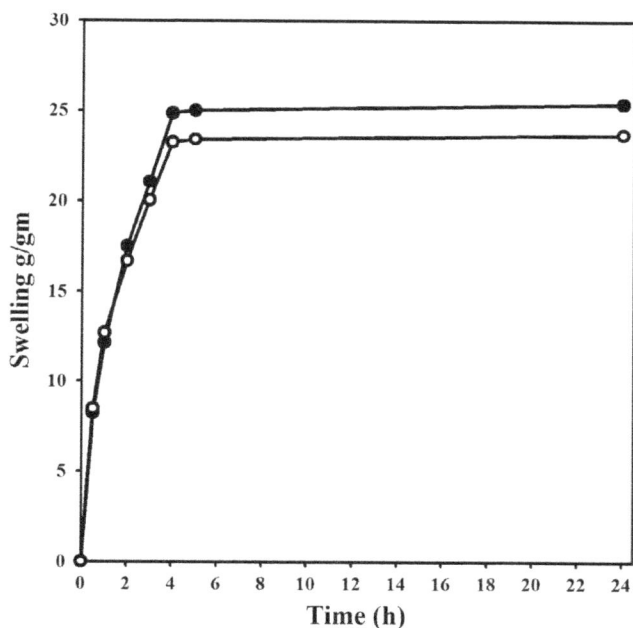

Figure 4. Swelling behavior of SBR/PVP/PMAA (●) and (SBR/PVP/PMAA)/Ag nanocomposite (○).

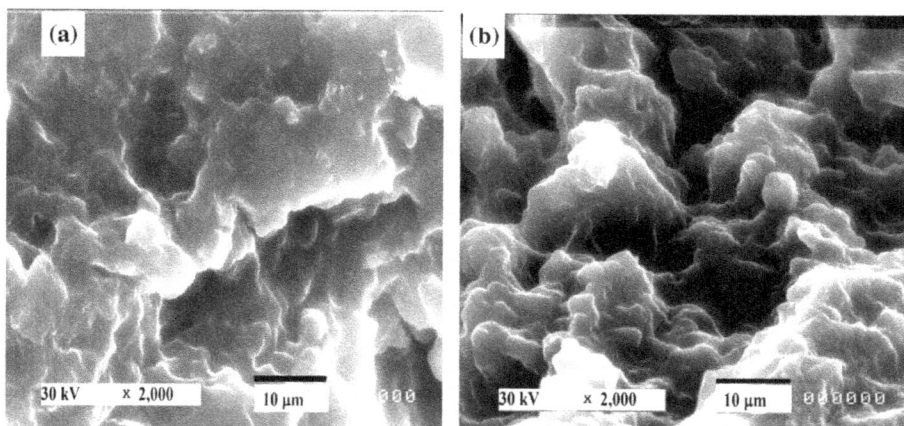

Figure 5. SEM micrographs of (a) SBR/PVP/PMAA and (b) (SBR/PVP/PMAA)/Ag nanocomposite.

Note: Magnitude 2000× and bar scale 10 μm.

The DSC of the prepared rubber hydrogel without AgNPs (Figure 6(c)) shows a pure phase suggesting that these are fully miscible between rubber-hydrogel phases with a homogeneous amorphous phase. Glassy temperature (T_g) for SBR at −56°C the glass transition temperature is shifted towards the higher temperature region upon −51°C due to introduction of crosslinked (Radhakrishnan, Sujith, & Unnikrishnan, 2007). The major endothermic peak centered at about 252°C is due to the complex formation of a link between PVP and PMAA with weak H-bond interaction (Polacco, Cascone, Petarca, & Peretti, 2000). DSC curve of (SBR/PVP/PMAA)/Ag nanocomposite shows an endothermic peak at 226°C which is attributed to Ag particles causing removal of H-bond interaction between chains of polymer.

3.7. Swelling in different pH

The swelling behavior of any hydrogel network depends upon the nature of the functional groups of the hydrogel. Figure 7 shows the equilibrium swelling behavior as a function of pH at room temperature. At pH 1 the swelling degree was 40 (g/g) which decreased to 13 (g/g) when AgNPs are embedded, this is due to polyvinylpyrrolidone coated AgNPs that are negative in nature (El Badawy et al.,

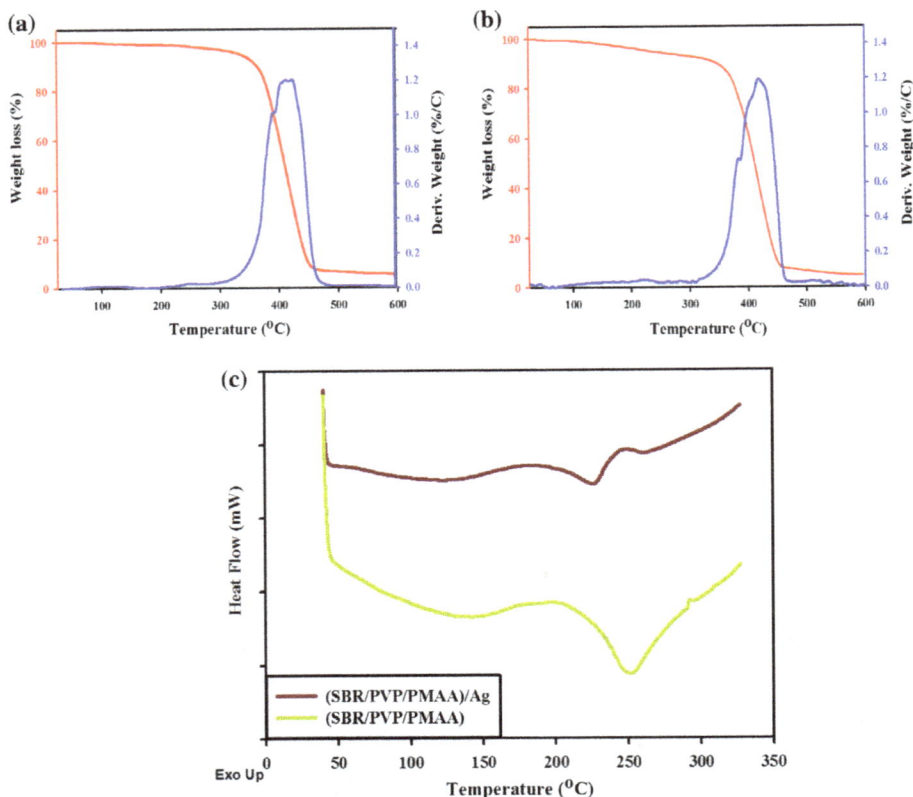

Figure 6. (a) TGA-DTG of SBR/PVP/PMAA, (b) TGA-DTG of (SBR/PVP/PMAA)/Ag nanocomposite and (c) DSC thermogram.

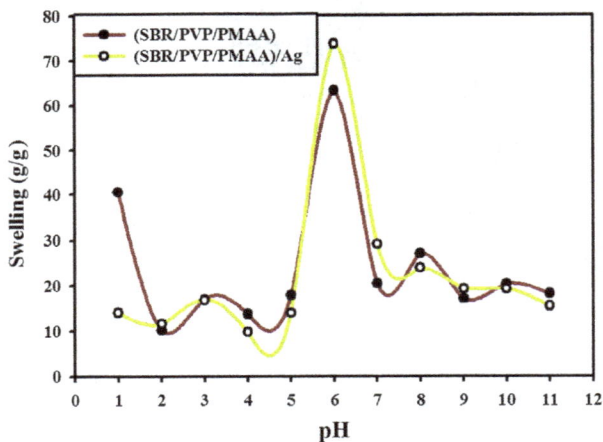

Figure 7. The swelling ratio as a function of the pH (●) (SBR/PVP/PMAA) and (○) (SBR/PVP/PMAA)/Ag nanocomposite.

2011). Hence, protonation on the surface causes neutralization of AgNPs which leads to a decrease in the repulsion force between particles and so the water uptake will be decreased. While in case (SBR/PVP/PMAA) at pH1 the COOH groups are neutralized which leads to absence of H-bond interaction with PVP. This causes an increase in swelling degree at 40 (g/g). At pH 2, 3, 4 and 5 the swelling degree was 10, 16, 13 and 17 (g/g) for (SBR/PVP/PMAA), respectively, and was 11, 13, 9 and 16 (g/g), respectively, in case (SBR/PVP/PMAA)/Ag. The results revealed that Ag caused a decrease in the swelling efficiency. The pKa value of PMMA was 5.7 (Dai, Ravi, Tam, Mao, & Gan, 2003) exceeding this value causes a dramatic increase in domains of COO– groups and swelling degree to 63 (g/g) for (SBR/PVP/PMAA) and in case (SBR/PVP/PMAA)/Ag, the swelling degree increases to 73 (g/g). This is attributed to the negative charge of AgNPs which repels with COO^- groups. While, the swelling of rubber-hydrogel samples decreases from pH 7 up to pH 11 as expected.

4. Conclusion

Preparation of pH sensitive rubber-hydrogel matrices using gamma irradiation gives a new material for appreciated applications related to pH effects. AgNPs (8–10 nm) are embedded into the prepared rubber-hydrogel to enhance its ability to be used as protecting agents to AgNPs with an average diameter of 8–10 nm. The DSC, TGA and SEM confirmed the one-phase formation of new matrices and ATR-FTIR confirmed the formation of complexes between PVP and PMAA. The new feature of SBR swelling improved in water and in different buffers solutions, this is due to incorporation with PVP/PMAA hydrogel.

Funding

The authors received no direct funding for this research.

Author details

Mohamed Mohamady Ghobashy[1]
E-mail: mohamed.ghobashy@eaea.org.eg
A. Awad[1]
E-mail: awadalaha@yahoo.com
Mohamed A. Elhady[1]
E-mail: mohamedelhady2000@yahoo.com
Ahmed M. Elbarbary[1]
E-mail: amelbarbary@yahoo.com

[1] Radiation Research of Polymer Department, National Center for Radiation Research and Technology (NCRRT), Atomic Energy Authority, P.O. Box 29, Nasr City, Cairo, Egypt.

References

Ahmad, A., & Al-Malaika, S. (2014). PET/EPR blends properties in the presence of compatibilisers containing glycidyl methacrylite. *Journal of Rubber Research, 17,* 219–232.

Ajji, Z., Mirjalili, G., Alkhatab, A., & Dada, H. (2008). Use of electron beam for the production of hydrogel dressings. *Radiation Physics and Chemistry, 77,* 200–202. https://doi.org/10.1016/j.radphyschem.2007.05.016

Akhavan, A., Sheikh, N., Khoylou, F., Naimian, F., & Ataeivarjovi, E. (2014). Synthesis of antimicrobial silver/hydroxyapatite nanocomposite by gamma irradiation. *Radiation Physics and Chemistry, 98,* 46–50. https://doi.org/10.1016/j.radphyschem.2014.01.004

Ali, A. E. H., El-Rehim, H. A., Hegazy, E.-S. A., & Ghobashy, M. M. (2006). Synthesis and electrical response of acrylic acid/vinyl sulfonic acid hydrogels prepared by γ-irradiation. *Radiation Physics and Chemistry, 75,* 1041-1046.

Ansarifar, A., Wang, L., Yoong, K. K., Osmani, M., & Pappu, A. (2010). Characterisation and use of glass fibre reinforced plastic waste Powder as filler in styrene-butadiene rubber. *Journal of Rubber Research, 13,* 162–174.

Bhattacharya, R., & Mukherjee, P. (2008). Biological properties of "naked" metal nanoparticles. *Advanced Drug Delivery Reviews, 60,* 1289–1306. https://doi.org/10.1016/j.addr.2008.03.013

Burkert, S., Schmidt, T., Gohs, U., Dorschner, H., & Arndt, K. F. (2007). Cross-linking of poly(N-vinyl pyrrolidone) films by electron beam irradiation. *Radiation Physics and Chemistry, 76,* 1324–1328. https://doi.org/10.1016/j.radphyschem.2007.02.024

Chahal, R. P., Mahendia, S., Tomar, A. K., & Kumar, S. (2010). Effect of ultraviolet irradiation on the optical and structural characteristics of *in situ* prepared PVP-Ag nanocomposites. *Digest Journal of Nanomaterials and Biostructures, 5,* 569–575.

Chang, C., Peng, N., He, M., Teramoto, Y., Nishio, Y., & Zhang, L. (2013). Fabrication and properties of chitin/hydroxyapatite hybrid hydrogels as scaffold nano-materials. *Carbohydrate Polymers, 91,* 7–13. https://doi.org/10.1016/j.carbpol.2012.07.070

Dai, S., Ravi, P., Tam, K. C., Mao, B. W., & Gan, L. H. (2003). Novel pH-responsive amphiphilic diblock copolymers with reversible micellization properties. *Langmuir, 19,* 5175–5177. https://doi.org/10.1021/la0340652

Ding, G., Kamulegeya, A., Chen, X., Chen, J., & Liu, Y. (2007). PVP magnetic nanospheres: Biocompatibility, *in vitro* and *in vivo* bleomycin release. *International Journal of Pharmaceutics, 328,* 78–85.

El Badawy, A. M., Silva, R. G., Morris, B., Scheckel, K. G., Suidan, M. T., & Tolaymat, T. M. (2011). Surface charge-dependent toxicity of silver nanoparticles. *Environmental Science & Technology, 45,* 283–287. https://doi.org/10.1021/es1034188

Elbarbary, A. M., & Ghobashy, M. M. (2017). Phosphorylation of chitosan/HEMA interpenetrating polymer network prepared by γ-radiation for metal ions removal from aqueous solutions. *Carbohydrate Polymers, 162,* 16–27. https://doi.org/10.1016/j.carbpol.2017.01.013

El-Hag Ali, A., El-Rehiem, H. A. A., Hegazy, E. S. A., & Ghobashy, M. M. (2007). Characterization and potential application of electro-active acrylamido-2-methyl propane sulfonic acid/acrylic acid copolymer prepared by ionizing radiation. *Journal of Macromolecular Science Part A: Pure and Applied Chemistry, 44,* 91–98. https://doi.org/10.1080/10601320601044559

Fong, H., & Reneker, D. H. (1999). Elastomeric nanofibers of styrene–butadiene–styrene triblock copolymer. *Journal of Polymer Science Part B: Polymer Physics, 37,* 3488–3493. https://doi.org/10.1002/(ISSN)1099-0488

Ghobashy, M. M. (2017). pH-sensitive wax emulsion copolymerization with acrylamide hydrogel using gamma irradiation for dye removal. *Radiation Physics and Chemistry, 134,* 47–55. https://doi.org/10.1016/j.radphyschem.2017.01.021

Ghobashy, M. M., & Abdeen, Z. I. (2016). Radiation crosslinking of polyurethanes: Characterization by FTIR, TGA, SEM, XRD, and Raman spectroscopy. *Journal of Polymers.* https://doi.org/10.1155/2016/9802514

Ghobashy, M. M., & Ehab, K. (2016) Sulfonated gamma-irradiated blend poly (styrene/ethylene-vinyl acetate) membrane and their electrical properties. *Advances in Polymer Technology.* doi:10.1002/adv.21781

Ghobashy, M. M., & Elhady, M. A. (2017). Radiation crosslinked magnetized wax (PE/Fe$_3$O$_4$) nano composite for selective oil adsorption. *Composites Communications, 3,* 18–22. https://doi.org/10.1016/j.coco.2016.12.001

Ghobashy, M. M., & Khafaga, M. R. (2016). Chemical modification of nano polyacrylonitrile prepared by emulsion polymerization induced by gamma radiation and their use for removal of some metal ions. *Journal of Polymers and the Environment, 25,* 343–348. doi:10.1007/s10924-016-0805-4

Gupta, N. V., & Shivakumar, H. G. (2012). Investigation of swelling behavior and mechanical properties of a pH-sensitive superporous hydrogel composite. *Iranian Journal of Pharmaceutical Research, 11,* 481–493.

Hadipour-Goudarzi, E., Montazer, M., Latifi, M., & Aghaji, A. A. G. (2014). Electrospinning of chitosan/sericin/PVA nanofibers incorporated with *in situ* synthesis of nano silver. *Carbohydrate Polymers, 113,* 231–239. https://doi.org/10.1016/j.carbpol.2014.06.082

He, C., Liu, L., Fang, Z., Li, J., Guo, J., & Wei, J. (2014). Formation and characterization of silver nanoparticles in aqueous solution via ultrasonic irradiation. *Ultrasonics Sonochemistry, 21,* 542–548. https://doi.org/10.1016/j.ultsonch.2013.09.003

Hu, Q., Wu, H., Zhang, L., Fong, H., & Tian, M. (2012). Rubber composite fibers containing silver nanoparticles prepared by electrospinning and *in situ* chemical crosslinking. *Express Polymer Letters, 6,* 258–265 https://doi.org/10.3144/expresspolymlett.2012.29

Jovanović, Ž., Krklješ, A., Stojkovska, J., Tomić, S., Obradović, B., Mišković-Stanković, V., & Kačarević-Popović, Z. (2011). Synthesis and characterization of silver/poly(N-vinyl-2-pyrrolidone) hydrogel nanocomposite obtained by *in situ* radiolytic method. *Radiation Physics and Chemistry, 80,* 1208–1215. https://doi.org/10.1016/j.radphyschem.2011.06.005

Kacarevic-Popovic, Z., Dragasevic, M., Krkljes, A., Popovic, S., Jovanovic, Z., Tomic, S., & Mišković-Stanković, V. (2010). On the use of radiation technology for nanoscale engineering of silver/hydrogel based nanocomposites for potential biomedical application. *The Open Conference Proceedings Journal, 1,* 200–206. https://doi.org/10.2174/22102892010010100200

Karadağ, E., Saraydin, D., & Guven, O. (2001). Radiation induced superabsorbent hydrogels. acrylamide/itaconic acid copolymers. *Macromolecular Materials and Engineering, 286,* 34–42. https://doi.org/10.1002/(ISSN)1439-2054

Khanna, P. K., Gokhale, R., Subbarao, V. V. V. S. (2004). Poly(vinyl pyrolidone) coated silver nano powder via displacement reaction. *Journal of Materials Science, 39,* 3773–3776. https://doi.org/10.1023/B:JMSC.0000030735.08903.a9

Kim, B. H., & Lee, J. S. (2015). One-pot photochemical synthesis of silver nanodisks using a conventional metal-halide lamp. *Materials Chemistry and Physics, 149,* 678–685. https://doi.org/10.1016/j.matchemphys.2014.11.026

Kim, B. S., & Peppas, N. A. (2003). In vitro release behavior and stability of insulin in complexation hydrogels as oral drug delivery carriers. *International Journal of Pharmaceutics, 266,* 29–37. https://doi.org/10.1016/S0378-5173(03)00378-8

Li, T., Park, H. G., & Choi, S. H. (2007). γ-Irradiation-induced preparation of Ag and Au nanoparticles and their characterizations. *Materials Chemistry and Physics, 105,* 325–330. https://doi.org/10.1016/j.matchemphys.2007.04.069

Li, Y., Maciel, D., Tomás, H., Rodrigues, J., Ma, H., & Shi, X. (2011). pH sensitive laponite/alginate hybrid hydrogels: Swelling behaviour and release mechanism. *Soft Matter, 7,* 6231–6238. https://doi.org/10.1039/c1sm05345k

Liu, Y. Y., Fan, X. D., Kang, T., & Sun L. (2004). A cyclodextrin microgel for controlled release driven by inclusion effects. *Macromolecular Rapid Communications, 25,* 1912–1916. https://doi.org/10.1002/(ISSN)1521-3927

Liu, Y. Y., Liu, W. Q., Chen, W. X., Sun, L., & Zhang, G. B. (2007). Investigation of swelling and controlled-release behaviors of hydrophobically modified poly(methacrylic acid) hydrogels. *Polymer, 48,* 2665–2671. https://doi.org/10.1016/j.polymer.2007.03.010

Lorestani, F., Shahnavaz, Z., Mn, P., Alias, Y., & Manan, N. S. A. (2015). One-step hydrothermal green synthesis of silver nanoparticle-carbon nanotube reduced-graphene oxide composite and its application as hydrogen peroxide sensor. *Sensors and Actuators B: Chemical, 208,* 389–398. https://doi.org/10.1016/j.snb.2014.11.074

Malina, D., Sobczak-Kupiec, A., Wzorek, Z., & Kowalski, Z. (2012). Silver nanoparticles synthesis with different concentrations of polyvinylpyrrolidone. *Digest Journal of Nanomaterials and Biostructures, 7,* 1527–1534.

Malkar, V. V., Mukherjee, T., & Kapoor, S. (2014). Synthesis of silver nanoparticles in aqueous aminopolycarboxylic acid solutions via γ-irradiation and hydrogen reduction. *Materials Science and Engineering: C, 44,* 87–91. https://doi.org/10.1016/j.msec.2014.08.002

Mullarney, M. P., Seery, T. A. P., & Weiss, R. A. (2006). Drug diffusion in hydrophobically modified N, N-dimethylacrylamide hydrogels. *Polymer, 47,* 3845–3855. https://doi.org/10.1016/j.polymer.2006.03.096

Peterson, D. S. (2014). pH Sensitive Hydrogel. In D. Li (Ed.), *Encyclopedia of Nanofluicis and Microfluidics* (2nd ed., 1–5). New York, NY: Springer.

Polacco, G., Cascone, M. G., Petarca, L., & Peretti, A. (2000). Thermal behaviour of poly(methacrylic acid)/poly(N-vinyl-2- pyrrolidone) complexes. *European Polymer Journal, 36,* 2541–2544 https://doi.org/10.1016/S0014-3057(00)00064-1

Radhakrishnan, C. K., Sujith, A., & Unnikrishnan, G. (2007). Thermal behaviour of styrene butadiene rubber/poly(ethylene-co-vinyl acetate) blends TG and DSC analysis. *Journal of Thermal Analysis and Calorimetry, 90,* 191–199. https://doi.org/10.1007/s10973-006-7559-5

Sadeghi, B., Sadjadi, M. A. S., & Pourahmad, A. (2008). Effects of protective agents (PVA & PVP) on the formation of silver nanoparticles. *International Journal of Nanoscience and Nanotechnology, 41,* 3–12.

Sarkawi, S. S., Kaewsakul, W., Sahakaro, K., Dierkes, W. K., & Noordermeer, J. W. M. (2016). A review on reinforcement of natural rubber by silica fillers for use in low-rolling resistance tires. *Journal of Rubber Research, 18,* 203–233.

Shin, H. S., Yang, H. J., Kim, S. B., & Lee, M. S. (2004). Mechanism of growth of colloidal silver nanoparticles stabilized by polyvinyl pyrrolidone in γ-irradiated silver nitrate solution. *Journal of Colloid and Interface Science, 274,* 89–94. https://doi.org/10.1016/j.jcis.2004.02.084

Silvert, P. Y., Herrera-Urbina, R., Duvauchelle, N., Vijayakrishnan, V., & Tekaia-Elhsissen, K. (1996). Preparation of colloidal silver dispersions by the polyol process. *Journal of Materials Chemistry, 6,* 573–577. https://doi.org/10.1039/JM9960600573

Sobczak-Kupiec, A., Malina, D., Wzorek, Z., & Zimowska, M. (2011). Influence of silver nitrate concentration on the properties of silver nanoparticles. *Micro & Nano Letters, 6,* 656–660. https://doi.org/10.1049/mnl.2011.0152

Sulaiman, G. M., Mohammad, A. A., Abdul-Wahed, H. E., & Ismail, M. M. (2013). January–March). Biosynthesis, antimicrobial and cytotoxic effects of silver nanoparticles using rosmarinus officinalis extract. *Digest Journal of Nanomaterials and Biostructures, 8,* 273–280.

Varkey, J. T., Augustine, S., & Thomas, S. (2000). Thermal degradation of natural rubber/styrene butadiene rubber latex blends by thermogravimetric method. *Polymer-Plastics Technology and Engineering, 39,* 415–435. https://doi.org/10.1081/PPT-100100038

Venkatarao, K., Datta, K. K. R., Eswaramoorthy, M., & George, S. J. (2011, February 1). Light-harvesting hybrid hydrogels: Energy-transfer-induced amplified fluorescence in noncovalently assembled chromophore-organoclay composites. *AngewandteChemie, 123*, 1211–1216.

Wan, L. S., Xu, Z. K., Huang, X. J., Huang, X. D., & Yao, K. (2007). Cytocompatibility of poly (acrylonitrile-co-N-vinyl-2-pyrrolidone) membranes with human endothelial cells and macrophages. *Acta Biomaterialia, 3*, 183–190. https://doi.org/10.1016/j.actbio.2006.09.007

Wang, H., Cai, L., Paul, A., Enejder, A., & Heilshorn, S. C. (2014). Hybrid elastin-like polypeptide–polyethylene glycol (ELP-PEG) hydrogels with improved transparency and independent control of matrix mechanics and cell ligand density. *Biomacromolecules, 15*, 3421–3428. https://doi.org/10.1021/bm500969d

Wang, Q., Wang, F., & Cheng, K. (2009). Effect of crosslink density on some properties of electron beam-irradiated styrene–butadiene rubber. *Radiation Physics and Chemistry, 78*, 1001–1005. https://doi.org/10.1016/j.radphyschem.2009.06.001

Wang, T., Huang, J., Yang, Y., Zhang, E., Sun, W., & Tong, Z. (2015). Bioinspired smart actuator based on graphene oxide-polymer hybrid hydrogels. *ACS Applied Materials & Interfaces, 7*, 23423–23430. https://doi.org/10.1021/acsami.5b08248

Wang, T., Turhan, M., & Gunasekaran, S. (2004). Selected properties of pH sensitive, biodegradable chitosan–poly (vinyl alcohol) hydrogel. *Polymer International, 53*, 911–918. https://doi.org/10.1002/(ISSN)1097-0126

Yang, Z., Zhai, D., Wang, X., & Wei, J. (2014). *In situ* synthesis of highly monodispersed nonaqueous small-sized silver nano-colloids and silver/polymer nanocomposites by ultraviolet photopolymerization. *Colloids and Surfaces A: Physicochemical and Engineering Aspects, 448*, 107–114. https://doi.org/10.1016/j.colsurfa.2014.02.017

Yong, K. C. (2015). Electrically conductive thermoplastic elastomer vulcanisate based on NBR and PP blends with polyaniline: Preparation and characterisation. *Journal of Rubber Research, 18*, 189–202.

Yong, K. C., & Mustafa, A. (2014). Natural rubber-rubberwood fibre laminated composites with enhanced stab resistance properties. *Journal of Rubber Research, 17*(1), 1–12.https://doi.org/10.1007/s11676-014-0428-3

Zhang, Z., Zhao, B., & Hu, L. (1996). PVP protective mechanism of ultrafine silver powder synthesized by chemical reduction processes. *Journal of Solid State Chemistry, 121*, 105–110. https://doi.org/10.1006/jssc.1996.0015

Zheng, Z., Sun, J., Yuan, Y., Duan, J., & Tian, X. (2016). An investigation into the compatibility and properties of natural rubber/poly(methylmethacrylate-co-laurylmethacrylate) latex blends. *Journal of Rubber Research, 17*, 219–232.

16

Crystal structure of bis(triphenylphosphonium) hexabromodigallate(II) in the correct space group: Conformational complexity in a heteroethane

Olivia N.J.M. Marasco[1], Sydney K. Wolny[1], Jackson P. Knott[1,2], Daniel Stuart[1,2], Tracey L. Roemmele[1,2] and René T. Boeré[1,2]*

*Corresponding author: René T. Boeré, Department of Chemistry and Biochemistry, University of Lethbridge, Lethbridge, AB, Canada T1K3M4; The Canadian Centre for Research in Advanced Fluorine Technologies, University of Lethbridge, Lethbridge, AB, Canada T1K3M4

E-mail: boere@uleth.ca

Reviewing editor: Massimiliano Arca, University of Cagliari, Italy

Abstract: The crystal structure of $[Ph_3PH]_2[Ga_2Br_6]$, previously described as having a disordered anion in the space group $R\bar{3}$, has been re-determined in the correct space group $P\bar{3}$, where it is fully ordered. Interestingly, two-thirds of the $[Ga_2Br_6]^{2-}$ dianions have an intermediate conformation with a Br–Ga–Ga–Br torsion angle of 36.91 (1)°, while the remaining is staggered as required from adopting a site with inversion symmetry. In the lattice, $[Ph_3PH]^+$ ions lie along the same threefold axes as the dianions and are oriented such that the P–H bond is directed towards a gallium atom. The phosphonium ions lie back-to-back and interact with relatively strong T-interactions between phenyl rings on adjacent cations. DFT calculations at the B3LYP/6–311+G(fd,) level have been used to determine the barriers to rotation in $[Ga_2X_6]^{2-}$ ions. For X = Cl and X = Br, the barriers are found to be very small, with values of 4.3 and 5.1 kJ mol^{-1} for the two halogens.

Subjects: **Computational and Theoretical Chemistry; Inorganic Chemistry; Crystallography**

Keywords: **hexahallodigallate; sextuple phenyl embrace; single bond rotation; conformation; X-ray crystal structure; density functional theory calculations**

1. Introduction

Gallium is most stable in the Ga(III) oxidation state while Ga(I) is also accessible (Lichtenthaler et al., 2015). By contrast, Ga(II) is rare except in dimers or oligomers (Evans & Taylor, 1969; Kloo, Rosdahl, & Taylor, 2002). The hexahalodigallates, $[Ga_2X_6]^{2-}$, have long been known as heavy analogues to substituted ethanes. Tuck (Windsor, Canada) and Taylor (Auckland, NZ) pioneered an elegant

ABOUT THE AUTHOR

René T. Boeré has taught in all areas of main group element chemistry and employs a research strategy of chemical synthesis underpinned by detailed spectroscopic and structural investigation. Solid-state structures determined by single-crystal X-ray diffraction methods are central to most of his research. He has a special interest in main-group element free radicals and hence in unusual oxidation states where radical species can be expected, such as gallium(II). He also seeks to place all his results into a broad theoretical framework for which he makes extensive use of computational methods—most often hybrid density-functional theory.

PUBLIC INTEREST STATEMENT

Data repositories have become of great significance in scientific research with the advent of affordable storage and the WWW for accessing information remotely. Of key importance in structural chemistry is the Cambridge Structure Database, which comprises over 850,000 entries from X-ray and neutron diffraction analyses and provides 3D geometrical parameters. Analyses of this data are only as good as the quality of the underlying information, and erroneous entries can impact the validity of conclusions based thereon. This article provides a corrective to an earlier result contained in the CSD that overemphasizes the prevalence of one particular form of hexahalogallate anions.

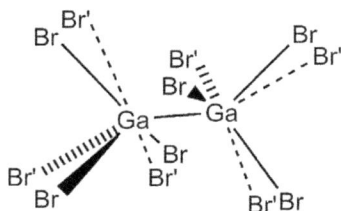

Figure 1. Disordered $[Ga_2Br_6]^{2-}$ anion in the refinement model for FUPSIS.

electrochemical preparative method for several salts thereof (Khan, Oldham, Taylor, & Tuck, 1980; Khan, Tuck, Taylor, & Rogers, 1986; Taylor & Tuck, 1983). The earliest crystal structure of a hexabromodigallate(II) to have entered the literature is the tetra-*n*-propylammonium salt (Cambridge Structure Database, CSD, refcode: PRAMBG; Cumming, Hall, & Wright, 1974; Groom & Allen, 2011). The ion has also been identified in a binary phase, Ga_2Br_3 (Hönle, Gerlach, Weppner, & Simon, 1986) and a closely-related ternary phase, $LiGaBr_3$ (Hönle & Simon, 1986). An unusual salt with *hexakis*-(N,N-dimethylformamide-O)-gallium(III) counter ions was isolated from a DMF extraction of a product obtained from oxidation of GaBr (refcode: FEGSUG) (Duan & Schnöckel, 2004). The crystal structure of the title compound was first reported in 1986 along with its chloride and iodide analogues (Khan et al., 1986). In this original report, the authors stated "beyond doubt ... a slight imperfection in the halogen atom positions caused a lowering of the symmetry to the space group $P\bar{3}$". However, the structure was unable to be solved in this lower symmetry space group due to insufficient measured reflections, and therefore the models were refined in the higher symmetry space group $R\bar{3}$. This structure (refcode: FUPSIS) is compiled in the CSD but the *apparent* 33% disorder in the bromine atom positions (Figure 1) has been supressed. The Br–Ga–Ga–Br′ torsion angle between the ordered and disordered halogens is about 44.2° whereas Br–Ga–Ga–Br is 60° (exact by symmetry).

Interest remains strong in the chemistry of various $[Ga_2X_6]^{2-}$ salts. $[N^nPr_4]_2[Ga_2Cl_6]$, refcode: TMAGAC (Brown & Hall, 1973) and $[Ph_3PH]_2[Ga_2Cl_6]$, refcode: FUPSIS (Khan et al., 1986) have been joined more recently by a salt of the mixed-halogen species $[Ga_2I_5Cl]^{2-}$, refcode: BAZTIH (Yurkerwich, Yurkerwich, & Parkin, 2011). Structures of various $[Ga_2I_6]^{2-}$ salts include FUPSEO (Khan et al., 1986), FASSOI (Tian, Pape, & Mitzel, 2004) and ISILEB (Baker, Jones, Kloth, & Mills, 2004). In the mixed-substituent anion $[Ga_2I_5(nhc)]^-$, refcode: OJAWEB, one iodide is replaced by a neutral N-heterocyclic carbene (Baker, Bettentrup, & Jones, 2003). There is also an extensive chemistry of $LX_2Ga–GaX_2L$ species with a wide range of neutral ligands L, all of which have Ga–Ga single bonds (selected references include: Baker, Bettentrup, & Jones, 2004; Ball, Cole, & McKay, 2012; Beagley et al., 1996; Duan & Schnöckel, 2004; Gordon et al., 1997; Nogai & Schmidbaur, 2002, 2004; Rickard, Taylor, & Kilner, 1999; Small & Worrall, 1982; Worrall & Small, 1982).

In the course of preparing the title compound, $[Ph_3PH]_2[Ga_2Br_6]$, we elected to obtain a low temperature crystal structure using a modern area-detector diffractometer to address the reported structural anomaly. We are now able to confirm the hypothesis of Khan et al. that the trigonal space group $P\bar{3}$ is correct. Furthermore, in this structural model, the bromine atom positions are fully ordered in two *independent* $[Ga_2Br_6]^{2-}$ ions which differ significantly in conformation about the Ga–Ga bond. One such molecule is perfectly staggered as required from its location at a site with crystallographic inversion symmetry, whereas a second molecule is in an intermediate conformation with a Br–Ga–Ga–Br torsion angle of 36.91 (1)°. This observation raises questions about the nature of conformational isomerism in hexahalodigallates(II), which is the subject of this report.

2. Results and discussion
The title complex, $[Ph_3PH]_2[Ga_2Br_6]$, was synthesized through direct electrochemical synthesis from elemental gallium by a variation on an optimized method (Taylor & Tuck, 1983). Under the acidic conditions, the added Ph_3P is protonated to provide an exceptional counterion for the trigonally symmetric anion and consequently crystals of $[Ph_3PH]_2[Ga_2Br_6]$ form during constant-current

electrolysis. Crystals filtered from the solution were found to include exemplars suitable for an X-ray diffraction analysis. Characterization was achieved by melting point and vibrational spectroscopy. Fourier transform infra-red (FTIR) and Raman vibrational spectra are in excellent agreement with the literature (Khan et al., 1986). In the Raman spectrum, the ν(P–H) band is found to be split into a doublet of 2,374 and 2,387 cm^{-1} (lit. 2,374 and 2,384 cm^{-1}). The deformation band is found at 869 cm^{-1} (lit. 870 cm^{-1}). The crucial, very intense A_{1g} combination ν(Ga–Br)/ν(Ga–Ga) band that is characteristic for a Ga–Ga single bond in all LX$_2$Ga–GaX$_2$L and [X$_3$Ga–GaX$_3$]$^{2-}$ species (Beamish, Boardman, Small, & Worrall, 1985) is found at 165.7 cm^{-1} (lit. 164 cm^{-1}).

2.1. Crystal structure determination

The crystal structure was successfully refined in the centrosymmetric trigonal space group $P\bar{3}$ (Figure 2 and Table 1). One [Ga$_2$Br$_2$]$^{2-}$ anion (Ga1;Br1) and one [Ph$_3$PH]$^+$ cation (P1) lie along the unique $\bar{3}$ axis; this anion is centred on Wyckoff position 1a with $\bar{3}$ site symmetry (so that the geometry is perfectly staggered), whereas the cation occupies Wyckoff position 2c with threefold rotational symmetry. The second [Ga$_2$Br$_2$]$^{2-}$ anion (Ga2,3;Br2,3) and two [Ph$_3$PH]$^+$ cations (P2,3) lie along the two threefold axes in the unit cell and all occupy Wyckoff positions 2d with site symmetries of 3. This anion has a smallest Br–Ga–Ga–Br torsion angle of 36.91 (1)°, so that it is almost half way between staggered and eclipsed. Symmetry related atoms with the same label are distinguished by colour coding in Figure 2. The P–H bonds of all the cations are directed towards Ga centres, whilst two Ph$_3$PH$^+$ moieties associate back-to-back, allowing their three phenyl rings to associate through phenyl carbon/ phenyl hydrogen T-interactions. This association is reminiscent of the supramolecular organization of Ph$_4$P$^+$ cations which has been dubbed the "sextuple phenyl embrace" with estimated attraction energy of 60–85 kJ mol^{-1} (Dance & Scudder, 1996). Similar contacts have been noted in dimers of SbPh$_3$ in a recent aromatic adduct structure, with CH$\cdots$C$_{aromatic}$ short contacts of 2.915–2.995 Å (Boeré, 2016). In the title structure, the T-interactions have lengths of 2.936 Å [C3$\cdots$H2' (−y, x − y, z)] on the $\bar{3}$ axis and on the 3 axis of 2.885 [C17$\cdots$H8], 2.995 [C8$\cdots$H18" (−x + y, 1 − x, z)] and 2.961 [C9$\cdots$H18# (−x + y, 1 − x, z)] Å. The T-interactions in the P1 cations and the two T-interactions in the P2,3 cations are replicated by the 3-fold rotational symmetry; these are not shown in Figure 2 to avoid additional

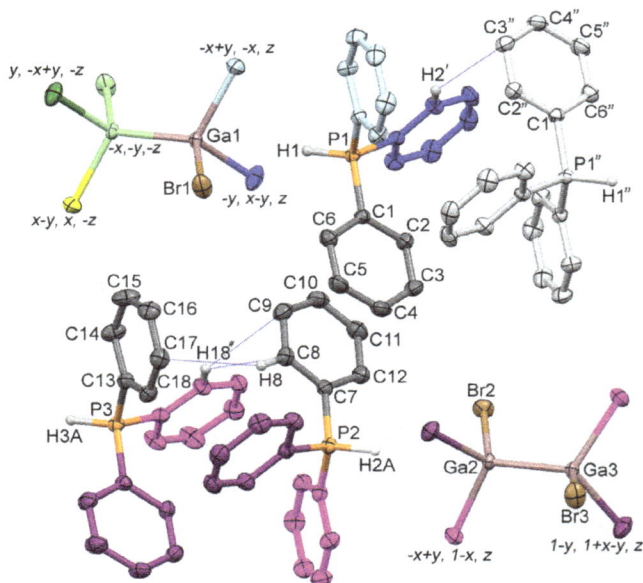

Figure 2. Displacement ellipsoids (50% probability) plot of the asymmetric unit depicting the molecular structures of the two independent [Ga$_2$Br$_2$]$^{2-}$ anions and the three independent [Ph$_3$PH]$^+$ cations (Ga1 with site symmetry $\bar{3}$; Ga2,3 with site symmetry 3). False colours indicate symmetry-related atoms (sym codes on diagram). A symmetry-equivalent P1 cation additional to the asymmetric unit is shown in light grey ("). Phenyl ring T-interactions from C–H to C are shown with dashed blue lines *once* for each symmetry- replicated interaction. H atoms on carbon are omitted, as are atom labels for symmetry related atoms.

Table 1. Selected interatomic distances, angles and torsions in the crystal structure of the title compound

Atoms	Distance (Å)	Atoms	Angle (°)	Atoms	Dihedral (°)
Ga1–Ga1^i	2.4124 (7)	Br1–Ga1–Ga1^i	115.115 (9)	Br1–Ga1–Ga^i–Br1^{vii}	60 (exact)
Br1–Ga1	2.3732 (3)	Br1–Ga1–Br1^{iv}	103.284 (11)	Br2–Ga2–Ga3–Br3 ^{viii}	36.91 (1)
Ga2–Ga3	2.3978 (5)	Br2–Ga2–Ga3	113.700 (9)	C1–P1···P^{ix}–C1^{ix}	60 (exact)
Br2–Ga2	2.3731 (3)	Br3–Ga3–Ga2	114.136 (9)	C7–P2···P3–C13^x	55.60 (1)
Br3–Ga3	2.3770 (3)	Br2–Ga2–Br2^v	104.931 (11)		
P1–C1	1.7794 (19)	Br3^{vi}–Ga3–Br3	104.431 (11)		
P1–H1	1.321 (18)	C1–P1–C1^{vii}	110.04 (6)		
P2–C7	1.785 (2)	C1–P1–H1	108.90 (7)		
P2–H2A	1.314 (18)	C7–P2–C7^v	111.09 (6)		
P3–C13	1.780 (2)	C7–P2–H2A	107.80 (7)		
P3–H3A	1.304 (18)	C13^{vi}–P3–C13	109.57 (7)		
Ga1···P1	4.518 (1)	C13–P3–H3A	109.37 (7)		
P1····P1^{ii}	6.156 (1)				
Ga2···P2	4.604 (1)				
Ga3···P3^{iii}	4.460 (2)				
P2···P3	6.145 (1)				

^{i}−x, −y, −z.

^{ii}−x, −y, 1 − z.

^{iii}x, y, 1 + z.

^{iv}−y, x − y, z.

^{v}−y + 1, x − y + 1, z.

^{vi}−x + y, 1 − x, z.

^{vii}y, −x + y, z. −x.

^{viii}−x + y, 1 − x, z.

^{ix}−x, −y, 1 − z.

^{x}−x + y, 1 − x, z.

clutter. The C1–P1···P1′–C1′ torsion angle between two such interlocked phosphonium ions is 60° (exact by lattice symmetry) for the P1 ions at Wyckoff position 2c, while the C7–P2···P3–C13 torsion is 55.60 (1)° for the pair of phosphonium ions occupying Wyckoff postion 2d.

The intriguing packing arrangement is well-suited to the cylindrical "dumbbell"-shaped $[Ga_2Br_6]^{2-}$ anions and the propeller-shaped phosphonium cations as depicted in a space filling diagram in Figure 3. The "vertical" columns of Ph_3PH^+—$[Ga_2Br_6]^{2-}$—$^+HPPh_3$ ions along the 3 and $\bar{3}$ axes are close-packed but alternate out of register by approximately half the unit cell c distances. Consequently, each $[Ga_2Br_6]^{2-}$ dimer is surrounded on six sides by phenyl ring C–H atoms of the two phosphonium ions.

In order to compare to the previously reported structure model (refcode: FUPSIS), especially to check whether the difference in refined models might have been caused by the change in temperature from ambient to 100 K in our data-set, we also refined our structure in space group $R\bar{3}$. Very close agreement is found for the anion to that in FUPSIS, and specifically the Br–Ga–Ga–Br′ torsion angle to the disordered site is about 40.0°. Thus, there is no evidence for a phase change having occurred over the temperature interval between the earlier structure and this determination. In Figure 4, *our* disordered model is superimposed on the true structure with the rhombohedral body diagonal oriented along the (⅔ ⅓ 0) 3 axis in the $P\bar{3}$ unit cell. Although the ratio of staggered $GaBr_3$ entities on the $\bar{3}$ axis and intermediate conformation entities on the 3 axis of the true structure is actually 1:2, there is a virtual coincidence of Br atom positions of the $R\bar{3}$ model with the Br2 atom in the $P\bar{3}$ unit cell, so that the apparent disorder population observable in the $R\bar{3}$ refinement is expected to be

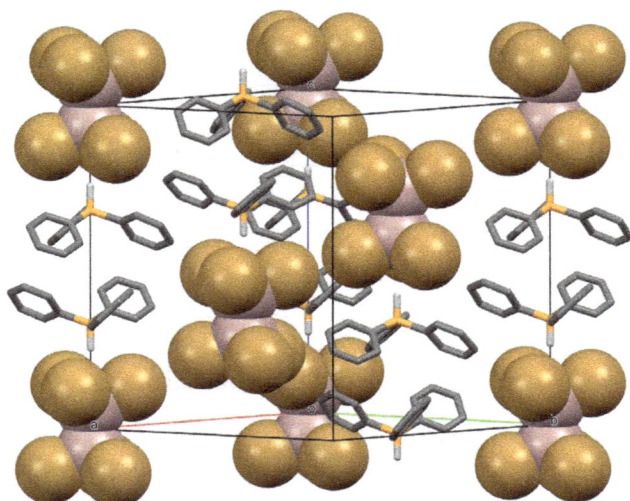

Figure 3. Packing diagram of the structure of the title compound in $P\bar{3}$. H atoms on the phenyl rings are omitted for clarity, as are all atoms along the front (1 1 0) unit cell edge. The phosphonium ions are rendered as rods, whereas the hexabromodigallate ions are rendered as "space filling" spheres.

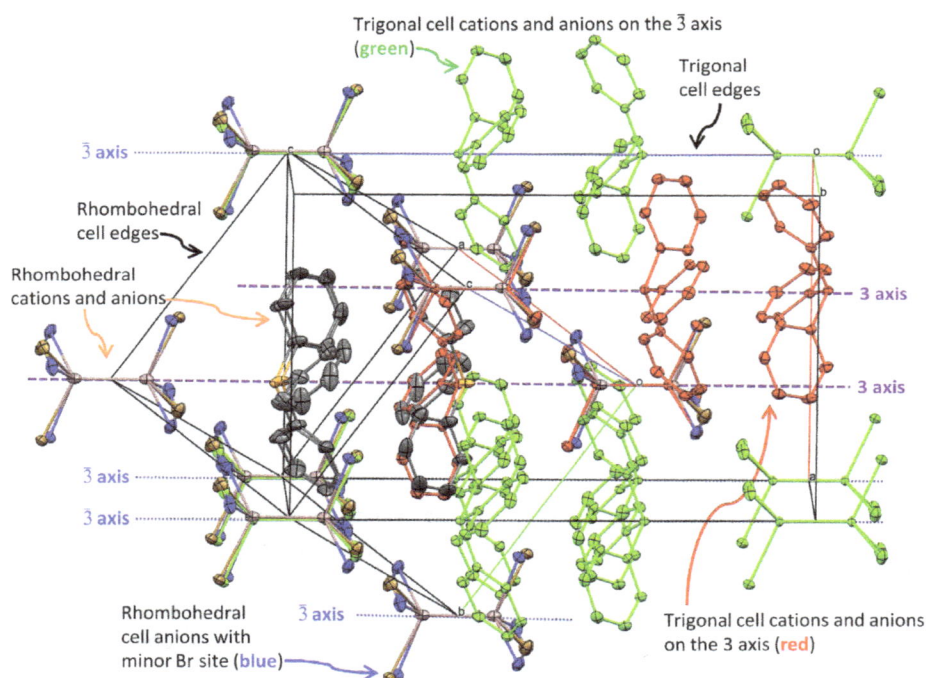

Figure 4. Overlaid unit cells of the title structure in $P\bar{3}$ with the alternate refinement in $R\bar{3}$ with apparent disorder as in (Khan et al., 1986), drawn as 30% displacement ellipsoids. The $\bar{3}$ body diagonal of the rhombohedral cell is aligned with the 3 axis of the trigonal cell along (⅔ ⅓ 0).

0.333. Freely refining the occupancies resulted in 0.375, in excellent agreement with this prediction (Khan et al., 1986). The disorder model they reported ignored the fates of the Ph_3PH^+ ions, implying that these ions are ordered. However, as can be seen in Figure 4, the displacement ellipsoids of the P and C atoms of this ion in the $R\bar{3}$ refinement are *all* severely distorted. In fact, both the positions of the atoms along the c axis *and* the rotational displacements around the 3 and $\bar{3}$ axes in the true structure are different for the independent phosphonium ions.

The small C7–P2···P3–C13 torsion angle of 55.60 (1)° seems to be associated with a shorter T-interaction in this pair than is observed for the ions on the $\bar{3}$ axis (Figure 4). This raises the distinct possibility that it is the stronger phosphonium ion intermolecular interactions that provide a driving force for adoption of the lower symmetry space group, thereby possibly inducing the intermediate conformation in the hexabromodigallate on the 3 axis through interactions between the phenyl rings and the digallate bromine atoms. Importantly, the *spacing* of the Ph_3PH^+—$[Ga_2Br_6]^{2-}$—$^+HPPh_3$ ions along the $\bar{3}$ axes are symmetrical as required by lattice symmetry, with Ga1···P1 = 4.518 (1) Å. However, along the 3 axis the spacing is long-short-long, with Ga2···P2 = 4.604 (1) and Ga3···P3′ = 4.460 (2) Å. It is this asymmetry that allows the P2···P3 distance to be 0.011 (1) Å shorter than P1···P1′, with the aforementioned lower C–P···P–C′ torsion angle and stronger T-interactions.

In summary, the crystal structure in the correct space group of the title compound holds several surprises. First, only one-third of the lattice occupancy of $[Ga_2Br_6]^{2-}$ ions is in the "expected" staggered geometry, whilst the remaining two-thirds are in an intermediate conformation. Secondly, the phosphonium ions on the 3 axis, also two-thirds of the total occupancy, relax by twisting so as to maximize the "sextuple phenyl embrace", providing a possible driving force for adoption of the lower symmetry crystal lattice.

In the crystal structure of $[N^nPr_4]_2[Ga_2Br_6]$, refcode: PRAMBG, the dianion has crystallographically required $\bar{1}$ site symmetry, rendering it staggered (Cumming et al., 1974). However, the $[Ga_2Br_7]^{2-}$ ions in binary and ternary solid phases, best formulated as $Ga_2[Ga_2Br_6]$ (Hönle et al., 1986) and $Li_2[Ga_2Br_6]$ (Hönle & Simon, 1986), surprisingly crystallize in eclipsed conformations. The structures of the two phases are different, but in both the M^+ counter ions are also strongly coordinated with the bromide ions, including bridges across the two Ga(II) atoms of the *same* ion. The point can be illustrated by considering the structure of $Li_2[Ga_2Br_6]$ (Figure 5). This shows that Li1 doubly bridges the eclipsed $[Ga_2Br_7]^{2-}$ anion, whereas Li2 is chelated by the $GaBr_3$ triad. These interactions could be the driving force for a potentially higher energy conformation of the anions. These authors commented that conformational changes in $[Ga_2Br_6]^{2-}$ are apparently quite small. In the more recently reported salt $[Ga(O-DMF)_6]_2[Ga_2Br_6]_3$ (refcode: FEGSUG) the cations are unable to interact strongly with the anion bromine atoms (Duan & Schnöckel, 2004). In this structure, one half of the $[Ga_2Br_6]^{2-}$ ions are staggered by imposed $\bar{1}$ site symmetry whilst the remainder have intermediate conformations, with a smallest torsion angle of 47.1°, so quite similar to what occurs in the structure of $[Ph_3PH]_2[Ga_2Br_6]$ in the correct space group.

2.2. Computed conformational energy profile

In order to address this structural diversity, we have undertaken DFT calculations at the B3LYP/6–311+G(fd,) level of theory. If the intermediate geometry from the title structure is geometry-optimized, it is found to reorganize during the calculations to a perfectly staggered conformation. This is, as expected for an ethane-like molecule, the most stable conformation. In order to ascertain approximate energies for such a transformation, a series of calculations were undertaken that scan the conformational coordinate from 63.3° to 183.3° in $[Ga_2X_6]^{2-}$ (X = Cl, Br). The results, shown in

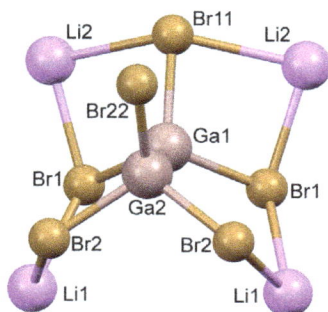

Figure 5. Representation of part of the crystal structure of $Li_2[Ga_2Br_6]$; there is a mirror plane through Br11, Ga1, Ga2, Br22. Li^+ ions coordinated to Br22 and all remaining bromine atoms completing the coordination spheres of Li1,2 have been omitted for clarity. The data is taken from Hönle and Simon (1986).

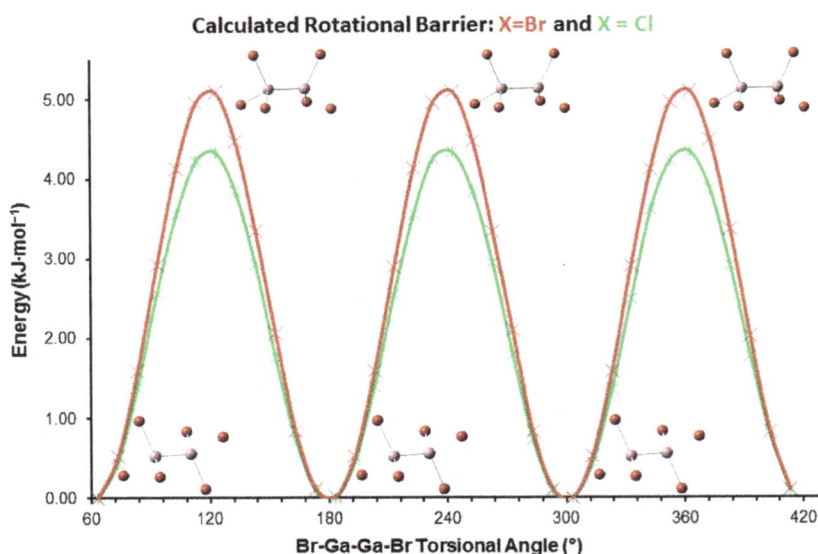

Figure 6. Conformational energy profile for the title anion computed at the B3LYP/6–311+G(fd,) level of theory for gas-phase [Ga$_2$Cl$_6$]$^{2-}$ and [Ga$_2$Br$_6$]$^{2-}$ ions. The calculations were undertaken in 10° steps in the X–Ga–Ga–X torsion angle starting from 63.3°.

Figure , indicate (1) that the staggered conformation is indeed the lower energy preferred geometry and (2) that the energetic cost of adopting the most hindered eclipsed conformation is very small, on the order of just 5 kJ mol^{-1}. The intermediate conformation encountered for two-thirds of the anions in the title crystal structure has no significance in the calculated profiles.

The rotational barrier in the title anion is surprisingly small, which is consistent with crystallographic characterization of eclipsed and intermediate forms in addition to the preferred staggered conformation amongst the small series of extant structurally characterized salts. To our knowledge, no experimental or calculated barriers to rotation have been reported for hexahalodigallate dianions. However, there is an intense interest in the barrier to rotation for the neutral Group 14 element ethane-like species (Cortés-Guzmán, Cuevas, Pendás, & Hernández-Trujillo, 2015; Johansson & Swart, 2013; Mo & Gao, 2007; Morino & Hirota, 1958; Quijano-Quiñones, Quesadas-Rojas, Cuevas, & Mena-Rejón, 2012). Table 2 contains data on some of these species. First of all, the barrier in ethane itself is known to be 12.8 kJ mol^{-1}, three times as high as in the title anion. However, upon progressing to the hydrides of the third and fouth period analogues, this value drops dramatically to only 3.1 kJ mol^{-1} for H$_3$GeGeH$_3$. Similarly, whilst the measured barrier heights in hexachloro- and hexabromoethanes rise dramatically to >73 and >83 kJ mol^{-1}, the barrier in Cl$_3$SiSiCl$_3$ is measured to be only 4.2 kJ mol^{-1}. Evidently, the much larger radius of Ga is more than enough to offset the large size of the six bromide substituents, so that the barrier height in [Ga$_2$Br$_6$]$^{2-}$ is quite similar to that in Cl$_3$SiSiCl$_3$.

Table 2. Experimental or calculated barriers to rotation in [X$_3$EEX$_3$] species			
Molecule	**Exp/calc**	**kJ/mol**	**References**
H$_3$CCH$_3$	c	12.8	Mo and Gao (2007)
H$_3$SiSiH$_3$	c	4.1	Mo and Gao (2007)
H$_3$GeGeH$_3$	c	3.1	Mo and Gao (2007)
H$_3$CSiH$_3$	c	6.1	Mo and Gao (2007)
Cl$_3$CCCl$_3$	e	73.2	Morino and Hirota (1958)
Br$_3$CCBr$_3$	e	83.7	Morino and Hirota (1958)
Cl$_3$SiSiCl$_3$	e	4.2	Morino and Hirota (1958)
[Ga$_2$Cl$_6$]$^{2-}$	c	4.34	This work
[Ga$_2$Br$_6$]$^{2-}$	c	5.09	This work

Ge_2Br_6 remains an under-investigated compound (Curtis & Wolber, 1972; Höfler & Brandstätter, 1975). Although it can be made as an air-sensitive colourless compound, it is unstable towards decomposition to $GeBr_2$ and $GeBr_4$ above 85°C (Curtis & Wolber, 1972). In the solid state, it may crystallize centrosymmetrically based on the non-coincidence of IR and Raman bands in its vibrational spectrum (Curtis & Wolber, 1972) but no crystal structure has been reported for this species. The ν(Ge–Ge) band was measured to be 196 cm^{-1} (Höfler & Brandstätter, 1975), implying a Ge–Ge bond that is somewhat stronger than the Ga–Ga bonds in digallates (ν(Ga–Ga) = 165.7 cm^{-1}). Both $Br_3SiSiBr_3$ (2/m site symmetry) and I_3SiSiI_3 ($\bar{3}$ site symmetry) have structures with crystallographically imposed staggered conformations (Berger, Auner, & Bolte, 2014). The crystal structures of Cl_3CCCl_3 in numerous polymorphs all crystallize with inversion site symmetries (Negrier, Tamarit, Barrio, & Mondieig, 2013; Sasada & Atoji, 1953). Br_3CCBr_3 crystallizes staggered with m site symmetry both at RT (Mandel & Donohue, 1972) and at 100 K (refcode: HEXBME); a recent study reports co-crystals of Br_3CCBr_3 with organometallic ruthenium complexes and here the molecules are found to be approximately staggered *without* crystallographic site symmetry in one case and in another to crystallize with $\bar{1}$ site symmetry (refcodes: FALYOI, FALYOI01) (Fuller et al., 2012). Consistent with the very high barrier energies for conformational changes in Br_3CCBr_3, the geometry in the former structure has a largest BrCCBr torsion angle of 179.4 (4)°, showing that it is effectively staggered even in the absence of lattice-imposed symmetry. The contrast with $[Ga_2Br_6]^{2-}$, which can adopt a wide range of conformations from eclipsed to staggered in different lattices, is dramatic.

3. Conclusions

Although it has been recognized since the discovery of the solid-state structures of $Li_2[Ga_2Br_6]$ and $Ga_2[Ga_2Br_6]$ that the barrier to Ga–Ga bond rotation in hexabromodigallate is likely to be small (Hönle et al., 1986; Hönle & Simon, 1986), no further investigation of this phenomenon appears to have been published. We have corrected the crystal structure of the phosphonium salt of this anion, $[Ph_3PH]_2[Ga_2Br_6]$, and demonstrated that here two out of three digallate anions have a distinctly intermediate conformation (smallest dihedral angle of 36.91 (1)°). Quantum calculations using standard DFT methods with large basis sets show that the barrier to rotation in the free, gas-phase anions is small, 4.3 and 5.1 kJ mol^{-1} for the hexachloro- and hexabromodigallates, respectively. In this regard, the hexahalodigallates (Cl, Br, I) are shown to be good heavy main-group analogues to ethane. Whereas in a dicarbon molecule, replacing H with such halogens radically increases their conformational barrier energies, in the digallates the naturally large size of gallium is very well matched to the size of these halogen atoms and consequently have very low barriers for rotation. It is almost certain that similar ordered structures with different conformations pertain in the structures of $[Ph_3PH]_2[Ga_2X_6]$ (X = Cl, I) (Khan et al., 1986). The solid phases $LiGaCl_3$ and $LiGaI_3$ show eclipsed geometry for Cl and staggered for I (Hönle, Miller, & Simon, 1988).

4. Experimental Section

4.1. Synthesis

A 100-mL electrolysis cell equipped with a gallium metal anode and a platinum wire cathode was flushed with dry N_2 gas, charged with 30 mL of acetonitrile, 0.50 g (1.5 mmol) Ph_3P, 2.0 mL (180 mmol) c. hydrobromic acid and cooled in an ice bath. A KEPCO ABC40 d.c. power supply and a Keithly 160B multimeter were used to maintain a steady-state current of 50 mA for the duration of the synthesis (~1.5 H). Yield: 0.20 g (0.23 mmol, 32% based on mass of gallium consumed) of colourless blocks (current efficiency = 0.53 mol F^{-1}). MP 159.7–161.9°C (lit. 160–162°C, Taylor & Tuck, 1983).

4.2. Crystal structure

Single crystals of $C_{36}H_{32}Br_6Ga_2P_2$ precipitated during electrochemical synthesis. A suitable crystal was selected and mounted using a fine glass capillary in frozen Paratone oil on a Bruker APEX-II CCD diffractometer. The crystal was kept at 173 (2) K during data collection. Using Olex2 (Dolomanov, Bourhis, Gildea, Howard, & Puschmann, 2009), the structure was solved with the ShelXT (Sheldrick, 2015a) structure solution programme using "intrinsic phasing" and refined with the ShelXL (Sheldrick, 2015b) refinement package using least squares minimization. Hydrogen atoms on the phenyl rings were treated as riding (C–H = 0.95 Å) with U_{iso} = 1.2 U of attached atom. The H atoms on P were re-strained to a bond distance of 1.32 (2) Å and also have U_{iso} = 1.2 U of P atoms. Restraining these distances was necessary for a stable refinement. Crystal Data for $C_{36}H_{32}Br_6Ga_2P_2$ (M = 1145.45 g/mol): trigonal, space group $P\bar{3}$ (No. 147), a = 13.9660 (5) Å, c = 17.6075 (7) Å, V = 2,974.2 (2) Å^3, Z = 3, T = 173 (2) K, μ(MoKα) = 7.510 mm^{-1}, D_{calc} = 1.919 g/cm^3, 37,168 reflections measured (3.368° ≤ 2θ ≤ 58.446°), 5,152 unique (R_{int} = 0.0246, R_{sigma} = 0.0175) which were used in all calcula-tions. The final R_1 was 0.0225 (I > 2σ(I)) and wR_2 was 0.0510 (all data). Several different crystals were investigated; some data sets displayed rather large residual electron density peaks approximately half-way between a Ga and a P atom along the 3 and $\bar{3}$-axes. The reported structure is the one that was least affected by this phenomenon; no satisfactory twin-law could be found for these features. Structure illustrations as well as geometric and symmetry analyses were undertaken using Mercury release 3.9 (Macrae et al., 2006). CCDC 1515022 contains the data deposition for this crystal struc-ture. These data can be obtained, free of charge, via http://www.ccdc.cam.ac.uk/products/csd/re-quest/ (or from the Cambridge Crystallographic Data Centre, 12 Union Road, Cambridge CB2 1EZ, U.K. (e-mail: deposit@ccdc.cam.ac.uk)).

4.3. Computation

DFT calculations were undertaken at the B3LYP/6–31G(fd,) and B3LYP/6–311+G(fd,) levels of theory using Gaussian W03 (Frisch et al., 2004). These methods have been used in closely related systems and verified against experimental data for rotational barriers (Mo & Gao, 2007). The geometry of the non-centrosymmetric $[Ga_2Br_6]^{2-}$ ion from the crystal structure was used as a starting point. Using ei-ther the smaller or larger basis sets, this geometry was found to be unstable towards the staggered conformation. Thereafter, the geometry was calculated at a series of fixed Br–Ga–Ga–Br torsion an-gles with full relaxation of all other variables. The torsions were stepped from 63° in 36 10-degree steps. With the smaller basis set, a metastable eclipsed form was detected with small local minima, but these disappear at the basis set limit. In no case was an intermediate conformation found to be stable. The chloro analogues were constructed from the original bromo anion by applying standard Ga–Cl single bond distances and then optimizing the same way as for the bromo species.

Funding
The diffractometer was funded by the Natural Sciences and Engineering Research Council of Canada and by the University of Lethbridge.

Author details
Olivia N.J.M. Marasco[1]
E-mail: o.marasco@uleth.ca
Sydney K. Wolny[1]
E-mail: sydney.wolny@uleth.ca
Jackson P. Knott[1,2]
E-mail: jackson.knott@uleth.ca
Daniel Stuart[1,2]
E-mail: stuart@uleth.ca
Tracey L. Roemmele[1,2]
E-mail: roemtl@uleth.ca
René T. Boeré[1,2]
E-mail: boere@uleth.ca
[1] Department of Chemistry and Biochemistry, University of Lethbridge, Lethbridge, AB, Canada T1K3M4.
[2] The Canadian Centre for Research in Advanced Fluorine Technologies, University of Lethbridge, Lethbridge, AB, Canada T1K3M4.

References
Baker, R. J., Bettentrup, H., & Jones, C. (2003). The reactivity of primary and secondary amines, secondary phosphanes and N-heterocyclic carbenes towards group-13 metal(I) halides. European Journal of Inorganic Chemistry, 2003, 2446–2451. doi:10.1002/ejic.200300068

Baker, R. J., Bettentrup, H., & Jones, C. (2004). The synthesis and structural characterisation of the first gallium(II) dialkylphosphide complex. Inorganic Chemistry Communications, 7, 1289–1291. doi:10.1016/j.inoche.2004.10.004

Baker, R. J., Jones, C., Kloth, M., & Mills, D. P. (2004). The reactivity of gallium(I) and indium(I) halides towards bipyridines, terpyridines, imino-substituted pyridines and bis(imino)acenaphthenes. New Journal of Chemistry, 28, 207–213. doi:10.1039/b310592j

Ball, G. E., Cole, M. L., & McKay, A. I. (2012). Low valent and hydride complexes of NHC coordinated gallium and indium. Dalton Transactions, 41, 946–952. doi:10.1039/c1dt11202c

Beagley, B., Godfrey, S. M., Kelly, K. J., Kungwankunakorn, S., McAuliffe, C. A., & Pritchard, R. G. (1996). The reactions of gallium metal with (p-MeOC$_6$H$_4$)$_3$AsI$_2$ and Et$_3$AsI$_2$; isolation of a novel gallium(II) arsine complex with a gallium-gallium bond and the X-ray crystal structures of GaI$_3$[(p-MeOC$_6$H$_4$)$_3$As] and Ga$_2$I$_4$(AsEt$_3$)$_2$. Chemical Communications, 2179–2180. doi:10.1039/CC9960002179

Beamish, J. C., Boardman, A., Small, R. W. H., & Worrall, I. J. (1985). Neutral complexes of Ga$_2$X$_4$ (X = Cl, Br) containing Ga–Ga bonds: The crystal and molecular structure of Ga2Cl4 2pyridine. Polyhedron, 4, 983–987. doi:10.1016/S0277-5387(00)84068-1

Berger, M., Auner, N., & Bolte, M. (2014). Hexabromo- and hexaiododisilane: Small and simple molecules showing completely different crystal structures. Acta Crystallographica Section C, Structural Chemistry, 70, 1088–1091. doi:10.1107/S2053229614022992

Boeré, R. T. (2016). Short contacts of the sulphur atoms of a

1,2,3,5-dithiadiazolyl dimer with triphenylstibine: First co-crystal with an aromatic compound. *CrystEngComm, 18,* 2748–2756. doi:10.1039/c6ce00351f

Brown, K. L., & Hall, D. (1973). Crystal structure of tetramethylammonium hexachlorodigallate(II). *Journal of the Chemical Society, Dalton Transactions,* 1843–1845. doi:10.1039/DT9730001843

Cortés-Guzmán, F., Cuevas, G., Pendás, Á. M., & Hernández-Trujillo, J. (2015). The rotational barrier of ethane and some of its hexasubstituted derivatives in terms of the forces acting on the electron distribution. *Physical Chemistry – Chemical Physics, 17,* 19021–19029. doi:10.1039/c5cp02774h

Cumming, H. J., Hall, D., & Wright, Carolyn E. (1974). Tetra-*n*-propylammonium hexabromodigallate(II). *Crystal Structure Communications, 3,* 107–109.

Curtis, M. D., & Wolber, P. (1972). Facile syntheses of germanium dibromide, hexabromodigermane, and tribromomethyltribromogermane. *Inorganic Chemistry, 11,* 431–433. doi:10.1021/ic50108a051

Dance, I., & Scudder, M. (1996). Supramolecular motifs: Concerted multiple phenyl embraces between Ph$_4$P$^+$ cations are attractive and ubiquitous. *Chemistry - A European Journal, 2,* 481–486. doi:10.1002/chem.19960020505

Dolomanov, O. V., Bourhis, L. J., Gildea, R. J., Howard, J. A. K., & Puschmann, H. (2009). OLEX2: A complete structure solution, refinement and analysis program. *Journal of Applied Crystallography, 42,* 339–341. doi:10.1107/S0021889808042726

Duan, T., & Schnöckel, H. (2004). Donor-stabilized galliumdihalides Ga$_2$X$_4$·2D (X = Cl, Br; D = Donor): An Experimental contribution on the variation of the Gallium-Gallium single bond. *Zeitschrift für Anorganische und Allgemeine Chemie, 630,* 2622–2626. doi:10.1002/zaac.200400321

Evans, C. A., & Taylor, M. J. (1969). Galliurn-gallium bonds in hexahalogenodigallate(II) ions. *Journal of the Chemical Society D,* 1201–1202. doi:10.1039/C29690001200

Frisch, M., Trucks, G. W., Schlegel, H., Scuseria, G. E., Robb, M. A., Cheeseman, J. R., ... Pople, J. A. (2004). *Gaussian 03, Revision C.02,* Wallingford CT: Gaussian, Inc.

Fuller, R. O., Griffith, C. S., Koutsantonis, G. A., Lapere, K. M., Skelton, B. W., Spackman, M. A., ... Wild, D. A. (2012). Supramolecular interactions between hexabromoethane and cyclopentadienyl ruthenium bromides: Halogen bonding or electrostatic organisation? *CrystEngComm, 14,* 804–811. doi:10.1039/c1ce05438d

Gordon, E. M., Hepp, A. F., Duraj, S. A., Habash, T. S., Fanwick, P. E., Schupp, J. D., ... Long, S. (1997). The preparation and structural characterization of three structural types of gallium compounds derived from gallium(II) chloride. *Inorganica Chimica Acta, 257,* 247–251. doi:10.1016/S0020-1693(96)05490-4

Groom, C. R., & Allen, F. H. (2011). The Cambridge Structural Database: Experimental three-dimensional information on small molecules is a vital resource for interdisciplinary research and learning. *WIREs Computational Molecular Science, 1,* 368–376. doi:10.1002/wcms.35

Höfler, F., & Brandstätter, E. (1975). Hydrohalogenation of hexaphenyldigermane and vibrational spectra of some Ge$_2$(C$_6$H$_5$)$_n$X$_{6-n}$ compounds. *Monatshefte für Chemie, 106,* 893–904. doi:10.1007/BF00900869

Hönle, W., Gerlach, G., Weppner, W., & Simon, A. (1986). Preparation, crystal structure, and ionic conductivity of digallium tribromide, Ga$_2$Br$_3$. *Journal of Solid State Chemistry, 61,* 171–180. doi:10.1016/0022-4596(86)90019-8

Hönle, W., Miller, G., & Simon, A. (1988). Preparation, crystal structure, and electronic properties of LiGaCl$_3$ and LiGaI *Journal of Solid State Chemistry, 75,* 147–155. doi:10.1016/0022-4596(88)90312-X

Hönle, W., & Simon, A. (1986). Preparation and crystal structures of LiGaBr$_4$ and LiGaBr$_3$. *Zeitschrift für Naturforschung, 41b,* 1391–1398. doi:10.1515/znb-1986-1113

Johansson, M. P., & Swart, M. (2013). Intramolecular halogen-halogen bonds?. *Physical Chemistry – Chemical Physics, 15,* 11543–11553. doi:10.1039/c3cp50962a

Khan, M. A., Tuck, D. G., Taylor, M. J., & Rogers, D. A. (1986). Crystal structures and vibrational spectra of triphenylphosphonium hexahalogenodigallates(II), (Ph$_3$PH)$_2$Ga$_2$X$_6$ (X = CI, Br, or I). *Journal of Crystallographic and Spectroscopic Research, 16,* 895–905. doi:10.1007/BF01188195

Khan, M., Oldham, C., Taylor, M. J., & Tuck, D. G. (1980). Preparative and structural studies of triphenylphosphonium salts. *Inorganic and Nuclear Chemistry Letters, 1980,* 469–474. doi:10.1016/0020-1650(80)80104-8

Kloo, L., Rosdahl, J., & Taylor, M. J. (2002). The nature of subvalent gallium and indium in aqueous media. *Polyhedron, 21,* 519–524. doi:10.1016/S0277-5387(01)01030-0

Lichtenthaler, M. R., Maurer, S., Mangan, R. J., Stahl, F., Mönkemeyer, F., Hamann, J., & Krossing, I. (2015). Univalent Gallium Complexes of Simple and *ansa*-Arene Ligands: Effects on the Polymerization of Isobutylene. *Chemistry: A European Journal, 21,* 157–165. doi:10.1002/chem.201404833

Macrae, C. F., Edgington, P. R., McCabe, P., Pidcock, E., Shields, G. P., Taylor, R., ... van de Streek, J. (2006). Mercury CSD 2.0— New features for the visualization and investigation of crystal structures. *Journal of Applied Crystallography, 39,* 453–457. doi:10.1107/S002188980600731X

Mandel, G., & Donohue, J. (1972). The refinement of the structure of hexabromoethane. *Acta Crystallographica, Section B: Structural Science, 28,* 1313–1316. doi:10.1107/S0567740872004182

Mo, Y., & Gao, J. (2007). Theoretical analysis of the rotational barrier of ethane. *Accounts of Chemical Research, 40,* 113–119. doi:10.1021/ar068073w

Morino, Y., & Hirota, E. (1958). Molecular structure and internal rotation of hexachloroethane, hexachlorodisilane, and trichloromethyl-trichlorosilane. *The Journal of Chemical Physics, 28,* 185–197. doi:10.1063/1.1744091

Negrier, P., Tamarit, J. L., Barrio, M., & Mondieig, D. (2013). Polymorphism in halogen-ethane derivatives: CCl$_3$–CCl$_3$ and ClF$_2$C–CF$_2$Cl. *Crystal Growth & Design, 13,* 782–791. doi:10.1021/cg301498f

Nogai, S., & Schmidbaur, H. (2002). Dichlorogallane Dehydrogenative Ga-Ga coupling and hydrogallation in gallium hydride complexes of 3,5-dimethylpyridine. *Organometallics, 23,* 5877–5880. doi:10.1021/om049393r

Nogai, S., & Schmidbaur, H. (2004). Dichlorogallane (HGaCl$_2$)$_2$: Its molecular structure and synthetic potential. *Inorganic Chemistry, 41,* 4770–4774. doi:10.1021/ic0203015

Quijano-Quiñones, R. F., Quesadas-Rojas, M., Cuevas, G., & Mena-Rejón, G. J. (2012). The rotational barrier in ethane: A molecular orbital study. *Molecules, 17,* 4661–4671. doi:10.3390/molecules17044661

Rickard, C. E. F., Taylor, M. J., & Kilner, M. (1999). Tetrachlorobis(hexamethylphosphoramide-O)digallium(II) (Ga Ga). *Acta Crystallographica, C55,* 1215–1216. doi:10.1107/S0108270199005843

Sasada, Y., & Atoji, M. (1953). Crystal structure and lattice energy of orthorhombic hexachloroethane. *The Journal of Chemical Physics, 21,* 145–152. doi:10.1063/1.1698566

Sheldrick, G. M. (2015). SHELXT—Integrated space-group and crystal-structure determination. *Acta Crystallographica Section A. Foundations and Advances, A71,* 3–8. doi:10.1107/S2053273314026370

Sheldrick, G. M. (2015). Crystal structure refinement with SHELXL. *Acta Crystallographica Section C. Structural Chemistry., C71,* 3–8. doi:10.1107/S2053229614026540

17

Solventless synthesis of new 4,5-disubstituted 1,2,3-selenadiazole derivatives and their antimicrobial studies

Aditi A. Jadhav[1], Vaishali P. Dhanwe[1], Prasad G. Joshi[1] and Pawan K. Khanna[1]*

*Corresponding author: Pawan K. Khanna, Nanochemistry Laboratory, Department of Applied Chemistry, Defence Institute of Advanced Technology (DIAT), Ministry of Defence, Govt. of India, Girinagar, Pune 411 025, India
E-mail: pawankhanna2002@yahoo.co.in
Reviewing editor: George Weaver, University of Loughborough, UK

Abstract: Two novel, namely 5-Phenyl-4-methyl-1, 2, 3-selenadiazole **(5h)** and 4-Phenyl-5-propyl-1, 2, 3-selenadiazole **(5i)** along with several other aliphatic and aromatic series of 1,2,3-selenadiazoles were synthesized at room temperature in one step under solventless conditions from the corresponding semicarbazones. All compounds were thoroughly characterized by various spectroscopic tools. The synthesized new and reported 1,2,3-selenadiazoles were found active against bacterial as well as fungal stains when screened for their antimicrobial activity against various pathogenic bacteria and fungi using agar disc diffusion as well as agar well diffusion method. Almost all selenadiazoles showed better antibacterial properties in comparison to established antibiotics like tetracycline. 4-ethyl-5-methyl-1,2,3-selenadiazole showed higher antimicrobial activity amongst the tested selenadiazoles.

Subjects: Biochemistry; Chemical Spectroscopy; Organic Chemistry

Keywords: organoselenium compounds; selenadiazoles; semicarbazones; solventless synthesis; antimicrobial studies

ABOUT THE AUTHOR

Pawan K. Khanna, the corresponding author, completed his PhD in Oganometallic Chemistry of Se & Te from IIT-Bombay in 1989–90. Subsequently he did postdoctoral research with Chris Morely at the Queens' University of Belfast and University of Wales at Swansea (UK). He worked in C-MET, Pune from 1993 to 1995. He was awarded the BOYSCAST fellowship of Govt of India during 1998–99 to work with David Cole-Hamilton at the Univesity of St. Andrews, Scotland. He moved to his current position of professor and head of Department of Applied Chemistry at Defence Institute of Advanced Technology (DIAT) in 2011 where he also served as dean of academics during 2011–2013. He has published over 150 research papers. *He was awarded MRSI medal in 2010 and Researcher of the Year award at DIAT in 2014.* Aditi Jadhav, Vaishali Dhanwe, Prasad Joshi are students of Pawan Khanna and are co-authors of the article. They are all developing organic and nanochemistry expertise.

PUBLIC INTEREST STATEMENT

This paper presents some novel compounds along with different series of heterocyclic organoselenium compounds i.e. 1, 2, 3-selenadiazoles synthesized at room temperature under solventless condition from corresponding semicarbazones and SeO_2. These one-step reaction methods are instant, environment-friendly, and convenient to work with. All compounds presented, are thoroughly characterized using different spectroscopic tools. The 1,2,3-selenadizoles are organoselenium compounds having many applications; which are useful as precursors for material chemistry and nanotechnology for the synthesis of various types of metal selenide nanoparticles. 1,2,3-selenadiazoles are biologically active compounds.

1. Introduction

The current interest in the chemistry of 1,2,3-selenadiazoles is mainly due to their chemical reactivity owing to their soft transformations upon thermolysis (Jadhav, Dhanwe, Joshi, & Khanna, 2015; Junling et al., 2004) and or photolysis that derive free selenium via elimination of a nitrogen molecule leading to formation of alkynes or new heterocycles (Jadhav, Dhanwe, et al., 2015; Meier & Voigt, 1972; Regitz & Krill, 1996; Jadhav & Khanna, 2015). Such new transformations make selenadiazoles and the end compounds thereof useful in pharmaceutical chemistry as well as precursors for coordination (Zhan, Liu, Fang, Pannecouque, & Clercq, 2009; Cervantes-Lee et al., 1998; Morley & Vaughan, 1993; Ford, Khanna, Morley & Vaira, 1999; Khanna & Morley, 1993) and materials chemistry including nanotechnology. (Khanna, 2005; Khanna, Gorte, & Morley, 2003; Khanna et al., 2009; Bhanoth, More, Jadhav, & Khanna, 2014; Jadhav, More, & Khanna, 2015).

The scope of 1,2,3-selenadiazole in pharmaceutical chemistry has widened due to resistance of micro-organisms against chemotherapeutic agents allowing infectious disease to become the second leading cause of death worldwide. Due to fast mutation process, many antibiotics become ineffective against the bacteria. As a result, bacterial resistance against antibiotic treatment is a common phenomenon. 1,2,3-selenadiazoles, in addition to their high-tech applications, have also been extensively studied for their pharmaceutical applications e.g. cytotoxicity and other biological activities. (Al-Smadi & Al-Momani, 2008; Pawar, Burungale, & Karale, 2009; Jalilian, Sattari, Bineshmarvasti, Daneshtalab, & Shafiee, 2003; El-Desoky, Badria, Abozeid, Kandeel, & Abdel-Rahman, 2013; Zhan et al., 2009; Xiao-Chun et al., 2012; Patil, Badami, & Puranik, 1994; Padmavathi, Sumathi, & Padmaja, 2002). Substituted 1,2,3-selenadiazoles and their derivatives have shown excellent antibacterial (Al-Smadi & Al-Momani, 2008; Pawar et al., 2009) antifungal (Jalilian et al., 2003), antitumor (El-Desoky et al., 2013; Xiao-Chun et al., 2012), and antiHIV properties (Zhan et al., 2009). Some of the 1,2,3-selenadiazole derivatives show antihaemostatic (Patil et al., 1994) and insecticidal activities (Padmavathi et al., 2002). Specially it is worth mentioning that thioacetanilides derivatives of 1,2,3-selenadiazoles showed antiHIV activity against HIV-I in MT-4 cells (Zhan et al., 2009). Antifungal study of 1,2,3-selenadiazoles e.g. sulphmoyl derivatives of 4,5-dihydronaphtho[1,2-d][1,2,3]selenadiazoles showed significant activity against *Cryptococcus neoformans.* Multiarm derivatives of 1,2,3-selenadiazoles were found highly active against *E. coli, S. aureus,* and *P. aurogenosa* bacteria (Al-Smadi & Al-Momani, 2008). Human melanoma cells (A375) growth was successfully inhibited by selenadiazole derivative i.e. 5-amino[1,2,5]selenadiazolo[3,4-*d*]pyrimidin-7-ol (El-Desoky et al., 2013). There are large number of such compounds useful as antimicrobial agents (Al-Smadi & Al-Momani, 2008; Pawar et al., 2009; Jalilian et al., 2003). Hence the researchers are continuously in the hunt for new molecules for controlling the bacterial infection timely and more effectively.

In view of excellent materials and biological applications of 1,2,3-selenadiazoles, it is warranted that meaningful studies should be conducted on such molecules to further enrich the knowledge bank so that their utility becomes more relevant. In order to tackle this issue, new 1,2,3-selenadiazole are required to be explored by conventional as well as non-conventional methods. We have recently reported solventless synthesis of cycloalkeno-1,2,3-selenadiazoles and tested their behavior towards several human pathogens (Jadhav, Dhanwe, et al., 2015). In our previous studies, a series of cyclic aliphatic 1,2,3-selenadiazoles were, for the first ever time, tested for their antibacterial activity and it was found that they act rather efficiently against number of microbes. To extend the feasibility of solventless synthesis and the effect of organic functionality via substitutions at 4 and 5 positions in the selenadiazole moiety on their antimicrobial activity, we herein report the synthesis of acyclic aliphatic and aromatic acetophenone derivatives of 1,2,3-selenadiazoles. Traditionally, 1,2,3-selandiazoles have been synthesized from the corresponding semicarbazones via ring closure due to mild oxidation of semicarbazones by selenium dioxide (selenious acid)(Regitz & Krill, 1996; bLabanauskas, Dudutiene, Matulis, & Urbelis, 2009; Lalezari, Shafiee, & Yalpani, 1971). Often the synthesis is based on solution method and except our last report (Jadhav, More, et al., 2015), there has not been any report described using solventless conditions for the preparation of 1,2,3-selenadiazoles. However, the solventless conditions have been occasionally mentioned for other types of selenadiazoles. (Junling et al., 2004). Photochemical sensitivity of reaction and the product alike, coupled with large solvent requirement and extended work-up process, warrants alternative approach to avoid pre-degradation of the compounds. Among the studied compounds, 5-Phenyl-4-methyl-1, 2, 3-selenadiazole **(5h)** and 4-Phenyl-5-propyl-1, 2, 3-selenadiazole **(5i)** have not been reported earlier. This article therefore deals with the two new selenadiazoles along with several other such compounds **(5a-f)** which have not been reported by solvelentless method. Additionally, their antibacterial properties against the chosen microbes are studied in the current work.

2. Experimental

2.1. Chemicals and methods

All chemicals and solvents (reagent or analytical grade) were purchased from Sigma–Aldrich Company and Merck India Ltd. and were used as received. The UV–visible spectra were recorded qualitatively at room temperature in the range of 200–800 nm using Analytik Jena SPECORD 210 PLUS UV spectrophotometer. FTIR spectra were recorded at room temperature in the range of 4000–800 cm^{-1} using FTIR Perkin Elmer spectrum two spectrometer. ^{1}H and ^{13}C NMR spectra (300 and 75 MHz, respectively) recorded on a Bruker DRX-300 instrument in CDCl$_3$, and the chemical shifts were reported relative to TMS as an internal standard. The high-resolution mass spectra were recorded on an Agilent Technologies 6540 UHD Accurate-Mass Q-TOF (LC/MS) spectrometer with electron spray ionization. The melting points were determined in open capillaries. The progress of reactions was monitored by TLC on Merck silica gel 60 F-254 aluminum sheets, eluent EtOAc, visualization UV light.

2.2. Synthesis

2-butanone semicarbazone (4a). The semicarbazone derivatives were synthesized using reported procedure (Al-Smadi & Ratrout, 2004). Yield 18.9 g (72%), white crystals, R_f 0.58, mp 132–134°C (EtOH); Reported 144°C (Ibrahim & Al-Difar, 2011). UV spectrum (EtOH), λ_{max} (nm): 273. IR spectrum, ν (cm^{-1}):3473 (secondary N–H), 3190 (amide N–H), 1684 (C=O), 1587 (C = N), 1111 (NC=O), 2888 (C-H). ^{1}H NMR spectrum, δ (ppm): 1.06–1.11 (3H, m, CH$_3$); 1.84 (3H, s, CH$_3$); 2.22–2.29 (2H, m, CH$_2$); 6.06 (2H, s, NH$_2$); 8.46 (1H, s, NH).^{13}C NMR spectrum, δ (ppm): 10.51 (CH$_3$); 15.22 (CH$_3$); 33.82 (CH$_2$); 151.69 (C = N); 158.36 (C=O). Found, m/z: 130.0997 [M]$^+$. C$_5$H$_{12}$N$_3$O. Calculated, m/z: 130.0995

3-pentanone semicarbazone (4b) was synthesized analogously to compound **4a** from 3-pentanone. Yield 21.20 g (85%), white crystals, R_f 0.54, mp 102–105°C (EtOH) Reported 113°C (Ibrahim & Al-Difar, 2011). UV spectrum (EtOH), λ_{max} (nm): 267. IR spectrum, ν (cm^{-1}): 3446 (secondary N–H), 3234 (amide N–H), 1673 (C=O), 1573 (C = N), 1112 (NC=O), 2955 (C–H). ^{1}H NMR spectrum, δ (ppm): 1.06–1.10 (6H, m, 2CH$_3$); 2.22–2.29 (4H, m, 2CH$_2$); 6.00 (2H, s, NH$_2$); 8.74 (1H, s, NH). ^{13}C NMR spectrum, δ (ppm): 9.60 (CH$_3$); 10.47 (CH$_3$); 22.39 (CH$_2$); 29.29 (CH$_2$); 155.95 (C = N); 158.86 (C=O). Found, m/z: 144.1131 [M + H]$^+$. C$_6$H$_{14}$N$_3$O. Calculated, m/z: 144.1136.

2-pentanone semicarbazone (4c) was synthesized analogously to compound **4a** from 2-pentanone. Yield 22.44 g (90%), white crystals, R_f 0.66, mp 106–108°C (EtOH). UV spectrum (EtOH), λ_{max} (nm): 257. IR spectrum, ν (cm^{-1}): 3472 (secondary N–H), 3219 (amide N–H), 1676 (C=O), 1590 (C = N), 1130 (NC=O), 2962 (C–H). ^{1}H NMR spectrum, δ (ppm): 0.90–0.95 (3H, m, CH$_3$); 1.49–1.64 (2H, m, CH$_2$), 1.84 (3H, m, CH$_3$), 2.17–2.22 (2H, m, CH$_2$); 5.94 (2H, br. s, NH$_2$); 8.69 (1H, s, NH). ^{13}C NMR spectrum, δ (ppm): 16.5 (CH$_3$); 20.21 (CH$_2$); 33.45 (CH$_2$); 19.45 (CH$_3$); 150.00 (C = N); 158.12 (C=O). Found, m/z: 144.1137 [M + H]$^+$. C$_6$H$_{14}$N$_3$O. Calculated, m/z: 144.1136

Isopropyl methyl semicarbazone (4d) was synthesized analogously to compound **4a** from 2-butanone. Yield 19.45 g (78%), white crystals, R_f 0.60, mp 110–115°C (EtOH). UV spectrum (EtOH), λ_{max} (nm): 262. IR spectrum, ν (cm^{-1}): 3472 (secondary N–H), 3179 (amide N–H), 1667 (C=O), 1574 (C = N), 1127 (NC=O), 2877 (C–H). ^{1}H NMR spectrum, δ (ppm): 1.06–1.11 (6H, m, 2CH$_3$); 1.84 (3H, S, CH$_3$); 2.21–2.32 (1H, m, CH); 6.18 (2H, s, NH$_2$); 8.35 (1H, s, NH). ^{13}C NMR spectrum, δ (ppm): 18.01 (2CH$_3$); 22.23

(CH$_3$); 38.01 (CH); 148.05 (C = N); 158.41 (C=O). Found, m/z: 144.1132 [M + H]$^+$. C$_6$H$_{14}$N$_3$O. Calculated, m/z: 144.1136.

Isobutyl methyl semicarbazone (4e) was synthesized analogously to compound **4a** from isobutyl methyl ketone. Yield 14.54 g (61.78%), white crystals, R_f 0.56, mp 130–135°C (EtOH). UV spectrum (EtOH), λ_{max} (nm): 259. IR spectrum, ν (cm^{-1}): 3424 (secondary N–H), 324 (amide N–H), 1675 (C=O), 1537 (C = N), 1134 (NC=O), 2871 (C–H). ^{1}H NMR spectrum, δ (ppm): 0.90–0.95 (6H, m, CH$_3$); 1.49–1.61 (1H, m, CH); 1.84 (3H, s, CH$_3$); 2.17–2.22 (2H, m, CH$_2$); 6.04 (2H, s, NH$_2$); 8.68 (1H, s, NH). ^{13}C NMR spectrum, δ (ppm): 18.23 (2CH$_3$); 22.54 (CH$_2$); 38.21 (CH); 20.12 (CH$_3$); 150.11 (C = N); 157.31 (C=O). Found, m/z: 158.1302 [M + H]$^+$. C$_7$H$_{16}$N$_3$O. Calculated, m/z: 158.1300.

Acetophnone semicarbazone (4f) was synthesized analogously to compound **4a** from acetophenone. Yield 18.14 g (82%), (Al-Smadi & Ratrout, 2004)

4-hydroxy acetophenone semicarbazone (4g) was synthesized analogously to compound **4a** from 4-hydroxy acetophenone. Yield 14.9 g (70%), (Al-Smadi & Ratrout, 2004)

Valerophenone semicarbazone (4h) was synthesized analogously to compound **3a** from 2-butanone. Yield 16.3 g (82%), white crystals, R_f 0.58, mp 163–165°C (EtOH). UV spectrum (EtOH), λ_{max} (nm): 270. IR spectrum, ν, cm^{-1}: 3453 (secondary N–H), 3188 (amide N–H), 3135 (Ar. H) 1678 (C=O), 1602 (Ar. C=C), 1575 (C = N), 1175 (NC=O), 2975 (C–H). ^{1}H NMR spectrum, δ (ppm): 0.91–0.93 (3H, m, CH$_3$); 1.21–1.26 (2H, m, CH$_2$); 1.58–1.64. (2H, m, CH$_2$); 2.70–2.73(2H, m, CH$_2$); 7.43–7.47(2H, m, Ar.CH); 7.72–7.75(3H, m, Ar.CH); 6.21(2H, s, NH$_2$); 8.90 (1H, s, NH). ^{13}C NMR spectrum, δ (ppm): 10.41 (CH$_3$); 27.2 (CH$_2$); 28.3 (CH$_2$); 29.78 (CH$_2$); 123.27, 127.03, 140.80 (Ar Cs); 144.54 (C = N); 158.65 (C=O). Found, m/z: 220.1469 [M + H]$^+$. C$_{12}$H$_{18}$N$_3$O. Calculated, m/z: 220.1467.

Propeophenone semicarbazone (4i) was synthesized analogously to compound **4a** from propeophenone. Yield 11 g (77.19%), white crystals, R_f 0.55, mp 165–168°C (EtOH). UV spectrum (EtOH), λ_{max} (nm): 260. IR spectrum, ν (cm^{-1}): 3451 (secondary N–H), 3199 (amide N–H), 3140 (Ar. H) 1689 (C=O), 1602 (Ar. C=C), 1575 (C = N), 1198 (NC=O), 2973 (C–H). ^{1}H NMR spectrum, δ (ppm): 1.18–1.22 (3H, m, CH$_3$); 2.69–2.72 (2H, m, CH$_2$); 7.48–7.52 (2H, m, Ar.CH); 7.70–7.74 (3H, m, Ar.CH); 6.2 (2H, s, NH$_2$); 8.60 (1H, s, NH). ^{13}C NMR spectrum, δ (ppm): 10.23 (CH$_3$); 20.12 (CH$_2$); 126.34, 128.43, 130.42 (ArCs); 150.16 (C = N); 158.40 (C=O). Found, m/z: 192.1134 [M + H]$^+$. C$_{10}$H$_{14}$N$_3$O. Calculated, m/z: 192.1132.

4-ethyl-1,2,3-selenadiazole (5a) (General Method). 2-butanone semicarbazone (**4a**) (5g, 0.029 mol) and selenium dioxide (3.2 g, 0.029 mol) were ground together in a mortar pestle at room temperature for around 20 min. The process was monitored by TLC using hexane–AcOEt, 7:3, as solvent system. The crude product was dissolved in 100 ml toluene and filtered. The filtrate was evaporated using a rotary evaporator. The product was purified by column chromatography on silica gel (60–120 mesh), using petroleum ether (bp 60–80°C)—toluene, 7:3, as eluent. Yellow liquid product was collected as a final product which is characterized by UV–visible, FTIR, ^{1}H and ^{13}C NMR spectroscopy, and mass spectrometry. Yield 3.12 g (50%), UV spectrum (toluene), λ_{max} (nm): 341. R_f 0.67. IR spectrum, ν (cm^{-1}): 2921–2847 (C–H), 1620 (C=C), 1454 (C–H), 1302 (C–N). ^{1}H NMR spectrum, δ (ppm), J(Hz): 1.45 (3H, J = 5, t, CH$_3$); 3.27 (2H, m, CH$_2$); 8.86 (1H, s, selenadiazole ring H). ^{13}C NMR spectrum, δ (ppm): 14.13 (CH$_3$); 29.99 (CH$_2$); 137.06 (C=C-N); 165.26 (C=C-Se). Found, m/z: 162.9776 [M + H]$^+$. C$_4$H$_7$N$_2$Se. Calculated, m/z: 162.9773.

4-ethyl-5-methyl-1, 2, 3-selenadiazole (5b) was synthesized analogously to compound **5a**. Yield 3.11 g (51%), yellow liquid, R_f 0.70. UV spectrum (toluene), λ_{max} (nm): 327. IR spectrum, ν (cm^{-1}): 2923–2870 (C–H), 1585 (C=C), 1445 (C–H δ), 1291 (C–N s). ^{1}H NMR spectrum, δ (ppm), J (Hz): 0.85–0.87 (3H, m, J = 5.0, CH$_3$); 2.04–2.07 (2H, m, J = 5.0, CH$_2$); 1.80 (3H, s, CH$_3$); ^{13}C NMR spectrum, δ (ppm): 16.16 (CH$_3$-CH2); 22.07 (CH$_2$); 17.86 (CH$_3$); 153.35 (C=C-N); 161.06 (C=C-Se). Found, m/z: 176.9927 [M + H]$^+$. C$_5$H$_9$N$_2$Se. Calculated, m/z: 176.9926.

4-propyl-1, 2, 3-selenadiazole (5c) was synthesized analogously to compound **5a**. Yield 2.9 g (48%), yellow liquid, R_f 0.78. UV spectrum (toluene), λ_{max} (nm): 341. IR spectrum, ν (cm^{-1}): 2917–2850 (C--H), 1645 (C=C), 1445 (C–H), 1300 (C–N). ^{1}H NMR spectrum, δ (ppm), J (Hz): 0.87 (3H, t, J = 5.0, CH$_3$); 1.71–1.77 (2H, m, J = 5, CH$_2$); 3.06 (2H, t, J = 5, CH$_2$); 8.80 (1H, s, selenadiazole ring H). ^{13}C NMR spectrum, δ (ppm): 13.32 (CH$_3$); 22.71 (CH$_2$-CH$_3$); 31.06 (CH$_2$-CH$_2$); 137.59 (C=C-N); 163.14 (C=C-Se). Found, m/z: 176.9928 [M + H]$^+$. C$_5$H$_9$N$_2$Se. Calculated, m/z: 176.9926.

4-iso-propyl-1, 2, 3-selenadiazole (5d) was synthesized analogously to compound **5a**. Yield 3.29 g (54%), yellow liquid, R_f 0.65. UV spectrum, λ_{max} (nm): 322. IR spectrum, ν, cm^{-1}: 2923–2869 (C–H), 1620 (C=C), 1462 (C–H), 1300 (C–N). ^{1}H NMR spectrum, δ (ppm), J (Hz): 1.35 (6H, m, CH$_3$); 3.45–3.53 (H, m, CH); (1H, S, heterocyclic ring H). ^{13}C NMR spectrum, δ (ppm): 22.58 (CH$_3$-CH-CH$_3$); 29.42 (CH); 135.87 (C=C-N); 169.65 (C=C-Se). Found, m/z: 176.9930 [M + H]$^+$. C$_5$H$_9$N$_2$Se. Calculated m/z: 176.9929.

4-isobutyl-1, 2, 3-selenadiazole (5e) was synthesized analogously to compound **5a**. Yield 3.12 g (52%), yellow liquid, R_f 0.78. UV spectrum (toluene), λ_{max} (nm): 325. IR spectrum, ν (cm^{-1}): 2923–2867 (C–H), 1633 (C=C), 1462 (C–H), 1292 (C–N). ^{1}H NMR spectrum, δ (ppm), J (Hz): 0.80 (6H, t, J = 5, CH$_3$); 1.99–2.06 (H, m, CH); 2.90 (2H, t, J = 5, CH$_2$); 8.71 (1H, s, selenadiazole ring H). ^{13}C NMR spectrum, δ (ppm): 21.75 (CH$_3$-CH-CH$_3$); 28.66 (CH); 37.86 (CH$_2$); 138.24 (C=C-N); 161.98 (C=C-Se). Found, m/z: 191.0089 [M + H]$^+$. C$_6$H$_{11}$N$_2$Se. Calculated, m/z: 191.0087.

4-(4-hydroxyphenyl) -1, 2, 3-selenadiazole (5f) was synthesized analogously to compound **5a**. Yield 2.61 g (45%), reddish solid, R_f 0.42. (Al-Smadi & Ratrout, 2004)

5-phenyl-1, 2, 3-selenadiazole (5g) was synthesized analogously to compound **5a**. Yield 2.81 g (48%), reddish solid, R_f 0.57. (Al-Smadi & Ratrout, 2004)

4-phenyl-5-propyl-1, 2, 3-selenadiazole (5h) was synthesized analogously to compound **5a**. Yield 3.3 g (48%), yellow viscous solid, R_f 0.62. UV spectrum (toluene), λ_{max} (nm): 327. IR spectrum, ν (cm^{-1}): 3027–3031(Ar C–H), 2853–2927 (C–H), 1600 (Ar C=C), 1660 (C=C), 1449 (C–H), 1227 (C–N). ^{1}H NMR spectrum, δ (ppm): 7.45–7.52 (3H, m, Ar CH); 7.68–7.70 (2H, m, Ar CH); 0.99–1.02 (3H, m, CH$_3$); 1.75–1.80 (2H, m, CH$_2$); 3.05–3.09 (2H, m, CH$_2$);. ^{13}C NMR spectrum, δ (ppm): 13.91 (CH3- CH$_2$- CH$_2$); 27.87 (CH$_3$- CH2- CH$_2$); 31.00 (CH$_3$- CH$_2$- CH2); 127.91, 128.35, 128.58, 129.41 (Ar Cs); 132.17 (C=C-N); 162.45 (C=C-Se). Found, m/z: 253.0244 [M + H]$^+$. C$_{11}$H$_{13}$N$_2$Se. Calculated, m/z: 253.0243.

5-phenyl-4-methyl-1, 2, 3-selenadiazole (5i) was synthesized analogously to compound **5a**. Yield 2.07 g (45%), pinkish solid, R_f 0.45. mp 130–132°C. UV spectrum (toluene), λ_{max} (nm): 300. IR spectrum, ν (cm^{-1}): 3078–2993 (Ar C–H), 2915 (C–H), 1595 (Ar C=C), 1658 (C=C), 1445 (C–H), 1285 (C–N). ^{1}H NMR spectrum, δ (ppm) : 2.59 (3H, s CH$_3$), 6.94–6.99 (3H, m, Ar CH); 7.87–7.90 (2H, m, Ar CH). ^{13}C NMR spectrum, δ (ppm): 31.58 (CH$_3$); 127.76, 128.35, 129.15, 132.39 (Ar Cs); 136.66 (C=C-N); 156.09 (C=C-Se). Found, m/z: 224.9934 [M + H]$^+$. C$_9$H$_9$N$_2$Se. Calculated, m/z: 224.9930.

2.3. Antimicrobial study

The antimicrobial study was carried out at Kulkarni Laboratory and Quality Management Services at Pune, India. Microbial strains used in the study were clinical isolates of bacteria *Staphylococcus aureus*, *Escherechia coli*, *Bacillus subtilis*, and *Pseudomonas aeruginosa*, as well as fungi *Aspergillus niger* and *Penicillium notatum*. Muller Hinton agar media (gm/litre) was used for media preparation. For 1000 ml Muller Hinton agar preparation peptone 5 g, sodium chloride 8 gm, beef infusion 3 gm, and agar 16 gm were weighed and dissolved in 1000 ml of distilled water and maintained pH 7.3–7.4 which was sterilized by autoclaving at 121°C for 15 min at 15 psi pressure.

All synthesized compounds **5a-i**, were screened for their antibacterial activity against Gram-positive and Gram-negative bacteria at concentration 0.0049 g/ml using two different methods involving agar disc diffusion and agar well method with Muller Hinton agar media (Table 1).The

Table 1. Sensitivity of human pathogenic microbes to 1,2,3-selenadiazoles 5a–i using the agar disc and agar well diffusion method

Sample codes	Conc. (g/ml)	Diameter of zone of inhibition (mm)									
		S. aureus		P. aeruginosa		E. coli		B. subtillis		A. nigar	P. notatum
		D.D.[1]	W.D.[2]	D.D.	WD	D.D.	W.D.	D.D.	W.D.	W.D.	W.D.
5a	0.0049	17	22	12	20	12	15	20	30	20	23
5b	0.0049	16	23	15	22	13	15	23	27	22	24
5c	0.0049	17	21	12	20	17	18	17	30	21	23
5d	0.0049	15	20	11	15	10	17	20	25	18	23
5e	0.0049	14	17	11	15	9	16	17	22	17	20
5f	0.0049	15	18	12	17	-	15	16	20	10	15
5g	0.0049	11	17	11	12	12	15	15	17	10	14
5h	0.0049	12	16	10	16	10	18	11	17	9	10
5i	0.0049	10	15	10	15	10	17	10	17	8	9
Tetracycline		20	12	12	23	-	-				

[1]D. D.: Disc diffusion method.

[2]W. D.: Well diffusion method; (–) not done.

selected bacterial suspension was spread on the surface of Muller Hinton agar plates respectively. In case of agar disc diffusion method, the synthesized compounds were impregnated on Whatman No. 1 filter paper disc (6 mm diameter) at a concentration of 0.0049 g/ml. Each disc is coded with the name of the agent. Discs were placed on solidified media and allowed to diffuse for 5 min; the plates were kept for incubation at 37°C for 24 h for bacteria. Dimethylsulfoxide (DMSO) was used as the control. At the end of incubation, antibacterial activity was determined by measuring zone of inhibition in mm around each of the disc and compared with standard DMSO.

To check the antimicrobial activity of the samples against test organisms by well diffusion method, freshly prepared nutrient agar was seeded with 1% inoculums of each test organism. The 8-mm diameter wells were cut with the help of cork borer. Each well was then filled with sample (approximately 100 μl). All the plates were incubated at 37°C for 24 h. The zone of inhibition around each disc was measured after completion of incubation time.

Antifungal activity of the samples: Similar procedure was used to check the antifungal activity of the test samples. Fungal strains used for these studies are *Aspergillus niger* and *Penicillium notatum*. Potato Dextrose agar was used to check the activity. The fungal spores was dispensed in sterile saline with 1% Tween 80 and seeded to PDA media. The 8-mm diameter wells were cut with the help of cork borer. Each well was then filled with sample. All the plates were incubated at room temperature for five days. The zone of inhibition around each disc was measured after completion of incubation time.

3. Results and discussion

In the current work, we present the synthesis of two novel 1,2,3-selenadiazoles (**5h,i**) and comparative study of antimicrobial behavior of different series of 1,2,3-selenadiazoles. The synthesized compounds were tested against two Gram-positive and Gram-negative bacteria as well as fungi species by two different methods of antimicrobial screening. We herein present a solventless method for preparation of 1,2,3-selenadiazoles from the respective semicarbazones. Semicarbazones **4a–i** can be easily synthesized by a reported procedure (Al-Smadi & Ratrout, 2004) from the respective ketones **3a–i** and semicarbazide hydrochloride **1**. We have earlier opined that a solventless cyclization process is a rapid method in comparison to the solution chemistry and requires lesser time with reduced or negligible amount of solvent for the isolation of product. In our experiments, the solventless reaction of semicarbazones **4a-i** with selenium dioxide requires hand grinding only and could be completed within 30–60 min producing 1,2,3-selenadiazoles **5a-i** in moderate yields. Additionally, during synthesis it was observed that solventless method is more feasible for the synthesis of aliphatic derivatives of 1,2,3-selenadiazoles **5a-e** compared to aromatic derivatives **5f-i**. It is further noted that compounds **5f-i** need longer grinding (reaction) time with low yield compared to compounds **5a-e**.

The structures of various semicarbazones **3a-i** and selenadiazoles **5a-i** were confirmed by UV–visible, FTIR, ^{1}H and ^{13}C NMR spectroscopy, and mass spectrometry. UV–visible spectra of 1,2,3-selenadiazoles **5a-i** showed absorption bands $\lambda_{(abs)}$ at ~300–350 nm due to π–π* electronic transition of C=C in conjugation with N = N in which the N and C are attached to the Se atom in heterocyclic ring. The longer wavelength absorption was observed due to extensive conjugation along with the presence of Se in heterocyclic ring.

FTIR spectra of compounds **5a-i** were recorded as KBr pellets. Figure SI 1 shows IR spectrum of 4-phenyl-5-propyl-1, 2, 3-selenadiazole. Generally, peaks at 2951–2848 cm^{-1} are due to C–H stretching mode of vibrations and at 1648–1627 cm^{-1} for C=C stretching mode of vibrations (due to Se-C=C–N moiety in 1,2,3-seleadiazoles).The IR transmission band at 1482–1436 cm^{-1} was assigned to C–H deformation mode vibrations, and the band at 1303–1286 cm^{-1} is assigned to C–N stretching mode of vibrations. Whereas, compounds having aromatic moieties showed additional bands at 3130–3080 cm^{-1} and 1580–1600 cm^{-1} which can be assigned to Ar C–H and C=C stretch, respectively. FTIR spectra show obvious variations in Ar C–H, C=C stretch along with other stretching modes of vibrations because of varying organic substituents at R and R' groups. The ^{1}H NMR spectra of 1,2,3-selenadiazoles **5a-i** in CDCl$_3$ showed different chemical shifts for CH, CH$_2$, and CH$_3$ protons in the range of δ 1.10–3.25 ppm for alkyl groups. For protons associated with the heterocyclic ring chemical shifts are observed in the range of δ 8.70–9.50 ppm. Similarly, compounds having aromatic moieties showed chemical shifts in the range of δ 6.90–8.3 ppm for aromatic ring protons. Likewise, the ^{13}C NMR spectrum of 1,2,3-selenadiazoles **5a-i** showed the expected number of signals due to different carbon atoms in the molecules. Chemical shift at around δ 130–137 ppm and δ 157–165 ppm are assigned to (C=C–N) and (C=C–Se) heterocyclic ring carbon atoms, respectively. For typical understanding the ^{1}H NMR and ^{13}C NMR spectrum of 4-phenyl-5-propyl-1, 2, 3-selenadiazole is shown in Figure SI2 and 3. Mass spectra of 1, 2, 3-selenadiazoles showed peaks with a set of isotopic components, characteristics of the presence of selenium which has 8 naturally occurring isotopes with atomic mass 72–82 out of which only 6 isotopes are stable. The m/z value of all the compounds

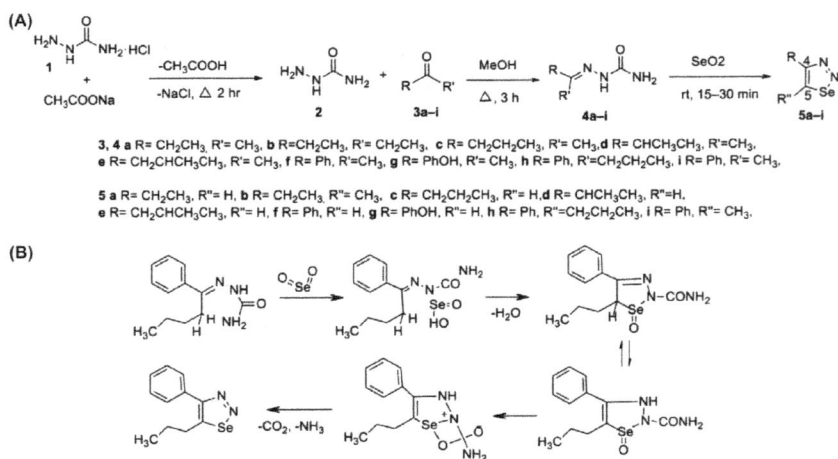

(A)

3, 4 a R= CH$_2$CH$_3$, R'= CH$_3$, b R=CH$_2$CH$_3$, R'= CH$_2$CH$_3$, c R= CH$_2$CH$_2$CH$_3$, R'= CH$_3$,d R= CHCH$_3$CH$_3$, R'=CH$_3$, e R= CH$_2$CHCH$_3$CH$_3$, R'= CH$_3$, f R= Ph, R'=CH$_3$, g R= PhOH, R'= CH$_3$, h R= Ph, R'=CH$_2$CH$_2$CH$_3$, i R= Ph, R'= CH$_3$.

5 a R= CH$_2$CH$_3$, R''= H, b R= CH$_2$CH$_3$, R''= CH$_3$, c R= CH$_2$CH$_2$CH$_3$, R''= H,d R= CHCH$_3$CH$_3$, R''=H. e R= CH$_2$CHCH$_3$CH$_3$, R''= H, f R= Ph, R''= H, g R= PhOH, R''= H, h R= Ph, R''=CH$_2$CH$_2$CH$_3$, i R= Ph, R''= CH$_3$.

(B)

Scheme 1. **(A) Synthesis of semicarbazones and their respective 1,2,3-selenadiazoles, (B) typical mechanism for synthesis of 4-phenyl-5-propyl-1, 2, 3-selenadiazole.**

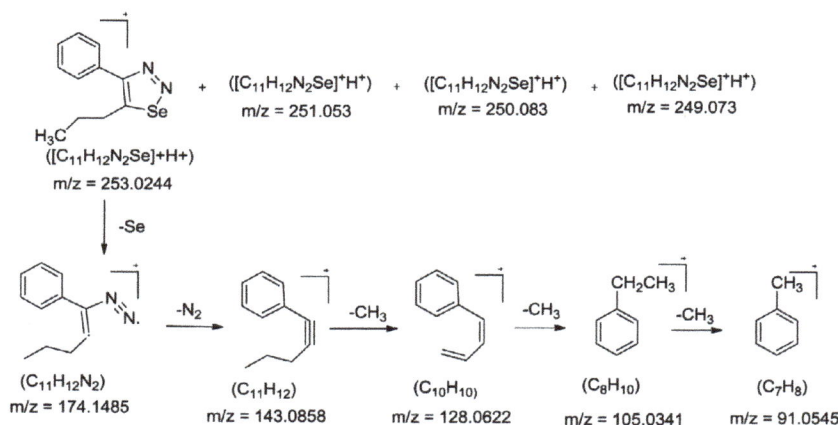

Scheme 2. Mass fragmentation pattern of 4-Phenyl-5-propyl-1, 2, 3-selenadiazole (5h).

corresponded to the respective protonated molecular ions. To study the fragmentation pattern through MS–MS, analysis of some samples were carried out. The mass fragmentation pattern of 4-Phenyl-5-propyl-1, 2, 3-selenadiazole is shown in scheme 2 (Figure SI 4a,b). In a typical fragmentation analysis, molecular ion peak undergoes initial breakdown by loss of selenium followed by elimination of a nitrogen molecule to form asymmetric alkynes. Alkyne so generated may further dissociate stepwise by loss of a methyl group to eventually result in the formation of toluene molecule.

3.1. Antimicrobial studies

The activity of synthesized selenadiazoles **5a–i** was tested against some human pathogenic microbes. For studying their activity, two Gram-negative (*Escherichia coli* and *Pseudomonas aeruginosa*) and two Gram-positive (*Staphylococcus aureus* and *Bacillus subtillis*) species were selected. For more authentication antibacterial activity of compounds were tested by two different methods i.e. agar disc diffusion method and agar well diffusion method. All the compounds described in this text were also screened for antifungal activity against *Aspergillus niger* and *Penicillium notatum*. From the results obtained by these methods, it was found that agar well method compared to agar disc diffusion method gives better results and thus can be considered more suitable for selenadiazoles. Encouraged with the suitability of agar well method for antibacterial screening, antifungal activity testing was also performed following the same method. Some of the tested compounds were highly active even at concentrations as low as 4.9 mg/ml (Table 1, Figure 1). For example, compound **5a** showed good inhibition against the highly resistant *Pseudomonas aeruginosa*. Generally, the

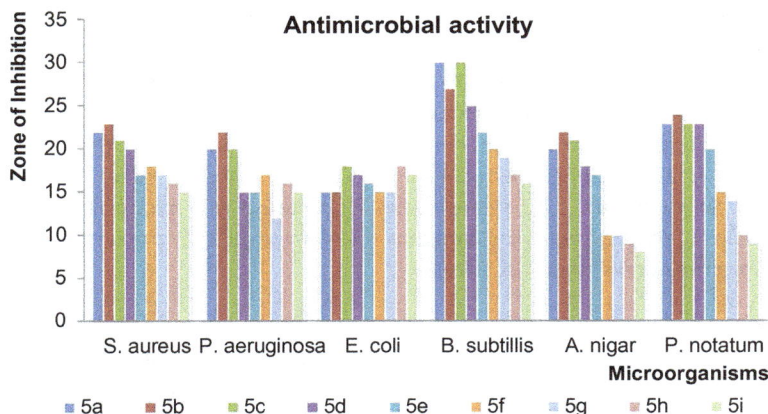

Figure 1. Sensitivity of 1,2,3-selenadiazoles 5a–i against human pathogenic microbes.

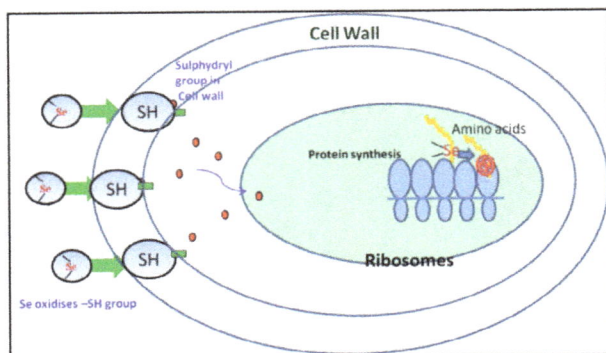

Figure 2. Proposed mechanism of bacterial cell growth inhibition by 1,2,3-selenadiazoles.

extracts of all selenadiazoles **5a–i** in dimethyl sulfoxide (DMSO) were active against all the tested pathogens, in the range of 8–25 mm diameter inhibition zone. DMSO used as control which showed no activity against any of the tested pathogens. On the basis of the presented data, it can be concluded that the tested 1,2,3- selenadiazoles **5a–i** possess good antimicrobial properties compared to tested antibiotic drugs such as tetracycline. Tetracycline is well-known antibiotic for the inhibition of microbial cell growth as it suppresses protein synthesis. Tetracycline showed zone of inhibition against tested bacteria in the range of (20–23 mm) where many of the selenadiazoles showed better zones inhibition than the drug. It is further observed that 1,2,3-selenadiazoles **5a–i** showed maximum activity against Gram-positive *Bacillus subtillis* and minimum activity against Gram-negative *E.Coli*. Amongst various selenadiazoles tested, compound **5b** showed maximum inhibition against all the pathogens. Selenadiazoles in which R and R' both groups are aliphatic showed better zone of inhibition in comparison to selenadiazoles in which R is an aromatic group. Within aliphatic series compounds, it was observed that compounds having branched chain alkyl groups showed lesser zone of inhibition compared to compounds having straight chain alkyl groups. The probable reason for variation in activity among selenadiazoles could be due to the variation in their stability and solubility. Since aliphatic series compounds are less stable, release of Se from such compounds is more efficient, thus making them more effective. However due to higher stability and less solubility aromatic series compounds in DMSO, they show marginally poor activity against the tested pathogens.

3.2. Mechanism of action
It has been reported that the cause of suppression of bacterial growth in presence of antibacterial agents could be due to several factors e.g. interference of compounds with cell wall synthesis, inhibition of protein synthesis by binding to various ribosomal subunits, suppression of nucleic acid synthesis, disturbances to metabolic pathway, and disruption of bacterial membrane structure. It is opined that selenadiazoles may show good antimicrobial activity due to the presence of diazole fragment along with selenium. There are a few selenium-containing organic compounds with a reported antibacterial activity (Radhakrishna et al., 2010). It can be postulated that selenium might partially replace sulfur in sulfur-containing amino acids present in the bacterial cells (Figure 2) and inhibit their growth due to toxicity of selenium compounds and metabolic disturbances to the microorganisms (Li, Liu, Wu, Liang, & Qu, 2002; Bemheim & Klein, 1941). In the present series of compounds, initial elimination of selenium from 1,2,3-selenadizoles may interact with microbes to alter their bioactivities. Such studies therefore, clearly indicate that selenium plays an important role for these compounds to enhance their antimicrobial activity.

4. Conclusions
In conclusion, synthesis and characterization of different series of other 1,2,3-selenadiazoles by solventless method are presented. Fragmentation pathway for some new molecules was ascertained by m/z spectroscopy. All the selenadiazoles were tested for their antimicrobial activity to highlight

that they showed excellent antimicrobial activity against wide range of bacteria and fungi. The instant formation of such organoselenium compounds will likely promote their application in pharmaceutical chemistry, biochemistry, microbiology.

Acknowledgements
Aditi A. Jadhav thanks Vice Chancellor (DIAT) for PhD fellowship. P. K. Khanna thanks DST, Govt. of India for research grant through project No. SR/S1/PC-39/2010.

Funding
The authors received no direct funding for this research.

Author details
Aditi A. Jadhav[1]
E-mail: aditijadhav@ymail.com
Vaishali P. Dhanwe[1]
E-mail: vdhanwe@gmail.com
Prasad G. Joshi[1]
E-mail: joshi.p13@gmail.com
Pawan K. Khanna[1]
E-mail: pawankhanna2002@yahoo.co.in
[1] Nanochemistry Laboratory, Department of Applied Chemistry, Defence Institute of Advanced Technology (DIAT), Ministry of Defence, Govt. of India, Girinagar, Pune 411 025, India.

References
Al-Smadi, M., & Al-Momani, F. (2008). Synthesis, characterization and antimicrobial activity of new 1,2,3-selenadiazoles. *Molecules, 13*, 2740–2749. doi:10.3390/molecules13112740

Al-Smadi, M., & Ratrout, S. (2004). New 1,2,3-selenadiazole and 1,2,3-thiadiazole derivatives. *Molecules, 9*, 957–967. doi:10.3390/91100957

Bemheim, F., & Klein, J. R. (1941). Action of sodium selenite on the oxidation of L-proline. *Journal of Biological Chemistry, 139*, 827–833.

Bhanoth, S., More, P. V., Jadhav, A., & Khanna, P. K. (2014). Core–shell ZnSe–CdSe quantum dots: A facile approach *via* decomposition of cyclohexeno-1,2,3-selenadiazole. *RSC Advances, 4*, 17526–677. doi:10.1039/C4RA00676C

Cervantes-Lee, F., Párkányi, L., Kapoor, R. N., Mayr, A. J., Pannell, K. H., Pang, Y., & Barton, T. J. (1998). De camethylpentasilacycloheptyne·Mo$_2$(CO)$_4$(η^5-C$_5$H$_5$)$_2$ and cycloheptyne·Mo$_2$(CO)$_4$(η^5-C$_5$H$_5$)$_2$. *Journal of Organometallic Chemistry, 562*, 29–33. doi:10.1016/S0022-328X(98)00367-2

El-Desoky, S. I., Badria, F. A., Abozeid, M. A., Kandeel, E. A., & Abdel-Rahman, A. H. (2013). Synthesis and antitumor studies of novel benzopyrano-1,2,3-selenadiazole and spiro[benzopyrano]-1,3,4-thiadiazoline derivatives. *Medicinal Chemistry Research, 22*, 2105–2114. doi:10.1007/s00044-012-0201-0

Ford, S., Khanna, P. K., Morley, C. P., & Vaira, M. Di (1999). Dinuclear diselenolenes derived from cycloalkeno-1,2,3-selenadiazoles and tetrakis(triphenylphosphine) palladium. *Journal of Chemical Society, Dalton Transaction*, 791–794, doi:10.1039/A807666I

Ibrahim, M. N., & Al-Difar, H. A. (2011). Synthesis and antibacterial activity of semicarbazone derivatives of some carbonyl compounds. *Der Chemica Sinica, 2*, 171–173. Retrieved from www.pelagiaresearchlibrary.com

Jadhav, A. A., Dhawne, V. P., Joshi, P. G., & Khanna, P. K. (2015). An efficient solventless synthesis of cycloalkeno-1,2,3-selenadiazoles, their antimicrobial studies, and comparison with parent semicarbazones. *Chemistry of Heterocyclic Compounds, 51*, 102–106.

Jadhav, A. A., & Khanna, P. K. (2015). Impact of microwave irradiation on cyclo-octeno-1,2,3-selenadiazole: Formation of selenium nanoparticles and their polymorphs. *RSC Adv, 5*, 44756–44763. http://dx.doi.org/10.1039/C5RA05701A

Jadhav, A. A., More, P. V., & Khanna, P. K. (2015). Rapid microwave synthesis of white light emitting magic sized clusters of CdSe: Role of oleic acid. *RSC Advances, 5*, 76733–76742. doi:10.1039/c5ra14625a

Jalilian, A. R., Sattari, S., Bineshmarvasti, M., Daneshtalab, M., & Shafiee, A. (2003). Synthesis and *in vitro* antifungal and cytotoxicity evaluation of substituted 4,5-dihydronaphtho[1,2-*d*][1,2,3]thia(or selena)diazoles. *Il Farmaco, 58*, 63–68. doi:10.1016/S0014-827X(02)00029-0

Junling, Z., Wenjie, Z., Jiahao, Z., Fang, Y., Yan, B., & Yiqun, L. (2004). Solid-state synthesis of heterocyclic aromatic selenium compounds at room temperature. *Chemical Journal on Internet (CJI), 6*,97. Retrieved from www.chemistrymag.org/cji/2004/06c097ne.htm

Khanna, P. K. (2005). Materials chemistry of 1,2,3-Selenadiazoles. *Phosphorus, Sulfur Silicon Related Elements, 180*, 951. doi:10.1080/10426500590906526

Khanna, P. K., & Morley, C. P. (1993). Synthesis and characterisation of new 2-bromovinyl selenides and their platinum group metal complexes. *Journal of Organometallic Chemistry, 450*, 109–114. doi:10.1016/0022-328X(93)80145-2

Khanna, P. K., Gorte, R. M., & Morley, C. P. (2003). Synthesis of cadmium selenide by use of selenadiazol in ethylene glycol—A greener source of selenium. *Materials Letters, 57*, 1464–1466. doi:10.1016/S0167-577X(02)01021-2

Khanna, P. K., More, P. V., Shewate, R., Beri, R. K., Viswanath, A. K., Singh, V., & Mehta, B. R. (2009). Preparation of CdSe quantum dots via thermolysis of a novel single source Cd/Se precursor derived from cyclohexeno-1,2,3-selenadiazole. Chemistry Letters, 38, 676. doi:10.1246/cl.2009.676

Labanauskas, L., Dudutiene, V., Matulis, D., & Urbelis, G. (2009). Synthesis of a new heterocyclic system: 3-phenylbenzimidazo[1,2-c]-[1,2,3]selenadiazole. *Chemistry of Heterocyclic Compounds, 45*, 1153–1154. doi:10.1007/s10593-009-0391-7 [Khim. Geterotsikl. Soedin. 2009, 1435.]

Lalezari, I., Shafiee, A., & Yalpani, M. (1971). 1,2,3-Selenadiazole and its derivatives. *Journal of Organic Chemistry, 36*, 2836–2838. doi:10.1021/jo00818a023

Li, X., Liu, Y., Wu, J., Liang, H., & Qu, S. (2002). Microcalorimetric study of *Staphylococcus aureus* growth affected by selenium compounds. *Thermochimica Acta, 387*, 57–61. doi:10.1016/S0040-6031(01)00825-5

Meier, H., & Voigt, E. (1972). Bildung und fragmentierung von cycloalkeno-1,2,3-selenadiazolen. *Tetrahedron, 28*, 187–198. http://dx.doi.org/10.1016/0040-4020(72)80068-1

Morley, C. P., & Vaughan, R. R. (1993). Synthesis and characterisation of cyclopentadienylcobalt and pentamethylcyclopentadienylcobalt diselenolenes. *Journal of Chemical Society Dalton Transaction, 5*, 703–707. doi:10.1039/DT9930000703

Padmavathi, V., Sumathi, R. P., & Padmaja, A. (2002). Effect of bifunctional heterocyclic sulfones on *Spodoptera litura*. *Journal of Ecobiology, 14*, 9–12.

Patil, B. M., Badami, B. V., & Puranik, G. S. (1994). *Indian Journal of heterocyclic chemistry 3*, 193.

Pawar, M. J., Burungale, A. B., & Karale, B. K. (2009). Synthesis and antimicrobial activity of spiro[chromeno[4,3-*d*][1,2,3] thiadiazole-4,1'-cyclohexane, spiro[chromeno[4,3*d*][1,2,3] selenadiazole-4,1'-cyclohexane and spiro[chroman-2,1'-cyclohexan]-4-one-5-spiro-4-acety2-(acetylamino)-Δ-1,3,4-thiadiazoline compounds *ARKIVOC,* (11), 97–107.

Radhakrishna, P. M., Sharadamma, K. C., Vagdevi, H. M., Abhilekha, P. M., Mubeen, S. R., & Nischal, K.

(2010). Synthesis and antibacterial activity of novel organoselenium compounds. *International Journal of Chemistry, 2,* 149–154. doi:10.5539/ijc.v2n2p149

Regitz, M., & Krill, S. (1996). 1,2,4-Selenadiphospholes—Novel heterocyclic compounds containing low-coordinated phosphorus atoms 1. *Phosphorus, Sulfur,*

and Silicon and the Related Elements, 115, 99–103. doi:10.1080/10426509608037957

Xiao-Chun, H., Jun-Sheng, Z., Tian-Feng, C., Yi-Bo, Z., Yi, L., & Wen-Jieand, Z. (2012). Synthesis, antioxidant and anticancer activities of 1,2,5-selenadiazole pyrimidine heterocyclic derivative ASPO. *Chemical Journal of Chinese Universities, 33,* 976–982. doi:10.3969/j. issn.0251-0790.2012.05.020

Zhan, P., Liu, X., Fang, Z., Pannecouque, C., & Clercq, E. (2009). 1,2,3-Selenadiazole thioacetanilides: Synthesis and anti-HIV activity evaluation. *Bioorganic Medicinal Chemistry, 17,* 6374–6379. doi:10.1016/j.bmc.2009.07.027

18

Nickel(II)-oxaloyldihydrazone complexes: Characterization, indirect band gap energy and antimicrobial evaluation

Ayman H. Ahmed[1]*, A.M. Hassan[1], Hosni A. Gumaa[1], Bassem H. Mohamed[1] and Ahmed M. Eraky[1]

*Corresponding author: Ayman H. Ahmed, Faculty of Science, Department of Chemistry, Al-Azhar University, Nasr City, Cairo, Egypt
E-mail: ayman_haf532@yahoo.com
Reviewing editor: Alexandra Martha Zoya Slawin, University of St. Andrews, UK

Abstract: A series of oxaloyldihydrazone ligands was prepared essentially by the usual condensation reaction between oxaloyldihydrazide and different aldehydes e.g. salicylaldehyde, 2-hydroxy-1-naphthaldehyde, 2-hydroxyacetophenone and 2-methoxy-benzaldehyde in 1:2 M ratio. The formed compounds were purified to give bis(salicylaldehyde)oxaloyldihydrazone (L_1), bis(2-hydroxy-1-naphthaldehyde) oxaloyldihydrazone (L_2), bis(2-hydroxyacetophenone)oxaloyldihydrazone(L_3) and bis(2-methoxy-benzaldehyde)oxaloyldihydrazone (L_4). All the oxaloyldihydrazones (L_1–L_4) and their relevant solid nickel(II) complexes have been prepared and structurally characterized on the basis of the elemental analyses, spectral (UV–vis, IR, mass and ^{1}H NMR), magnetism and thermal (TG) measurements. The dihydrazones coordinate to the metal center forming mononuclear complexes with L_1, L_3 and L_4 in addition to binuclear complex with L_2. The metal center prefers tetrahedral stereochemistry upon chelation. The optical indirect band gap energy for all compounds underlies the range of semiconductor materials. The prepared ligands and their metal complexes have been assayed for their antimicrobial activity against fungi as well as Gram-positive and Gram-negative bacteria. The resulting data indicate the ability of the investigated compounds to inhibit the growth of some micro-organisms, where L_2 showed the highest activity among all the compounds. Minimum inhibitory concentration (MIC) of L_2 against the growth of five micro-organisms was

ABOUT THE AUTHOR

Ayman H. Ahmed was born in Benha, Egypt. He completed his undergraduate work in chemistry at Zagazig University (1991). He obtained his MSc degree (1998) and PhD (2002) at Al-Azhar University. Assistant professor in 2007, where he is now a full professor of inorganic chemistry. His work is relevant to coordination compounds and their biological studies in addition to synthesis of organic compounds as corrosion inhibitors. His current research is in the area of synthesis and characterization of solid complexes especially zeolite-encapsulated complexes. As a part of his ongoing studies, herein, he contributed in reporting the nickel(II)-oxaloyldihydrazone complexes: characterization, indirect band gap energy and antimicrobial evaluation

PUBLIC INTEREST STATEMENT

In view of the biological importance of oxaloyldihydrazones and their chelates, four types of dihydrazone ligands and their Ni(II) complexes have been synthesized and structurally characterized by elemental analyses, spectral (UV–vis., IR, mass and ^{1}H NMR), magnetism and thermal (TG) measurements. The metal center prefers tetrahedral stereochemistry upon chelation. The optical indirect band gap energy for all compounds lays in the range of semiconductor materials. The metal complexes showed lower optical band gap energy than the original ligands indicating better conduction of complexes compared with the free ligands. All the investigated compounds have been assayed for their antimicrobial activity against fungi as well as Gram-positive and Gram-negative bacteria. The isolated solids may be considered as antimicrobial agents towards some micro-organisms.

determined which gives better response against *Aspergillus fumigatus* and *Bacillis subtilis* compared with some selected standard drugs.

Subjects: Human Biology; Inorganic Chemistry; Optoelectronics

Keywords: hydrazone complexes; synthesis; stereochemistry; biological activity

1. Introduction

Hydrazones and their derivatives constitute a versatile class of organic compounds that are important for drug design, organocatalysis and also for the syntheses of heterocyclic compounds (Barbazan et al., 2008). In addition, they find use as plasticizers, polymerization inhibitors and antioxidants. Despite hydrazones have been under study for a long time owing to their easy preparation, much of their basic chemistry remains unexplored. In fact, hydrazones played a central role in the development of coordination chemistry. Hydrazones obtained by the condensation of 2-hydroxy or methoxy aldehydes and ketones with hydrazides are considered potential polynucleating ligands because they have amide, azomethine and phenol/methoxy functions, thus offering a variety of bonding possibilities in metal complexes (Sherif & Ahmed, 2010). Extensive studies have revealed that the lone pair on trigonally hybridized nitrogen atom of the azomethine group is responsible for the chemical and biological activity (Ahmed & Ewais, 2012; Ghasemian et al., 2015; Hassan et al., 2015), compounds of this type have a great biological activity as antitumor and antiviral agents (Rollas & Küçükgüzel, 2007; Yuan, Lovejoy, & Richardson, 2004). Nevertheless, a lot of complexes derived from hydrazones [M = Cu(II), Ni(II), Pd(II), Co(II), V(IV) and Ru(II)] have been studied (Hassan et al., 2015), little complexes of oxaloyldihydrazone ligands have been identified (Hassan et al., 2015; Lal, Adhikari, Kumar, Chakraborty, & Bhaumik, 2002; Lal, Basumatary, Adhikari, & Kumar, 2008; Salavati-Niasari & Sobhani, 2008). This is because oxaloyldihydrazones are soluble only in high polar solvents such as DMF and DMSO which requires much effort to isolate them or their complexes in pure form. Of worthy mention, some hydrazone complexes have been incorporated in zeolite-Y by analogous methods and the resulting materials were inferred by various physicochemical characterization techniques (Ahmed, 2014; Ahmed & Mostafa, 2009; Ahmed & Thabet, 2011, 2015).

Actually, investigation on new antimicrobial agents is important due to the resistance acquired by several pathogenic micro-organisms. The synthesized chemical compounds, which are used for the treatment of infectious diseases, are known as chemotherapeutic agents. Every year thousands of compounds are synthesized to find out potential chemotherapeutic agent to combat pathogenic micro-organisms. In this regard, mixed ligand metal complexes of ampicillin and chloramphenicol prepared using Ni(II), Co(II) and Fe(III) chlorides hexahydrate disclosed that the activity of the metal complexes had more potent antibacterial activity than the parent drugs (Ahmed & Ewais, 2012; Ajani, 2008). Indeed, hydrazones and their coordination compounds including nickel metal have wide applications as antitumour (Yuan, Lovejoy, & Richardson, 2004), antimicrobial activity (Al-Sha'alan, 2007), antimalarial (Walcovrt, Loyevsky, Lovejoy, Gordeuk, & Richardson, 2004) and antiviral (Abdel-Aal, El-Sayed, & El-Ashry, 2006). The biological activity of hydrazone may be attributed to the formation of stable chelates with transition metals present in the cell, thus many vital enzymatic reaction cannot take place in the presence of hydrazone.

In view of the importance of oxaloyldihydrazones and their chelates, we have undertaken the synthesis and structural characterization studies on bivalent metal (Ni) complexes with some types of oxaloyldihydrazones (L_1-L_4). The optical band gap of all isolated compounds has been determined to describe their optical and electronic properties. Moreover, the biological activity of the ligands and their complexes has been investigated to show the possibility for their uses in pharmaceutical industry as antibiotics to treat some diseases in medicine area.

2. Experimental

2.1. Materials and methods

The selected metal salts, diethyl oxalate, hydrazine monohydrate were purchased from Sigma–Aldrich. The employed aldehydes were of E-Merck grade. Oxalicdihydrazide was prepared by the recipe described in Ref. (Hassan et al., 2015), (Exp/Lit. m.p = 240°C). Other chemicals and solvents were of highest purity and used without further purification. The ^{1}H NMR spectra were recorded on a Jeol-FX-90Q Fourier NMR spectrometer at 25°C using DMSO solvent and TMS as an internal standard. Mass spectra of the ligands were performed by a Shimadzu-GC-MS-QP1000 EX using the direct inlet system. Metal contents (%wt) were estimated complexometrically by EDTA using xylenol orange as indicator and solid hexamine as buffer (pH 6). Elemental analyses (CHN), spectral (UV–vis., FTIR), μ_{eff} and thermal (TG) measurements were carried out as reported (Salama, Ahmed, & El-Bahy, 2006). The optical band gap energy (E_g) of product compounds was calculated from Tauc's equations (Rashad, Turky, & Kandil, 2013; Tauc, 1968).

2.2. Preparations

2.2.1. Preparation of oxaloyldihydrazone ligands

The dihydrazone ligands; bis(salicylaldehyde)oxaloyldihydrazone (L_1), bis(2-hydroxy-1-naphthaldehyde)oxaloyldihydrazone (L_2), bis(2-hydroxyacetophenone) oxaloyldihydrazone (L_3) and bis(2-methoxybenzaldehyde)oxaloyldihydrazone (L_4); were prepared by general condensation procedure. Oxalicdihydrazide (0.01 mol) dissolved first in hot water (20 cm^3) followed by adding methanol (40 cm^3) was mixed with the appropriate aldehyde [salicylaldehyde, 2-hydroxy-1-naphthaldehyde, 2-hydroxyacetophenone and 2-methoxybenzaldehyde] (0.02 mol) in absolute methanol. The resulting mixture was refluxed for 3 h under constant stirring. The product separated out on concentrating the solution to half of its volume and cooling. The crystals of the desired ligand was collected by filtration through a Buchner funnel and dried in the oven at 50°C for 2 h. After that, the ligand was recrystallized from DMF–MeOH$_{aq}$ mixed solvent, collected, washed thoroughly on filter paper by acetone to remove any excess of DMF and then dried in an electric oven at 50°C for 2 h. The authenticity of the ligands was proved by elemental analyses, (IR, mass and ^{1}H NMR) spectroscopy (Tables 1 and 2).

2.2.2. Preparation of solid complexes

The dihydrazone ligand (1 mmol) was dissolved in a minimum amount of DMF (20 ml) and then 50 ml methanol was added. The resulting solution was added slowly to a methanolic solution of Ni(II) acetate. The resulting mixture was heated under reflux for 3 h and then reduced to 15 cm^3 by evaporation on hot plate. The resulting reaction mixture was cooled down to room temperature and the colored solid complexes were filtered off, washed several times with successive portions of hot solvents [DMF, 10–15 ml, methanol, 30–40 ml and acetone, 20–30 ml], respectively, to remove any excess of unreacted ligand and finally dried in an electric furnace at 80°C for 5 h.

2.3. Biological activity

2.3.1. Agar diffusion well method to determine antimicrobial activity

The synthesized compounds were screened for their antimicrobial activity against seven different test organisms having environmental and clinically importance. The micro-organism inoculums were uniformly spread using sterile cotton swab on a sterile Petri dish containing malt agar (for fungi) and nutrient agar (for bacteria). Ten mg/ml of each sample was added to each well (10 mm diameter holes cut in the agar gel, 20 mm apart from one another). The systems were incubated for 24–48 h at 37°C (for bacteria) and at 28°C (for fungi). After incubation, micro-organism growth was observed. Inhibition of the bacterial and fungal growth were measured in mm. Tests were performed in triplicate (Smânia, Monache, Smânia, & Cuneo, 1999). Minimal inhibitory concentration (MICs) of bis(2-hydroxy-1-naphthaldehyde)oxaloyldihydrazone (L_2) was determined by dissolving the

Compound	Symbol	M.p (°C) Color	C	H	N	M	¹H-NMR Chemical shift (δppm)	μ_{eff}	M⁺ Found/ calcd	E_g (eV)
$C_{16}H_{14}N_4O_4$	L_1	>300	59.1	5.4	16.2	–	12.6(NH, s), 10.98(OH, s), 8.75(CH=N, s),	–	326.0/ 326.0	2.73
		Yellow	58.9	4.3	17.2	–	6.6–8.40 (aromatic protons, m)			
$C_{24}H_{18}N_4O_4$	L_2	>300	68.1	5.3	12.9	–	12.76(NH, s), 12.57(OH, s), 9.74(CH=N, s),	–	427.1/426.4	2.46
		Yellow	67.7	4.3	13.2	–	7.0–8.8 (aromatic protons, m)			
$C_{18}H_{18}N_4O_4$	L_3	>300	59.5	6.1	14.7	–	12.85(NH, s), 11.85(OH, s), 6.6–8.0(aromatic	–	356.4/354.2	2.78
		Pale yellow	61	5.1	15.8	–	protons, m), 2.48(CH₃, s),			
$C_{18}H_{18}N_4O_4$	L_4	>300	59.2	5.8	15.7	–	12.3(NH, s), 11.9(OH, s), 8.95(CH=N, s)		355.0/354.4	3.05
		White	61	5.1	15.8	–	6.8–8.5(aromatic protons, m), 3.93(OCH₃, s)	–		
$[Ni(L_1)(OAc)_2].2H_2O$	1	>300	44.0	4.2	10.1	11.9	–	3.8		2.28
		Orange	44.6	4.5	10.4	10.9	–			
$[Ni_2(L_2-2H)(H_2O)_2].3H_2O$	2	>300	44.9	3.4	8.2	17.9		3.9		1.87
		Break red	45.8	3.8	8.9	18.6	–			
$[Ni(L_3-2H)].CH_3OH$	3	>300	51.0	4.3	12.2	12.4		3.6		2.08
		Red	51.5	4.6	12.6	13.3	–			
$[Ni(L_4-H)(OAc)].0.5H_2O$	4	>300	51.0	4.2	11.5	12.3		3.4		2.42
		Green	50.0	4.4	11.7	12.2	–			

Table 1. Analytical, physical and spectroscopic data of the oxaloyldihydrazones and their related metal complexes

specimen in DMSO and using different concentrations (0.001–0.078 mg/l). The minimal inhibitory concentration (MICs) was determined after incubation period.

3. Results and discussion

Oxaloyldihydrazones ligands (L_1-L_4) and their solid complexes with Ni(II) ion have been isolated in a pure form. Physical, analytical and spectroscopic data of the hydrazones and their related metal complexes are given in Tables 1 and 2. The complexes are air stable for long time, insoluble in MeOH/ EtOH, Et₂O, CHCl₃, acetone, CCl₄ as well as benzene. Complexes 1–3 are soluble in DMF and DMSO while complex 4 is partially soluble. Comparison of the elemental analysis for both the calculated

Table 2. Significant IR and electronic absorption data of oxaloyldihydrazones and their nickel complexes

Symbol	ν(OH) phenolic (enolic)	ν(OH) H₂O/ MeOH	ν(NH)	ν(C–H) aromatic (aliphatic)	ν(C=O)	ν(C=N)	ν(C–O) phenolic	ν(C–OMe)	ν(C=C)	δ(C–H) aromatic out of plane	ν(M-O) phenolic/ enolic (carbonyl/ methoxy)	λmax,nm (assignments)
L_1	3149 (3278)	-/-	3204	3062 (2924)	1666	1620	1275		1403 1458 1486	756		387(n-π*, C=N), 372(n-π*, C=O), 340(π -π*, C=N), 300 (π -π*, C=O), 288(π -π*, Phenyl)
L_2	3476 (-)	-/-	3166	3043 (2926)	1705 1660	1621	1287		1465 1541 1574	741		395(n-π*, C=N), 390(n-π*, C=O) 370(π -π*, C=N), 348(π -π* C=O), 320(π -π*, Phenyl)
L_3	3448 (-) -/-	-/-	3293	3048 (2923)	1689 1651	1607	1246		1449 1486 1512	746		378(n-π*, C=N), 348(n-π* C=O), 322(π -π*, C=N), 300 (π-π*, C=O), 280(π -π*, Phenyl)
L_4	3227		3202	3041 (2940)	1653	1600		964	1464 1486 1572	760		371(n-π*, C=N), 345(n-π* C=O), 320(π -π*, C=N), 290 (π -π*, C=O), 270(π -π*, Phenyl)
1	3362 (-)	/3400	3204	3063 (2924)	1666	1605	1276		1458 1486	757		595 (³T₁→³T₁(P)) 450(L→ MCT)
2	(-)	3377		3054 (2925)		1617 1602	1302		1457 1538	747	547/458	666 (³T₁→³T₁(P)) 470(L→ MCT)
3	(-)	3400	3283	3013 (2927)	1678	1611	1273		1445 1489 1536	748	540/ /495	660 (³T₁→³T₁(P)) 445(L→ MCT)
4	(-)	3385	3202	3055 (2926)	1653	1602 1580		970	1464 1488	756	/525	600 (³T₁→³T₁(P)) 450(L→MCT)

and found percentages indicates that the compositions of the synthesized complexes coincide well with the proposed formulae.

3.1. Chemical structures of dihydrazones and Ni$_{II}$-complexes

3.1.1. IR spectra and bonding
The positions of the significant IR bands of dihydrazones and their nickel complexes are registered in Table 2.

Ligands: oxloyldihydrazones (L_1-L_4) can exist either in the trans (staggered) configuration or cis-configuration, Figure 1, (Hassan et al., 2015; Lal, Adhikari, & Kumar, 1993). In cis-configuration, the dihydrazone can adopt either anti-cis-configuration or syn-cis-configuration. Infrared spectra showed strong bands at 1602–1617 cm^{-1} assignable to the azomethine group ($\nu_{C=N}$). The observation of these bands confirmed the interaction of dihydrazides with aldehydes forming azomethine link-ages. The bands of $\nu(OH)_{phenolic/naphthoic}$, $\nu(NH)$ and $\nu(C=O)$ for L_1-L_3 were noticed at (3149, 3204 and 1666), (3476, 3166 and 1705(m)+1660(v.s)) and (3448, 3293 and 1689(m)+1651(v.s)) cm^{-1}, respec-tively. Meanwhile, L_4 which does not contain o-hydroxy group revealed $\nu(NH)$ at 3202 cm^{-1} and $\nu(C=O)$ at 1653 cm^{-1}. The appearance of both $\nu(C=O)$ and $\nu(NH)$ simultaneously in the IR spectra of

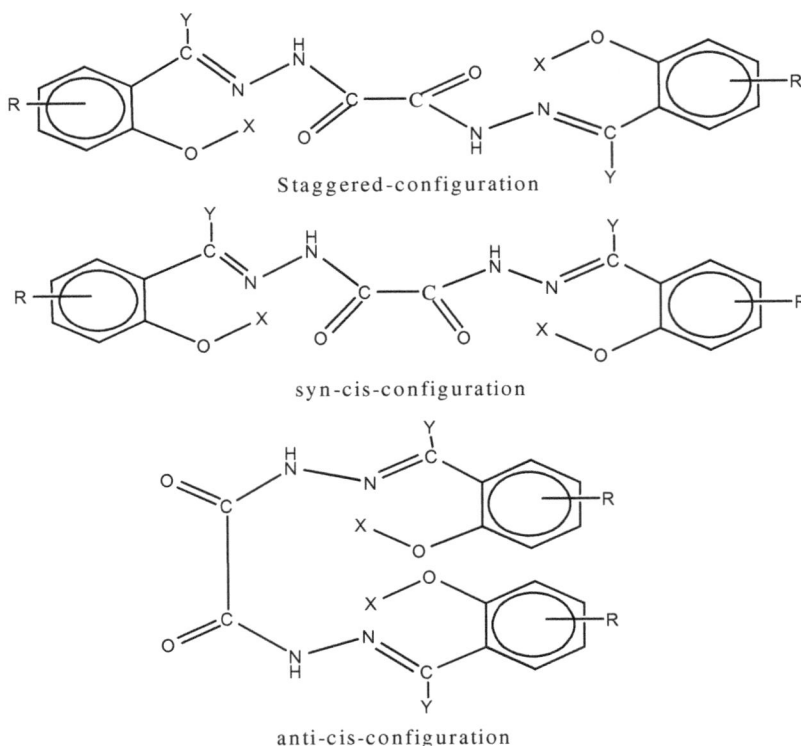

Staggered-configuration

syn-cis-configuration

anti-cis-configuration

Figure 1. Proposed structures of oxaloyldihydrazone ligands, where L_1 (R = H, X = H, Y = H), L_2 (R = ph, X = H, Y = H), L_3 (R = H, X = H, Y = CH$_3$) and L_4 (R = H, X = CH$_3$, Y = H).

L_1-L_4 indicates the presence of keto forms. The appearance of carbonyl groups (C=O) near 1660 cm^{-1} as one band in case of L_1 and L_4 and pair of bands near (1660, 1700 cm^{-1}) for each L_2 and L_3 indicates the trans configuration (staggered, Figure 1) for (L_1 and L_4), while mixture of [cis(syn/anti-cis-structure) + trans (staggered structure), Figure 1] isomers for (L_2 and L_3). This suggestion results from the field effect phenomenon which can be elucidated as follows: when the two carbonyl (C=O::) groups are in the same direction (cis configuration), the non-bonding electrons present on oxygen atoms cause electrostatic repulsion. This causes a change in the state of hybridization of C=O group and also make it to go out of the plane of the double bond. Thus, the configuration diminished and absorption occurs at higher wave number. Subsequently, cis is absorbed (due to the field effect) at higher frequency compared to trans isomer. Thereby, L_2 and L_3 showed two bands for C=O, one of them (at high frequency) associated with syn/anti-cis-structure, while that at lower frequency is related to staggered configuration. The low intensity of high frequency bands (near 1700 cm^{-1}) compared with that at low frequency (near 1660 cm^{-1}) indicates the domination of staggered structure. Of worthy mention, NMR spectroscopy is in good agreement with this suggestion. The observation of OH (phenolic) group in L_1 at lower position (3149 cm^{-1}) is taken as evidence to the persistence of interamolecular H-bonding between the phenolic-OH and azomethine group (O–H···N) (Burger, Ruff, & Ruff, 1965). The proposition of intermolecular H-bond (O–H···O–H) between ligand molecules is excluded owing to the sharpness of this band. The remarkable downward frequency shift indicates the strength of this bond. In fact, ^{1}H NMR did not distinguish this H-bond in its spectra because execution of NMR analysis in high polar solvent (DMSO) led to break these bonds. The phenolic-OH group did not form H-bonding in case of ligands L_2 and L_3 where it is observed after 3400 cm^{-1}. Further, the existence of two bands at 3278 cm^{-1} (L_1) and 3227 cm^{-1} (L_4) may be attributed to ν(OH)$_{enolic}$ in H-bond bonding with azomethine group. The lower shift for these two bands than the normal value asserts the weakness of this bond. The obscure of enolic OH in NMR spectra of L_1 and L_4 may be attributed to factors: (1) its lower concentrations. Minute molecules may change one or two of its C=O groups into enolized configuration but the keto form is still dominated, (2) dilution by high polar DMSO disrupted these H-bonds returning the modified molecules into keto forms. The significant ν(C–O) groups associated to the aromatic ring of the dihydrazones L_1-L_4 were observed at 1275, 1287, 1246 and 964 cm^{-1}, respectively (Burger et al., 1965; Chanu, Kumar, Ahmed, & Lal, 2012). All of the investigated hydrazones ligands showed 3–4 bands in the range 1400–1600 cm^{-1} related to ν(C = C) of the aromatic ring. Also, each ligand exhibited strong band in the region 700–800 cm^{-1} corresponding to the out-of-plane deformation of the aromatic ring. The positions of other bands assigned to υ(CH)$_{aliphatic}$ near 2900 cm^{-1} and υ(CH)$_{aromatic}$ near 3050 cm^{-1} are demonstrated in Table 2.

Ni(II) complexes: In these complexes, the dihydrazones (L_1-L_4) acts as bi-, tri-, tetra-dentate ligands towards the centered nickel(II) ion. Appearance of both ν(NH) and ν(C=O) after chelation (in case of **1**) at the same position in addition to a negative shift of ν(C=N) indicated that its corresponding ligand coordinated in keto form via the azomethine group only. This negative shift of ν(C=N) arises from a decrease in the π-bond character of the azomethine (–C=N) group as a result of nitrogen to metal coordination (Ali, Ahmed, Mohamed, & Mohamed, 2007). The shift of ν(OH) phenolic of L_1 to higher frequency upon complexation (3149 →3362 cm^{-1}) is due to the cleavage of H-bond of the type CH = N.....OH$_{phenolic}$. Obscure of both ν(NH) and ν(C=O) after chelation (in case of **2**) with splitting of ν(C=N) and positive shift of ν(C–O)$_{naphthoic}$ evidenced that C=O and NH groups are not contributed in coordination, whereas the corresponding ligand binds to the metal in enol form via the azomethine and deprotonated o-hydroxyl groups. The splitting of C=N occurred due to the enolization of two C=O groups in dihydrazone molecule assuming enol–enol skeleton. For **3** and in comparison to free L_3, the negative shift of ν(NH) (3293→3283 cm^{-1}), positive shift of υ(C–O$_{phenolic}$, 1246→1273 cm^{-1}), obscure ν(OH)-phenolic as well as observation of ν(C=O) at its normal position (1678 cm^{-1}) ascertained the coordination of L_3 with Ni(II) ion through NH and deprotonated o-hydroxy groups. Despite the formation seven-membered ring around the nickel ion may be discouraged and rare, remaining of υ(CH = N) at nearly the same location ruled out its anticipation in construction of complex geometry. The observation of C=O as one band in **3** instead of two bands in L_3 with slight change in its location may be due to the presence of the two oxygen atom related to (O = C–C=O) part, Figure 2, in trans or cis symmetry. Perhaps the absorption of this group (C=O) at somewhat higher frequency (1678 cm^{-1}) supports the first suggestion i.e., cis structure. Unnoticeable ν(OH) of o-hydroxyl group in complexes **2** and **3** species supports the deprotonation of this group during the coordination of L_2 and L_3. In case of complex **4**, half of the L_4 molecule was altered to enol form and coordination takes place by one side (one OMe, one C=N and one deprotonated enolized C–OH group). This proposition depends on the weakness in intensity of [ν(NH), ν(C=O) and ν(C–OMe)] and splitting of C=N. The splitting in the vibrational stretching C=N band observed in **4** at (1602, 1580 cm^{-1}) substantiated the presence of dissimilar azomethine groups (participated and unparticipated) (Salapathy & Sahoo, 1970). This result emphasizes the contribution of one azomethine group in bonding, Figure 2. The broadening of C=N band (at 1580 cm^{-1}) may be due to the absorption interference of C=N group generated by enolization.

Undoubtedly, the appearance of new M-O bands in all above complexes (Table 2) manifested the connection between the ligand and the central metal ion through the oxygen atom of ortho group

Table 3. The MIC of bis(2-hydroxy-naphthaldehyde)oxaloyldihydrazone (L_2)		
Tested micro-organism	**L_2/MICS (µg/ml)**	**Standard**
FUNGI		**Amphotericin B**
Aspergillus fumigatus	15.6	23.7
Synccephalastrum racemosum	31.25	19.7
GRAM-POSITIVE BACTERIA		**Ampicillin**
Streptococcus pneumoniae	125	23.8
Bacillis subtilis	15.6	32.4
GRAM-NEGATIVE BACTERIA		**Gentamicin**
Escherichia coli	500	19.9

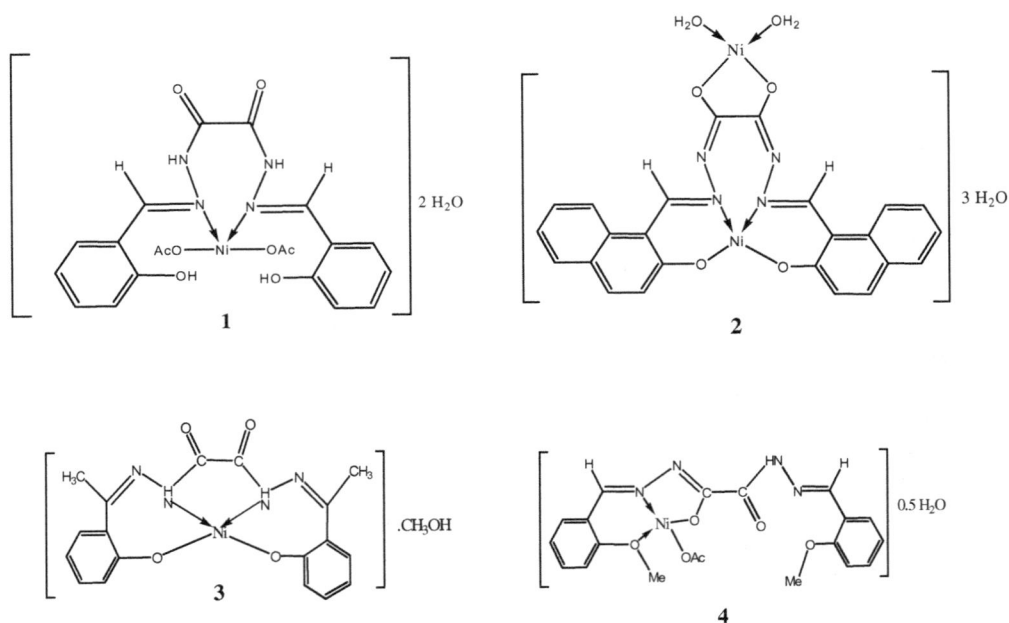

Figure 2. Suggested structures of the Ni^{II}-oxaloyldihydrazone complexes.

(Kumar et al., 2011). Furthermore, crystalline and coordinated water molecules in nickel specimens were supposed to be based on the appearance of the broad bands within the range 3300–3500 cm^{-1}. Other bands assignable to acetate ion (OAc) were detected in complex **1** at 1466 and 1359 cm^{-1}, while in complex **4** at 1354 and 1437 cm^{-1} suggesting ν_{as} and ν_s carboxylic modes, respectively. The larger difference between the ν_{as} and ν_s frequencies confirmed the coordination of acetate as a uni-dentate anion through the C–O moiety of the carboxylic group (Nakamoto, 1970). The slight changes in the positions of $\nu(C = C)_{ph}$ upon chelation arise from metal–ligand interaction.

3.1.2. ^{1}H NMR Spectra

The assignment of the main signals in ^{1}H NMR spectra of all ligands (Hassan et al., 2015) is given in Table 2, where nickel complexes are not scanned owing to their paramagnetic nature. ^{1}H NMR spectra of all ligands exhibited multiple signals of the aromatic protons in the 6.5–8.5 ppm region. The signals of equal integration observed in L$_1$, L$_2$ and L$_3$ at δ (12.6, 11), (12.8, 12.6) and (12.9, 11.8) ppm downfield of TMS have been assigned to NH and ortho-OH protons, respectively. On the other hand, L$_4$ revealed a signal at 12.3 ppm attributed to secondary NH group. Upon the addition of D$_2$O, the OH and NH signals were obscured. Further, the existence of the δOH (phenolic/naphthoic) at its normal frequency excluded any intramolecular hydrogen bonding operating between ortho-OH and CH = N group (CH = N.....H-O). The azomethine signals, δ(CH = N), observed only in L$_1$, L$_2$ and L$_4$ have been assigned at 8.75, 9.73 and 8.95 ppm, respectively. As reported in the literature (Kumar et al., 2011), if the dihydrazone exists in the syn-cis-configuration or staggered configuration, the δOH, δNH and δCH = N, resonances, each should appear as a singlet. However, the appearance of these signals in the form of six signals (doublet of doublet) indicates anti-cis (chair) configuration. Actually, the features of the ^{1}H-NMR spectra of the dihydrazones are in consistent with the syn-cis- or staggered configuration. According to IR interpretation mentioned above, the staggered configuration is well-defined/dominated for all ligands (L$_1$-L$_4$). ^{1}H NMR spectra of L$_3$ and L$_4$ differ from other ligands spectra where they showed signals at 2.55 and 3.9 ppm downfield of TMS due to the methyl (CH$_3$) and methoxy (OCH$_3$) protons, respectively (Abd El-Wahab, Abd El-Fattah, Ahmed, Elhenawy, & Alian, 2015; Hassan et al., 2015; Sherif & Ahmed, 2010).

3.1.3. Electronic spectra

The assignments of the observed electronic absorption bands of the oxaloylhydrazones and their metal complexes (Ali et al., 2007; Lever, 1984; Sutton, 1968) as well as the magnetic data of the formed chelates are shown in Table 1.

The electronic data of the studied ligands in paraffin oil exhibited five absorption bands at λ_{max} (nm) equals 371–395 (n-π*, C=N), 345–390 (n-π*, C=O), 320–370 (π-π*, C=N), 290–348 (π-π*, C=O) and 270–320 (π-π*, aromatic ring). All of nickel complexes revealed a broad band within the range 595–666 nm assigned to $^3T_1 \rightarrow {}^3T_1$ (P) transition. The observation of this band suggested a tetrahedral geometry around the Ni(II) ion. In spite of the complexes 2 and 3 having red color which is a characteristic for the square planar Ni(II) complexes, the paramagnetic nature of these complexes excluded this suggestion (square planner Ni(II) complexes are diamagnetic). Another band was observed for each complex in the range 445–470 nm characteristic to L-M charge transfer. This ligand–metal charge transfer (L→MCT) transition may be associated to, most probably, an electronic excitation from the HOMO of ortho-phenolate/naphtholate/methoxy oxygen to the LUMO of Ni(II) ion (McCollum et al., 1994).

3.1.4. Magnetic behavior

From the magnetic studies, the values of μ_{eff} in case of Ni^{II}-complexes (**1, 3 and 4**) which contain one metal ion are close to spin only values and may be considered in good consistent with the proposed structures. For other consequential binuclear complexes (**2**), the magnetic moment values are less than that expected per molecular formula regarding the existence of two nickel ions in the proposed structure will enhance the magnetization. This can be explained on the basis of metal–metal interaction. It is suggested that these complexes have high spin but the presence of two metals near to each other in the same molecule may cause partial quenching of the spin moments of the metal ions (spin coupling) decreasing the magnetism (Sutton, 1968).

3.1.5. TG studies of Ni(II)-complexes

To examine the thermal stabilities of Ni^{II}-complexes, thermogravimetric analysis (TG) for all complexes was carried out. The thermogram (TGA) data of complex **(1)** exhibited three stages of

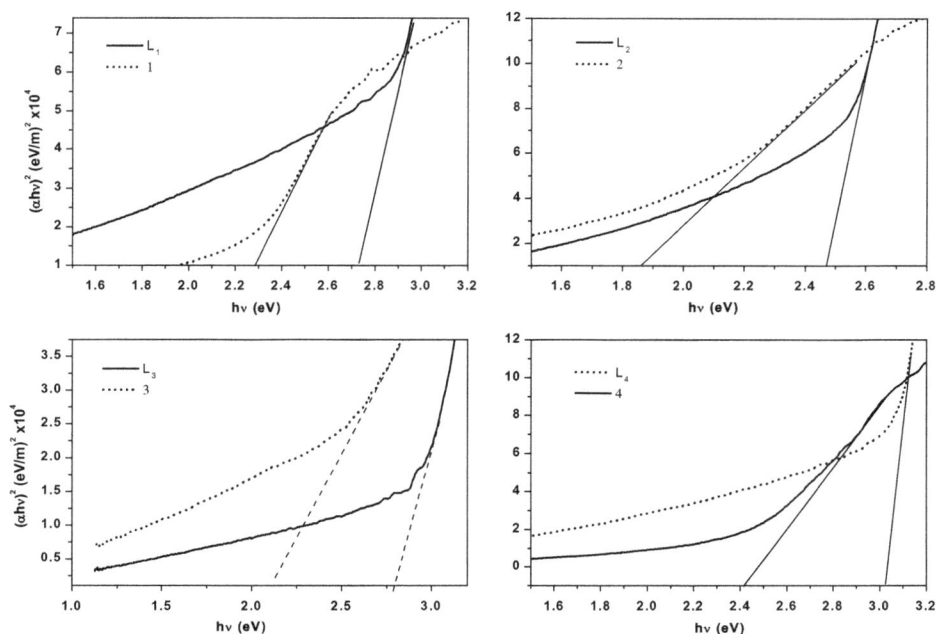

Figure 3. Plots of $(\alpha h\nu)^2$ vs. $h\nu$ of oxaloyldihydrazones and their Ni^{II}-complexes.

decomposition at different temperatures. The first stage in temperature range (50–160°C) corresponds to the loss of two crystalline water molecules (found/calculated = 6.5/6.7)% while the other two stages correspond to the gradual decomposition of the complex forming a mixture of NiO and $NiCO_3$ at the last stage (found/calculated = 17.9/18.0)%. The thermogram (TGA) data of complex (**2**) revealed five stages of decomposition. The first stage (50–155°C) corresponds to the loss of three crystalline water molecules (found/calculated = 17.9/18.0)%. The second stage corresponds to the loss of (found/calculated = 6.7/5.7)% of the weight of complex at temperature range from (155–250°C) corresponding to the loss of two coordinated water molecules. The other three stages correspond to the gradual decomposition of the complex with the formation of $2NiCO_3$ at the last stage (found/calculated = 37.4/37.7)%. TGA curve of complex (**3**) gave four stages of decomposition started at 50, 195, 250 and 400°C. The first stage (50–140°C) corresponds to the loss of one methanol molecule (found/calculated = 6.3/7.2)%. The other three stages correspond to the gradual decomposition of the complex with the formation of NiO + $NiCO_3$ mixture at the last stage (found/calculated = 18.3/17.3)%. TG curve of complex (**4**) provided four stages of decomposition. The first one (50–250°C) is assigned to the loss of 0.5 crystalline water molecule beside one acetate group (found/calculated = 12.8/13.7)%. The other three stages are attributed to the gradual decomposition of the complex with the formation of NiO and $NiCO_3$ mixture at the last stage (found/calculated = 20.8/20.2)%.

In view of data presented and discussed above, structures of the nickel complexes can be represented by Figure 2.

3.1.6. Optical properties

To clarify the conductivity of the isolated complexes, the optical band gap energy (E_g) of oxaloyldihydrazones and their Ni(II) complexes have been calculated from the following equations:

The measured transmittance (T) was used to calculate approximately the absorption coefficient (α) using the relation

$$\alpha = 1/d \ln(1/T)$$

where d is the width of the cell and T is the measured transmittance. The optical band gap was estimated using Tauc's equation (Rashad et al., 2013; Tauc, 1968):

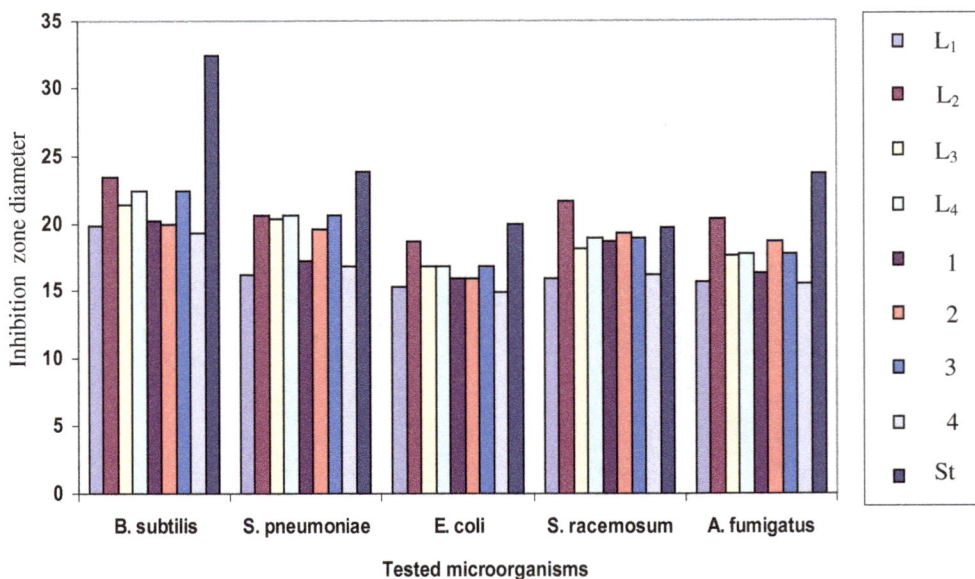

Figure 4. Antimicrobial activity for oxaloyldihydrazone ligands and their nickel(II) complexes.

$$\alpha h\nu = A(h\nu - Eg)^m$$

where m is equal to 1/2 and 2 for direct and indirect transition, respectively, A is an energy-independent constant.

The values of α calculated from the first equation were used to plot $(\alpha h\nu)^2$ vs. $h\nu$ (Figure 3) from which an indirect band gap was found by extrapolating the linear portion of the curve to $(\alpha h\nu)^2 = 0$. The values of indirect optical band gap E_g were determined and given in Table 1. The E_g values of oxaloyldihydrazones (L_1-L_4) and Ni(II)-L complexes were found to be at 3.06–3.42 and 1.87–2.42 eV, respectively as indicated in Figure 3 and Table 1. Inspection of Table 1 revealed higher E_g values of ligands compared with their corresponding complexes. As reported in the literature (Karipcin et al., 2007), it is suggested that after complexation, metal leads to raise mobilization of the ligand electrons by accepting them in its shell. It can be evaluated that after formation of the complex, the chemical structure of the ligands is changed, the width of the localized levels is expanded and in turn, the band gap is smaller. This result is very significant in applications of electronic and optoelectronic devices because of the lower optical band gap of the materials (Mott & Davis, 1979). Of worthy mention, small band gap facilitates electronic transitions between the HOMO–LUMO energy levels and makes the molecule more electroconductive (Sengupta, Pandey, Srivastava, & Sharma, 1998). The obtained band gap values suggest that these complexes are semiconductors and lay in the same range of highly efficient photovoltaic materials. So, the present compounds could be considered potential materials for harvesting solar radiation in solar cell applications (Fu, Guo, Liu, Cai, & Huang, 2005; Karipcin et al., 2007; Mott & Davis, 1979; Sengupta et al., 1998).

3.1.7. Antimicrobial investigations

The results revealed a variable sensitivity of the **1–4** compounds towards the micro-organisms as shown in Figure 4. The data showed that all tested compounds have an appropriate activity against all the tested micro-organisms except for *P. aeruginosa* as Gram-negative bacteria and for *C. albicans* as fungi. The oxaloyldihydrazone ligand L_2 exhibited the highest activity against all the tested micro-organisms, fungi, Gram-positive bacteria and Gram-negative bacteria. Complex **3** showed higher inhibition zones against *B. subtilis* and *S. pneumoniae* as Gram-positive bacteria as well as *E. coli* as Gram-negative bacteria, while the metal complex **2** exhibited the highest inhibition zone against *S. racemosum* and *Aspergillus fumigatus*. Complex **4** showed lower inhibition zones against all the tested micro-organisms except for *S. pneumoniae* as Gram-positive bacteria and *S. racemosum* as fungi. The dihydrazone ligand (L_1) showed lower inhibition zones against *S. pneumoniae* as Gram-positive bacteria and *S. racemosum* as fungi.

In sum, the most active compound among all the tested compounds against the seven tested micro-organisms was L_2 due to its wide spectrum of activity (Figure 4). The enhancement in antibacterial activity of the nickel(II) complexes **(2, 3)** can be explained based on the chelation theory (Mishra & Singh, 1997) where the activity of hydrazone increases upon coordination. This enhancement in activity may be rationalized on the basis that their structures mainly possess an additional C=N bond. Hydrazone ligands and their complexes act as antimicrobial agents against Gram-negative and Gram-positive bacteria, fungi and yeast (Loncle, Brunel, Vidal, Dherbomez, & Letourneux, 2004; Vicini, Zani, Cozzini, & Doytchinova, 2002). Chelation reduces the polarity of the metal ion because of partial sharing of its positive charge with the donor groups and possibly the π-electron delocalization within the completely chelate ring system that is formed during coordination (Efthimiadou, Psomas, Sanakis, Katsaros, & Karaliota, 2007). These factors increase the lypophilic nature of the central metal atom and hence increase the hydrophobic character and liposolubility of the molecule favoring its permeation through the lipid layer of the bacterial membrane. Consequently, the metal complexes can easily penetrate into the lipid membranes and block the metal-binding sites of enzymes of the micro-organisms, thus destroying them more aggressively. These metal complexes also affect the respiration process of the cell and thus block the synthesis of proteins, which restrict further growth of the organism.

3.1.7.1. Minimum inhibitory concentration (MIC) of bis(2-hydroxy-1-naphthaldehyde)oxaloyldihydrazone (L₂): The antimicrobial screening concentrations of the compounds to be used were estimated from minimum inhibitory concentration (MIC) value, i.e. the lowest concentration of compounds which completely inhibits bacterial and fungal growth. The MIC of the synthesized L$_2$ against the growth of micro-organisms was determined by standard serial dilution method (Cappucino & Sherman, 1999) and compared with that of standard drug (Amphotericin B for fungi, Ampicillin for Gram-positive bacteria and Gentamicin as Gram-negative bacteria), Table 3. The results showed that bis(2-hydroxy-1-naphthaldehyde)oxaloyldihydrazone (L$_2$) has higher activity against *S. racemosum* than *A. fumigatus* and no activity against *C. albicans* (as fungi), higher activity against *S. pneumoniae* than *B. subtilis* (as Gram-positive bacteria), lower activity against *E. coli* and finally no activity against *P. aeruginosa* (as Gram-negative bacteria). Some factors are responsible for the high activity of L$_2$ towards several micro-organisms. (i) The high stability of L$_2$ and this can be indicated from its high melting points (mp < 300°C), (ii) It has also been proposed that concentration plays a vital role in increasing the degree of inhibition, as the concentration increases the activity increases, (iii) the high ability of L$_2$ to form hydrogen bond with the active centers of the cell constituents via the azomethine nitrogen atom, resulting in interference with the normal cell process and finally, (iv) it has been suggested that the hydrazones might inhibit enzyme production owing to the presence of nonbonding electron pairs on the N and O atoms. The existence of hydroxyl, carbonyl and imine (C=N) groups enhances antimicrobial activity.

4. Conclusion

A series of nickel(II)-hydrazone complexes derived from oxaloyldihydrazones (L$_1$-L$_4$) obtained by usual condensation of oxaloyldihydrazide and some selected aldehydes, namely salicylaldehyde (L$_1$), 2-hydroxynaphthaldehyde (L$_2$), 2-hydroxyacetophenone (L$_3$) and 2-methoxybenzaldehyde (L$_4$) have been isolated in pure form. All the ligands and their four solid nickel(II) complexes can be prepared by traditional reflux method. Comparison of the elemental analysis for both calculated and found percentages indicates that the compositions of the isolated solid complexes coincide well with the proposed formulae. Elemental analyses (C, H, N and M%), UV–Vis, FTIR spectroscopy and magnetism besides the change in ligand color upon complexation evidenced the formation of the desired nickel–hydrazone complexes. The dihydrazones give mononuclear complexes with L$_1$, L$_3$ and L$_4$ but binuclear complex is formed with L$_2$. The nickel in all complexes has tetrahedral geometry. The indirect band gap energy (E$_g$) of ligands and their related complexes lay in the range of semiconductor materials. The investigated compounds were assayed for their antimicrobial activities against seven test micro-organisms. The results suggested that the hydrazones and their complexes might retard enzyme production of some micro-organisms. The existence of the hydroxyl, carbonyl and azomethine (C=N) groups can enhance antimicrobial sensitivity.

Funding
The authors received no direct funding for this research.

Author details
Ayman H. Ahmed[1]
E-mail: ayman_haf532@yahoo.com
A.M. Hassan[1]
E-mail: alimelbraky@hotmail.com
Hosni A. Gumaa[1]
E-mail: hosniag@yahoo.com
Bassem H. Mohamed[1]
E-mail: bassem_shb2000@yahoo.com
Ahmed M. Eraky[1]
E-mail: eraky_83@yahoo.com
[1] Faculty of Science, Department of Chemistry, Al-Azhar University, Nasr City, Cairo, Egypt.

References
Abd El-Wahab, H., Abd El-Fattah, M., Ahmed, A. H., Elhenawy, A. A., & Alian, N. A. (2015). Synthesis and characterization of some arylhydrazone ligand and its metal complexes and their potential application as flame retardant and antimicrobial additives in polyurethane for surface coating. *Journal of Organometallic Chemistry, 791,* 99–106.

Abdel-Aal, M. T., El-Sayed, W. A., & El-Ashry, E. H. (2006). Synthesis and antiviral evaluation of some sugar arylglycinoylhydrazones and their oxadiazoline derivatives. *Archiv der Pharmazie, 339,* 656–663. http://dx.doi.org/10.1002/(ISSN)1521-4184

Ahmed, A. H. (2014). Zeolite-encapsulated transition metal chelates: Synthesis and characterization. *Review Inorganic Chemical, 34,* 153–175.

Ahmed, A. H., & Ewais, E. (2012). Physicochemical and antimicrobial investigation on some selected arylhydrazone complexes. *Journal of Chemical*

Pharmaceutical Research, 4, 3349–3360.

Ahmed, A. H., & Mostafa, A. G. (2009). Synthesis and identification of zeolite-encapsulated iron (II), iron (III)-hydrazone complexes. *Materials Science and Engineering: C, 29*, 877–883.
http://dx.doi.org/10.1016/j.msec.2008.07.023

Ahmed, A. H., & Thabet, M. S. (2011). Metallo-hydrazone complexes immobilized in zeolite Y: Synthesis, identification and acid violet-1 degradation. Journal of Molecular Structure, *1006*, 527–535.
http://dx.doi.org/10.1016/j.molstruc.2011.09.061

Ahmed, A. H., & Thabet, M. S. (2015). Physicochemical studies and heterogeneous hydroxylation of benzene on Fe^{II}, Fe^{III}-semicarbazone/zeolite-Y clathrates. *Synthesis and Reactions in Inorganic and Metal-Organic Chemistry, 45*, 1632–1641
http://dx.doi.org/10.1080/15533174.2015.1031038

Ajani, O. (2008). Synthesis, characterization, antimicrobial activity and toxicology study of some metal complexes of mixed antibiotics. *African Journal Pure Applied Chemistry, 2*, 69–74.

Al-Sha'alan, N. H. (2007). Antimicrobial activity and spectral, magnetic and thermal studies of some transition metal complexes of a Schiff base hydrazone containing a quinoline moiety. *Molecules, 12*, 1080–1091.

Ali, A. M., Ahmed, A. H., Mohamed, T. A., & Mohamed, B. H. (2007). Chelates and corrosion inhibition of newly synthesized Schiff bases derived from o-tolidine. *Transition Metal Chemistry, 32*, 461–467.
http://dx.doi.org/10.1007/s11243-007-0184-8

Barbazan, P., Carballo, R., Covelo, B., Lodeiro, C., Lima, J. C., & Vazquez-Lopez, E. M. (2008). Synthesis, characterization, and photophysical properties of 2-hydroxybenzaldehyde [(1E)-1-pyridin-2-ylethylidene]hydrazone and its rhenium(I) complexes. *European Journal of Inorganic Chemistry, 2008*, 2713–2720.
http://dx.doi.org/10.1002/(ISSN)1099-0682

Burger, K., Ruff, I., & Ruff, F. (1965). Some theoretical and practical problems in the use of organic reagents in chemical analysis—IV. *Journal of Inorganic and Nuclear Chemistry, 27*, 179–190.
http://dx.doi.org/10.1016/0022-1902(65)80208-1

Cappucino, J. G., & Sherman, N. (1999). *Microbiology a laboratory manual*. Addison: Wesley California.

Chanu, O. B., Kumar, A., Ahmed, A., & Lal, R. A. (2012). Synthesis and characterisation of heterometallic trinuclear copper(II) and zinc(II) complexes derived from bis(2-hydroxy-1-naphthaldehyde)oxaloyldihydrazone. *Journal of Molecular Structure, 1007*, 257–274.
http://dx.doi.org/10.1016/j.molstruc.2011.11.001

Efthimiadou, E. K., Psomas, G., Sanakis, Y., Katsaros, N., & Karaliota, A. (2007). Metal complexes with the quinolone antibacterial agent N-propyl-norfloxacin: Synthesis, structure and bioactivity. *Journal of Inorganic BioChemistry, 101*, 525–535.
http://dx.doi.org/10.1016/S0223-5234(02)01378-8

Fu, M. L., Guo, G. C., Liu, X., Cai, L. Z., & Huang, J. S. (2005). Syntheses, structures and properties of three selenoarsenates templated by transition metal complexes. *Inorganic Chemistry Communications, 8*, 18–21.
http://dx.doi.org/10.1016/j.inoche.2004.10.021

Ghasemian, M., Kakanejadifard, A., Azarbani, F., Zabardasti, A., Shirali, S., Saki, Z., & Kakanejadifard, S. (2015). The triazine-based azo-azomethine dyes; synthesis, characterization, spectroscopy, solvatochromism and biological properties of 2,2'-(((6-methoxy-1,3,5-triazine-2,4-diyl)bis(sulfanediyl)bis(2,1-phenylene)) bis(azanylylidene)bis(methanylylidene))bis(4-(phenyldiazenyl)phenol). Spectrochimica Acta, *38*, 643–647.

http://dx.doi.org/10.1016/j.saa.2014.11.048

Hassan, A. M., Ahmed, A. H., Gumaa, H. A., Mohamed, B. H., Eraky, A. M., & Chem, J. (2015). Manganese(II) complexes of N,O-chelating dihydrazone: Synthesis, characterization, optical properties and corrosion inhibition on aluminum in HCl solution. *Journal of Chemical Pharmaceutical Research, 7*, 91–104.

Karipcin, F., Dede, B., Caglar, Y., Hur, D., Ilican, S., Caglar, M., & Sahin, Y. (2007). A new dioxime ligand and its trinuclear copper(II) complex: Synthesis, characterization and optical properties. *Optics Communications, 272*, 131–137.
http://dx.doi.org/10.1016/j.optcom.2006.10.079

Kumar, A., Lal, R. A., Chanu, O. B., Borthakur, R., Koch, A., Lemtur, A., ... Choudhury, S. (2011). Synthesis and characterization of a binuclear copper(II) complex [Cu(H$_2$ slox)]$_2$ from polyfunctional disalicylaldehyde oxaloyldihydrazone and its heterobinuclear copper(II) and molybdenum(VI) complexes. *Journal of Coordination Chemistry, 64*, 1729–1742.
http://dx.doi.org/10.1080/00958972.2011.580845

Lal, R. A., Adhikari, S., Kumar, A., Chakraborty, J., & Bhaumik, S. (2002). Synthesis and characterization of manganese(IV) complexes derived from the direct reaction of manganese(II) acetate tetrahydrate with oxaloyldihydrazide and 2-hydroxy-1-naphthaldehyde. *Synthesis and Reactivity in Inorganic and Metal-Organic Chemistry, 32*, 81–96.
http://dx.doi.org/10.1081/SIM-120013148

Lal, R. A., Basumatary, D., Adhikari, S., & Kumar, A. (2008). Synthesis and properties of mononuclear and binuclear molybdenum complexes derived from bis(2-hydroxy-1-naphthaldehyde)oxaloyldihydrazone. *Spectrochimica Acta Part A: Molecular and Biomolecular Spectroscopy, 69*, 706–714.
http://dx.doi.org/10.1016/j.saa.2007.05.023

Lal, R. A., Adhikari, S., & Kumar, A. (1993). Spectroscopic studies of isomerization of coordinated dihydrazone in tetranuclear dioxouranium(VI) complexes derived from bis(o-hydroxynaphthaldehyde) oxaloyldihydrazone. *Indian Journal of Chemical, 36A*, 1063–1067.

Lever, A. B. P. (1984). *Inorganic electronic spectroscopy*. Amsterdam: Elsevier.

Loncle, C., Brunel, J. M., Vidal, N., Dherbomez, M., & Letourneux, Y. (2004). Synthesis and antifungal activity of cholesterol-hydrazone derivatives. *European Journal of Medicinal Chemistry, 39*, 1067–1071.
http://dx.doi.org/10.1016/j.ejmech.2004.07.005

McCollum, D. G., Hall, L., White, C., Ostrandor, R., Rheingold, A. L., Whelan, J., & Bosnich, B. (1994). Bimetallic reactivity. Preparation and characterization of symmetrical bimetallic complexes of a binucleating macrocyclic ligand, cytim, containing 6- and 4-coordinate sites. *Inorganic Chemistry, 33*, 924–933.
http://dx.doi.org/10.1021/ic00083a016

Mishra, L., & Singh, V. K. (1997). Co(ll), Ni(ll) and Cu(II) and Zn(II) complexes with Schiff bases derived from 2-aminobenzimidazoles and pyrazolycarboxaldehyde. *Indian Journal of Chemical Technology, 32A*, 446.

Mott, N. F., & Davis, E. A. (1979). *Electronic process in non-crystalline materials*. Oxford: Calendron Press.

Nakamoto, K. (1970). *Inorganic spectra of inorganic and coordination compounds* (2nd ed.). New York, NY: Wiley.

Rashad, M. M., Turky, A. O., & Kandil, A. T. (2013). Optical and electrical properties of Ba1-xSr(x)TiO$_3$ nanopowders at different Sr$_{2+}$ ion content. *Journal of Materials Science - Materials in Electronics, 24*, 3284–3291.

Rollas, S., & Küçükgüzel, S. G. (2007). Biological activities of hydrazone derivatives. *Molecules, 12*, 1910–1039.
http://dx.doi.org/10.3390/12081910

Salama, T. M., Ahmed, A. H., & El-Bahy, Z. M. (2006). Y-type zeolite-encapsulated copper(II) salicylidenep-aminobenzoic Schiff base complex: Synthesis, characterization and carbon monoxide adsorption. *Microporous and Mesoporous Materials, 89,* 251–259. http://dx.doi.org/10.1016/j.micromeso.2005.10.036

Salapathy, S., & Sahoo, B. (1970). Salicylaldazinate metal chelates and their I.R. spectra. *Journal of Inorganic and Nuclear Chemistry, 32,* 2223–2227. http://dx.doi.org/10.1016/0022-1902(70)80500-0

Salavati-Niasari, M., & Sobhani, A. (2008). Ship-in-a-bottle synthesis, characterization and catalytic oxidation of cyclohexane by host (nanopores of zeolite-Y)/ guest (Mn(II), Co(II), Ni(II) and Cu(II) complexes of bis(salicylaldehyde)oxaloyldihydrazone) nanocomposite materials. *Journal of Molecular Catalysis A: Chemical, 285,* 58–67. http://dx.doi.org/10.1016/j.molcata.2008.01.030

Sengupta, S. K., Pandey, O. P., Srivastava, B. K., & Sharma, V. (1998). Synthesis, structural and biochemical aspects of titanocene and zirconocene chelates of acetylferrocenyl thiosemicarbazones. *Transition Metal Chemistry, 23,* 349–353. http://dx.doi.org/10.1023/A:1006986131435

Sherif, E. M., & Ahmed, A. H. (2010). Synthesizing new hydrazone derivatives and studying their effects on the inhibition of copper corrosion in sodium chloride solution. *Synthesis and Reactivity in Inorganic and Metal-Organic Chemistry, 40,* 365–372.

Smânia, A., Monache, F. D., Smânia, E. F. A., & Cuneo, R. S. (1999). Antibacterial activity of steroidal compounds isolated from *Ganoderma applanatum* (Pers.) Pat. (Aphyllophoromycetideae) fruit body. *International Journal of Medicinal Mushrooms, 1,* 325–330. http://dx.doi.org/10.1615/IntJMedMushr.v1.i4

Sutton, D. (1968). *Electronic spectra of transition metal complexes.* London: McGraw Hill.

Tauc, J. (1968). Optical properties and electronic structure of amorphous Ge and Si. *Materials Research Bulletin, 3,* 37–46. http://dx.doi.org/10.1016/0025-5408(68)90023-8

Vicini, P., Zani, F., Cozzini, P., & Doytchinova, I. (2002). Hydrazones of 1,2-benzisothiazole hydrazides: Synthesis, antimicrobial activity and QSAR investigations. *European Journal of Medicinal Chemistry, 37,* 553–564. http://dx.doi.org/10.1016/S0223-5234(02)01378-8

Walcovrt, A., Loyevsky, M., Lovejoy, D. B., Gordeuk, V. R., & Richardson, D. R. (2004). Novel aroylhydrazone and thiosemicarbazone iron chelators with anti-malarial activity against chloroquine-resistant and -sensitive parasites. *International Journal of Biomedical Cell Biology, 36,* 401–407. http://dx.doi.org/10.1016/S1357-2725(03)00248-6

Yuan, J., Lovejoy, D. B., & Richardson, D. R. (2004). Novel di-2-pyridyl-derived iron chelators with marked and selective antitumor activity: In vitro and in vivo assessment. *Blood, 104,* 1450–1458. http://dx.doi.org/10.1182/blood-2004-03-0868

19

Morphological changes in giant vesicles comprised of amphiphilic block copolymers by incorporation of ionic segments into the hydrophilic block chain

Eri Yoshida[1]*

*Corresponding author: Eri Yoshida, Department of Environmental and Life Sciences, Toyohashi University of Technology, 1–1 Hibarigaoka, Tempaku- cho, Toyohashi 441–8580, Aichi, Japan
E-mail: eyoshida@ens.tut.ac.jp
Reviewing editor: George Weaver, University of Loughborough, UK

Abstract: The morphological changes in giant spherical vesicles comprised of am-philhilic poly(methacrylic acid)-*block*-poly(methyl methacrylate-*random*-methacrylic acid), PMAA-*b*-P(MMA-*r*-MAA), by incorporation of the 3-sulfopropyl methacrylate potassium salt (SpMA) into the hydrophilic PMAA block chains were investigated. The vesicles were reduced in size as the SpMA ratio increased due to the variation in the critical packing shape by the expansion of the hydrophilic surface area by the incorporation of the more hydrophilic ionic segments. The increase in the SpMA ratio also delayed the transition from the spherical vesicles to a film-like bilayer. The giant vesicles containing the SpMA segments were disrupted into a nonspecific form by the electrostatic interaction with poly(allylamine hydrochloride) at a 2.5 M ratio of the unit for the allylamine hydrochloride (AH) to the SpMA. A large excess of the AH/SpMA above 5.0 caused no disruption, but partial fusion of the vesicles accompanied by a decrease in size.

Subjects: Biochemistry; Materials Chemistry; Materials Science; Nanoscience & Nanotechnology; Organic Chemistry; Physical Chemistry; Polymer Science; Polymer Technology; Polymers & Plastics

Keywords: morphological changes; amphiphilic diblock copolymers; giant vesicles; film-like bilayer; ionic segments

ABOUT THE AUTHOR

Dr Eri Yoshida is an Associate Professor at Department of Environmental and Life Sciences, Toyohashi University of Technology, Japan. She is the author of over 100 refereed articles and 20 reviews and book chapters and obtained over 20 patents. She is also an Editorial Board member of 7 international peer reviewed journals. Dr Yoshida research interests include: (a) Molecular self-assembly of polymer surfactants, (b) Controlled/living radical polymerization, (c) Molecular design of functional polymers, and (d) Polymer syntheses using supercritical carbon dioxide. The study in this paper concerns the molecular self-assembly of polymers prepared using the photo-controlled/living radical polymerization technique established by the author. She is a member of several professional societies.

PUBLIC INTEREST STATEMENT

A great variety of living things inhabit the natural world. They are quite different in form, size, and lifestyle, however, share something in common in a basic unit of their bodies. It's a cell. A cell is a container enclosed with a thin two-layer membrane formed by surfactants of lipids. The functions of lipid bilayer membrane are closely related not only to essential life activities of living things but also to the origin of life.

A giant vesicle is a micro-sized container with a closed bilayer structure formed by surfactant and is regarded as an artificial model of biomembrane for cells based on similarities in its size and structure. This study demonstrated the maintenance of a giant vesicle structure imitating the way in living cells with the aim of creating biomimetic functions for the giant vesicles comprised of non-natural synthetic polymer surfactants.

1. Introduction

Cytoplasmic membranes have many important roles in essential life processes, such as structural maintenance (Branton, Cohen, & Tyler, 1981), selective transport (Elston, Wang, & Oster, 1998; Nishi & Forgac, 2002), endocytosis (Anderson, Brown, & Goldstein, 1977; Ford et al., 2002; Itoh, Kigawa, Kikuchi, Yokoyama, & Takenawa, 2001), and signal transduction (Oancea & Meyer, 1998; Rizo & Sudhof, 1998). These functions are fulfilled by the membrane proteins embedded or anchored in the lipid bilayer matrix, or attached on the bilayer surface. The integral membrane proteins and lipid anchored membrane proteins are fixed in the lipid bilayer through the van der Waals interaction between their hydrophobic domains and the hydrocarbon chains of the lipids (Johnson & Cornell, 1999; Popot & Engelman, 2000). On the other hand, the peripheral membrane proteins located on the bilayer surface are electrostatically bound to the polar head groups of the lipids often via the integral membrane proteins (Branton et al., 1981; Hyvonen et al., 1995; Kim, Mosior, Chung, Wu, & McLaughlin, 1991). While the hydrophilic head groups of the lipids serve as the binders of the membrane proteins to induce their own functions during specific situations on the bilayer, they determine the membrane curvature and cell morphology along with the hydrophobic hydrocarbon chains based on the individual critical packing shape dependent on the optimal hydrophilic surface area and hydrocarbon chain volume (Bigay & Antonny, 2012; Cullis & De Kruijff, 1979; Israelachvili, Mitchell, & Ninham, 1976).

Giant vesicles are regarded as artificial models for cytoplasmic membranes of cells and organelles, such as erythrocytes (Elgsaeter, Stokke, Mikkelsen, & Branton, 1986), mitochondria (Frey & Mannella, 2000), and chloroplasts (Sadava, 1993) based on the similarities in their size and structure. In recent years, it has been demonstrated that the micrometer-sized giant vesicles comprised of amphiphilic poly(methacrylic acid)-*block*-poly(methyl methacrylate-*random*-methacrylic acid) diblock copolymer, PMAA-*b*-P(MMA-*r*-MAA), provided new biomembrane models based on some similarities concerning the structural properties (Yoshida, 2015a) and response to temperature and pH (Yoshida, 2013, 2015b). The similarities also included the pore formation for ionic compounds passing across the hydrophobic phase of the vesicle bilayer (Yoshida, 2015c). It has also been reported that the morphology of the giant vesicles was effectively controlled by manipulating the critical packing shape of the copolymer using the block length (Yoshida, 2014a), molar ratio of the monomer units (Yoshida, 2014b, 2014c), and mixed composition (Yoshida, 2015d, 2015e), coupled with the physical conditions, such as the copolymer concentration, solvent affinity, and stirring speed during the photopolymerization-induced self-assembly of the copolymers into the vesicles (Yoshida, 2015f). The morphologies were varied to spherical, elliptical, worm-like, film-like, and villus-like forms by these factors (Yoshida, 2015d, 2015e, 2015g).

In the present study, the morphological changes in the giant vesicles by incorporation of ionic segments in the hydrophilic PMAA block chain and by the electrostatic interaction between the ionic segments and a polyelectrolyte were investigated. This paper describes the morphology of the giant vesicles was evaluated by incorporation of 3-sulfopropyl methacrylate potassium salt (SpMA) units into the PMAA block chain and by the electrostatic interaction between the SpMA units in the hydrophilic surface of the vesicles and a polyelectrolyte of poly(allylamine hydrochloride) (PAH).

2. Experimental

2.1. Instrumentation

The photopolymerization-induced self-assembly was performed using an Ushio optical modulex BA-H502, an illuminator OPM2–502H with a high-illumination lens UI-OP2SL, and a 500 W super high-pressure UV lamp (USH-500SC2, Ushio Co. Ltd.). ^1H NMR measurements were conducted using Jeol ECS400 and ECS500 FT NMR spectrometers. Gel permeation chromatography (GPC) was performed using a Tosoh GPC-8020 instrument equipped with a DP-8020 dual pump, a CO-8020 column oven, and a RI-8020 refractometer. Two gel columns, Tosoh TSK-GEL α-M were used with

N,N-dimethylformamide (30 mM LiBr and 60 mM H_3PO_4) as the eluent at 40°C. Field emission scanning electron microscopy (FE-SEM) measurements were performed using a Hitachi SU8000 scanning electron microscope.

2.2. Materials

Methyl methacrylate (MMA) was passed through a column packed with activated alumina to remove an inhibitor and distilled over calcium hydride. Methacrylic acid (MAA) was distilled under reduced pressure. These purified monomers were degassed with Ar for 15 min with stirring just before use. Methanol (MeOH) was refluxed over magnesium with a small amount of iodine for several hours, then distilled. 4-Methoxy-2,2,6,6-tetramethylpiperidine-1-oxyl (MTEMPO) was prepared as reported previously (Miyazawa, Endo, Shiihashi, & Okawara, 1985). SpMA and (4-$tert$-butylphenyl)diphenylsulfonium triflate (^{t}BuS) were purchased from Sigma-Aldrich and used as received. PAH (Mn = 150,000) and 2,2′–azobis[2-(2-imidazolin-2-yl)propane] (V-61) were purchased from Nittobo and Wako Pure Chemical Industries, respectively and used without further purification. Extrapure N_2 gas with over 99.9995 vol% purity and Ar gas with over 99.999 vol% purity were obtained from Taiyo Nippon Sanso Corporation.

2.3. Preparation of P(MAA-r-SpMA) end-capped with MTEMPO

V-61 (22.8 mg, 0.0911 mmol), MTEMPO (18.0 mg, 0.0966 mmol), ^{t}BuS (24.0 mg, 0.0512 mmol), MAA (2.010 g, 23.3 mmol), SpMA (58.0 mg, 0.235 mmol), and MeOH (4 mL) were placed in a 30-mL test tube joined to a high vacuum valve. The contents were degassed several times using a freeze-pump-thaw cycle and then charged with N_2. The polymerization was carried out at room temperature for 5.5 h with irradiation at 9.3 ampere by reflective light using a mirror with a 500 W high-pressure mercury lamp to avoid any thermal polymerization caused by the direct irradiation (Yoshida, 2012). MeOH (11 mL) and distilled water (5 mL) that were degassed by bubbling Ar for 15 min were added to the product under a flow of Ar. After the product was completely dissolved in the aqueous MeOH solution, part of the mixture (ca. 1 mL) was withdrawn to determine monomer conversions and molecular weight of the P(MAA-r-SpMA) prepolymer. The solution was poured into ether (50 mL) to precipitate a polymer. The precipitate was collected by filtration and dried in vacuo for several hours to obtain a polymer (67.2 mg). The conversions of MAA and SpMA were 75% and 82%, respectively based on ^{1}H NMR. The molecular weight and its distribution of the prepolymer were estimated to be Mn = 21600 and Mw/Mn = 1.73 by GPC based on PMAA standards. The degree of polymerization was DP = 242 based on the molecular weight. The residual P(MAA-r-SpMA) prepolymer solution was subjected to the following polymerization-induced self-assembly.

2.4. Photopolymerization-induced self-assembly of P(MAA-r-SpMA)-b-P(MMA-r-MAA)

The solution of the prepolymer (4 mL containing the MTEMPO-capped P(MAA-r-SpMA) (0.01933 mmol), unreacted MAA (1.163 mmol), and unreacted SpMA (8.317×10^{-3} mmol)), MMA (664.6 mg, 6.638 mmol), and MAA (42.6 mg, 0.495 mmol) were placed in a 30-mL test tube joined to a high vacuum valve under a flow of Ar. The contents were degassed several times using a freeze-pump-thaw cycle and finally charged with N_2. The polymerization was carried out for 7 h at room temperature and a 600-rpm stirring speed by UV irradiation at 9.3 ampere using a reflective light from a mirror. Part of the solution (ca. 0.5 mL) was withdrawn to determine the monomer conversions by ^{1}H NMR. A mixed solvent (MeOH/water = 3/1 v/v, 20 mL) was added to the dispersion solution to precipitate aggregates. The aggregates were cleaned with the mixed solvent by a repeated sedimentation-redispersion process. The resulting aggregates were stored in the presence of a small amount of the mixed solvent.

2.5. Reaction of the giant vesicles and PAH

The giant vesicles comprised of P(MAA$_{0.990}$-r-SpMA$_{0.010}$)$_{242}$-b-P(MMA$_{0.841}$-r-MAA$_{0.159}$)$_{270}$ (3.1 mg containing 1.565×10^{-4} mmol of the SpMA unit) were dispersed in an aqueous MeOH solution (0.5 mL, MeOH/water = 3/1 v/v). PAH (14.6 mg, 0.1561 mmol of the allylamine hydrochloride (AH) unit) was dissolved in an aqueous MeOH solution (1 mL, MeOH/water = 3/1 v/v). A PAH solution (10 μL, 1.561×10^{-3} mmol of the AH unit) was added to the vesicle dispersion at room temperature. The mixture was vigorously shaken and stored at room temperature for 7 days to precipitate the vesicles. The precipitates were isolated by decantation and dried in air to subject to FE-SEM.

2.6. SEM observations

The aggregates were dried in air and subjected to the FE-SEM measurements at 1.0 kV without coating. The size distribution of vesicles was estimated as reported previously (Kobayashi, Uyama, Yamamoto, & Matsumoto, 1990).

3. Results and discussion

In order to evaluate the morphologies of giant vesicles containing ionic segments in the hydrophilic surface, amphiphilic diblock copolymers consisting of the hydrophilic PMAA block supporting the SpMA ionic units and the hydrophobic P(MMA-r-MAA) block were prepared by photopolymerization-induced self-assembly using the photo-NMP technique (Scheme 1). Prior to the synthesis of the diblock copolymer, the P(MAA-r-SpMA) prepolymers end-capped with MTEMPO were prepared by the photo-NMP method in methanol using the V-61 initiator, the MTEMPO mediator, and the ᵗBuS accelerator. The prepolymers with different SpMA contents are listed in Table 1. The monomer conversions were estimated by ¹H NMR (Figure 1(a)); the SpMA conversion was determined using the signal intensity based on the protons at 4.27 ppm for the ester methylene attached to the oxygen of the unreacted monomer vs. the protons at 4.03–4.19 ppm of the polymerized unit, while the MAA conversion was based on the α-methyl protons at 1.90 ppm of the unreacted monomer vs. the protons at 0.79–2.28 ppm of the polymerized main chains. The molar ratios of the monomer units were calculated using the monomer conversions. The prepolymers contained ca. 1–10 mol% of the SpMA units. The degree of polymerization (DP) was determined on the basis of the molecular weight estimated by GPC based on PMAA standards. A decrease in the molecular weight of the copolymer with an increase in the SpMA content is due to the GPC estimation for the prepolymers with the different SpMA contents. The amphiphilic P(MAA-r-SpMA)-b-P(MMA-r-MAA) diblock copolymers were prepared by the block copolymerization of MMA and MAA employing the MTEMPO-capped P(MAA-r-SpMA) prepolymers. The resulting diblock copolymers are listed in Table 2. The MMA conversion was estimated by ¹H NMR using the ester methyl protons at 3.74 ppm for the unreacted MMA and those at 3.54–3.70 ppm for the polymerized units in the block copolymer. The MAA conversion was determined using the α-methyl protons at 1.90 ppm for the unreacted MAA and 1.92 ppm for MMA on the basis of the MMA conversion (Figure 1(b)). Some copolymers slightly contained the SpMA units in the P(MMA-r-MAA) blocks due to the remaining unreacted SpMA monomers in the prepolymer solution. The DP of the P(MMA-r-MAA) block was calculated using the monomer conversions and the prepolymer concentration equal to the initial concentration of MTEMPO (Yoshida, 2009). The amphiphilic diblock copolymers simultaneously self-assembled into giant vesicles as the block

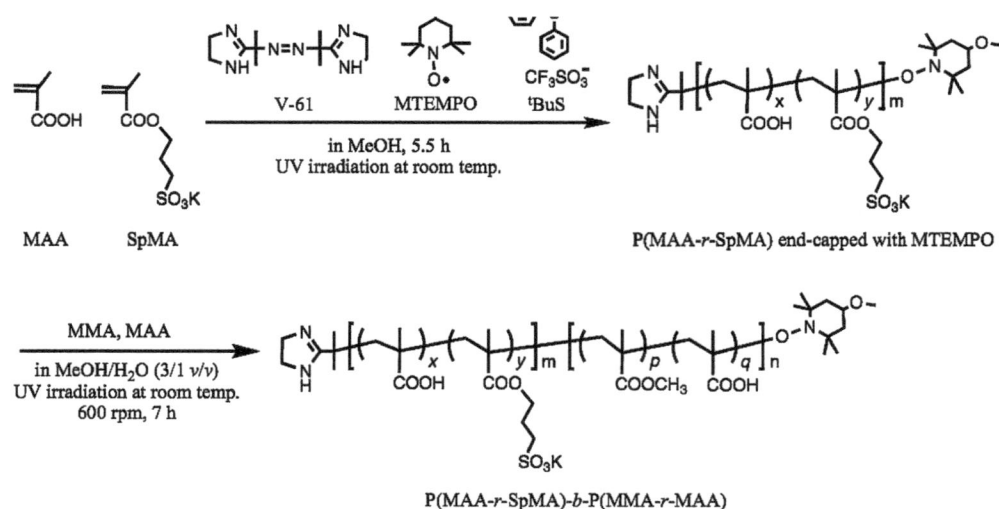

Scheme 1. Preparation of the P(MAA-r-SpMA)-b-P(MMA-r-MAA) diblock copolymers by the photopolymerization-induced self-assembly.

Table 1. The P(MAA-r-SpMA) prepolymers end-capped with MTEMPO

[MAA]$_0$(M)	[SpMA]$_0$(M)	Conversion (%)		Molar ratio		DP	Mn^a	Mw/Mn^a	Prepolymer
		MAA	SpMA	MAA	SpMA				
5.84	0.0589	75	82	0.989	0.011	242	21600	1.73	Pre-1
5.72	0.177	76	92	0.964	0.036	230	21400	1.69	Pre-2
5.60	0.295	79	90	0.943	0.057	200	19300	1.63	Pre-3
5.31	0.589	82	95	0.886	0.114	163	17300	1.65	Pre-4

Notes: [V-61]$_0$ = 0.0228 M, [MTEMPO]$_0$ = 0.0242 M, [^{t}BuS]$_0$ = 0.0128 M. Polymerization: 5.5 h.
^{a}Estimated by GPC based on PMAA standards.

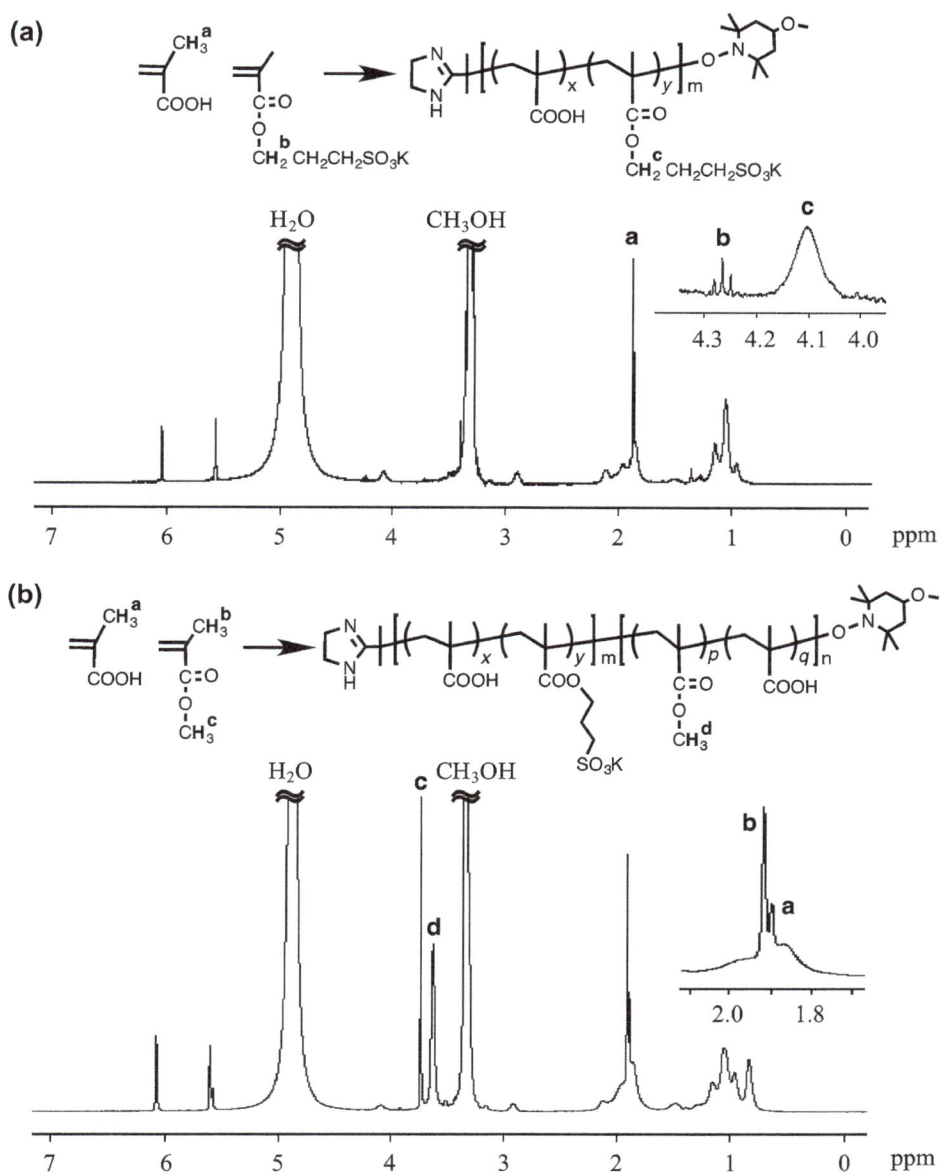

Figure 1. ^{1}H NMR spectra of the P(MAA-r-SpMA) prepolymer (a) Pre-4 and the diblock copolymer (b) BC-41. Solvent: CD$_3$OD-d$_4$/CDCl$_3$ = 3/1 (v/v).

Table 2. The P(MAA-*r*-SpMA)-*b*-P(MMA-*r*-MAA) diblock copolymers

Prepolymer	$[MMA]_0$ (M)	$[MAA]_0$ (M)	$[SpMA]_0$ (mM)	Conversion (%)			Molar ratio[a]			DP^b	Mn^c	Mw/Mn^c	Diblock copolymer
				MMA	MAA	SpMA	MMA	MAA	SpMA				
Pre-1	1.66	0.417	2.08	66	50	41	0.841	0.159	0	270	43600	1.55	BC-11
Pre-1	1.54	0.517	2.08	70	56	18	0.791	0.209	0	284	47900	1.46	BC-12
Pre-1	1.45	0.617	2.08	69	57	12	0.738	0.262	0	279	56100	1.48	BC-13
Pre-2	1.66	0.412	2.69	68	50	30	0.846	0.154	0	277	42300	1.53	BC-21
Pre-2	1.54	0.516	2.69	67	55	58	0.785	0.214	0.001	274	49100	1.48	BC-22
Pre-2	1.45	0.598	2.69	75	65	36	0.738	0.262	0	305	55300	1.50	BC-23
Pre-3	1.66	0.412	5.66	72	56	62	0.837	0.160	0.003	297	54700	1.64	BC-31
Pre-3	1.54	0.518	5.66	70	56	62	0.786	0.211	0.003	283	54800	1.55	BC-32
Pre-3	1.45	0.586	5.66	69	55	64	0.753	0.244	0.003	273	69900	1.50	BC-33
Pre-4	1.66	0.408	5.89	68	44	37	0.861	0.137	0.002	270	49100	1.81	BC-41
Pre-4	1.59	0.473	5.89	69	53	48	0.813	0.185	0.002	277	53200	1.80	BC-42
Pre-4	1.54	0.517	5.89	69	51	24	0.799	0.200	0.001	274	66900	2.22	BC-43
Pre-4	1.45	0.413	5.89	69	57	38	0.742	0.255	0.002	279	99100	1.84	BC-44

Notes: $[Prepolymer]_0$ = 4.83 mM. Polymerization: 7 h.

[a]Molar ratio of the monomer units in the hydrophobic copolymer block.

[b]DP for the hydrophobic copolymer block.

[c]Estimated by GPC based on PMAA standards for the diblock copolymer.

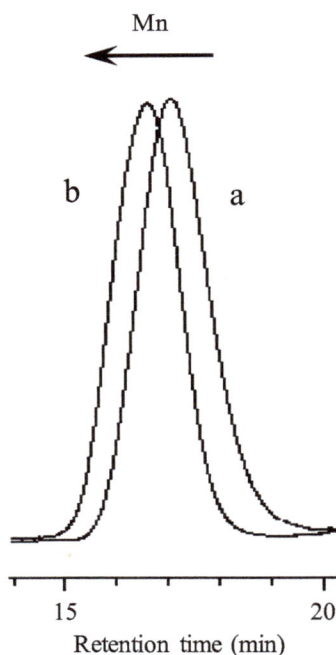

Figure 2. GPC profiles of the P(MAA-*r*-SpMA) prepolymer (a) Pre-1 and the diblock copolymer (b) BC-11.

copolymerization proceeded in an aqueous methanol solution (MeOH/water = 3/1 *v/v*). As shown in the GPC profiles of the block copolymer and prepolymer, the giant vesicles obtained by the photopolymerization-induced self-assembly included no prepolymer due to purification by a repeated sedimentation-redispersion process (Figure 2).

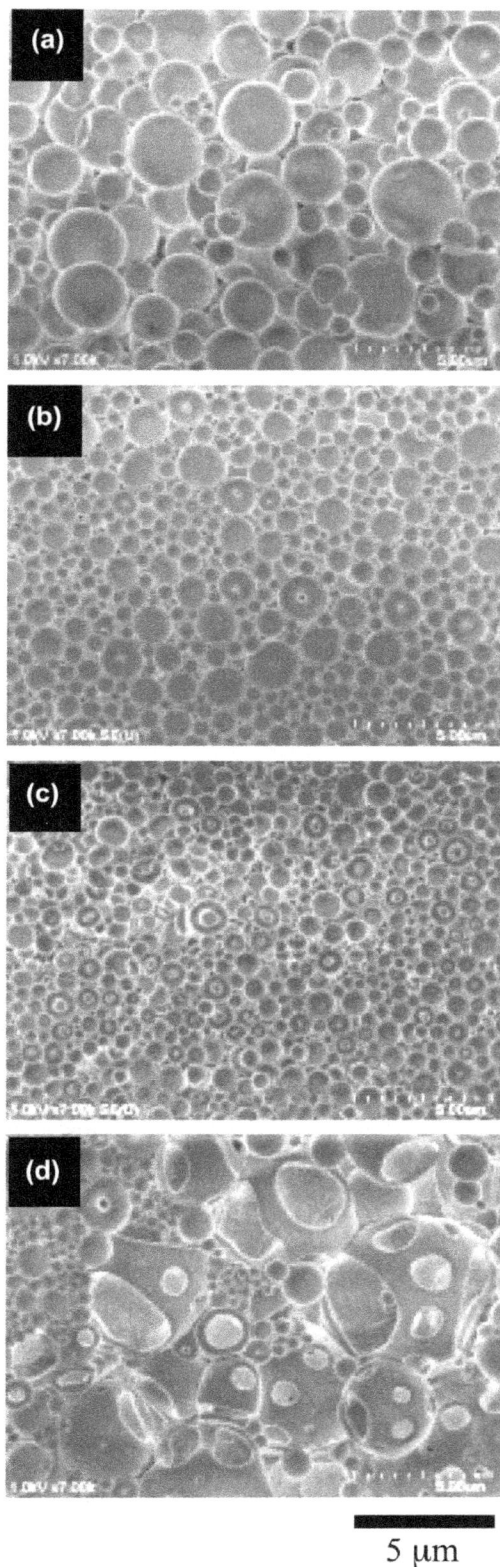

5 μm

Figure 3. FE-SEM images of the giant vesicles obtained by the diblock copolymers with different SpMA contents; (a) BC-11, (b) BC-21, (c) BC-31, and (d) BC-41.

Table 3. Morphologies of the diblock copolymers			
Diblock copolymer	Morphology	Dn (μm)	Dw/Dn
BC-11	S	1.51	1.56
BC-12	S + F	7.46[a]	1.70[a]
BC-13	F	–	–
BC-21	S	0.809	1.61
BC-22	S + F	8.14[a]	2.38[a]
BC-23	F	–	–
BC-31	S	0.787	1.25
BC-32	S + F	1.54[a]	3.09[a]
BC-33	F	–	–
BC-41	S	0.619	2.21
BC-42	S	0.549	3.79
BC-43	S + F	0.577[a]	1.58[a]
BC-44	F	–	–

Notes: S spherical vesicle, F Film-like bilayer.

[a]Estimated for spherical vesicles.

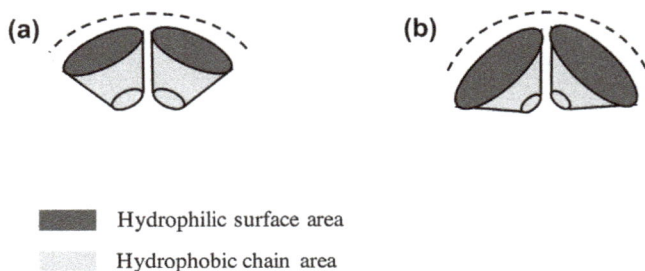

(a)   (b)

Hydrophilic surface area

Hydrophobic chain area

Scheme 2. Variation in bilayer curvature based on different critical packing shapes; (a) a truncated cone-like shape and (b) a more sharply truncated cone-like shape.

The SEM observations revealed that the morphology of the giant vesicles was dependent on the SpMA contents. Figure 3 shows the FE-SEM images of the giant vesicles formed by the diblock copolymers. Based on the negligible differences in the DP and MMA ratio, the vesicle size (Dn) decreased with an increase in the SpMA content (Table 3). This size decrease is accounted for by the variation in the critical packing shape of the copolymer; a truncated cone-like critical packing shape forming the spherical vesicles changed into a more sharply truncated cone due to the expansion of the optimal hydrophilic surface area by the incorporation of the more hydrophilic SpMA units into the hydrophilic block chain (Scheme 2). This more sharply truncated cone-like shape increased the curvature of the bilayer, resulting in a decreased vesicle size. It was found that the copolymer with the SpMA units over 10 mol% in the hydrophilic block chain produced large unstable vesicles having some holes along with much smaller spherical vesicles. The copolymer with a still more sharply truncated cone-like critical packing shape formed the rims of the holes that require a much higher curvature.

The diblock copolymers produced a difference in the morphological transition from spherical vesicles to a film-like bilayer based on their SpMA contents. Figure 4 shows the variation in the morphology of the copolymer with a 1.1-mol% SpMA content by a decrease in the MMA ratio of the hydrophobic block. The spherical vesicles were changed into very large spherical vesicles by a 5-mol% decrease in the MMA ratio. By a further 5-mol% decrease in the MMA ratio, the large spherical vesicles were transformed into a film-like bilayer. These morphological changes were based on the variation in the critical packing shape from a truncated cone into a cylinder by the expansion of the hydrophobic volume due to a decrease in its hydrophobicity of the P(MMA-r-MAA) block. A similar transformation was observed for the vesicles of the 3.6-mol% SpMA copolymer (Figure 5). On the

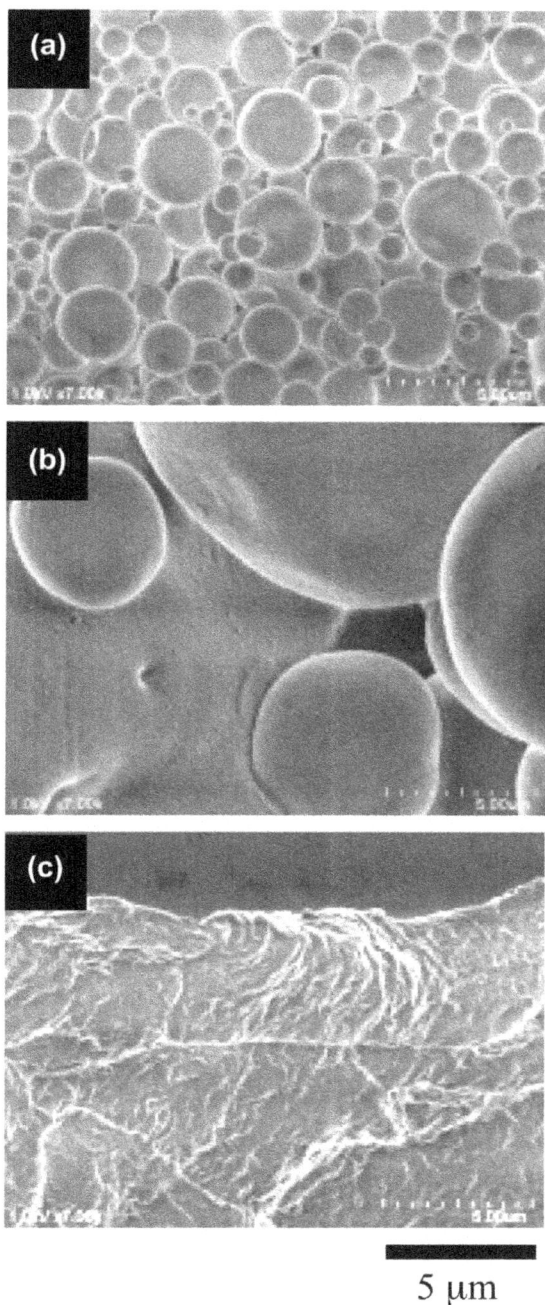

5 μm

Figure 4. The morphologies of the 1.1-mol% SpMA copolymers with different MMA ratios; (a) BC-11, (b) BC-12, and (c) BC-13.

other hand, the vesicles formed by the 5.7-mol% copolymer were still spherical by a 5-mol% decrease in the MMA units, although a flat bilayer was partly observed (Figure 6). The vesicles were transformed into a flexible film-like bilayer by a further decrease in the MMA ratio. Furthermore, the vesicles with some holes for the 11.4-mol% SpMA copolymer were changed into many small spherical vesicles along with the much larger vesicles by a 5-mol% decrease in the MMA ratio (Figure 7). An ca. 10-mol% decrease in the MMA ratio produced a flexible bilayer bearing the signs of fusion of the vesicles. It was deduced that an increase in the ionic segments in the hydrophilic block chains prevented the transition from the spherical vesicles to a film-like bilayer.

(a)

(b)

(c)

5 μm

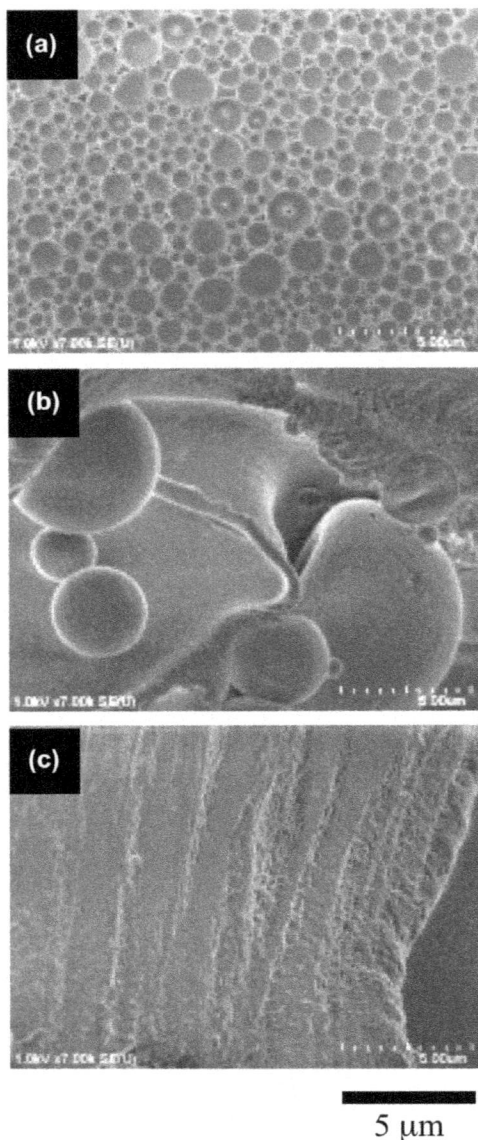

Figure 5. The morphologies of the 3.6-mol% SpMA copolymers with different MMA ratios; (a) BC-21, (b) BC-22, and (c) BC-23.

The human erythrocyte membrane maintains its shape by the cytoskeleton of the spectrin network via the peripheral proteins bound to the integral membrane protein and to the polar head groups of the lipids, like phosphatidylinositol and phosphatidylserine through an electrostatic interaction (Branton et al., 1981). The morphological changes in the spherical vesicles were investigated by the electrostatic interaction between a PAH polyelectrolyte and the SpMA units in the hydrophilic surface of the vesicles. The spherical vesicles formed by the 1.1-mol% SpMA copolymer were soaked in PAH solutions of different concentrations. The morphological changes in the vesicles by the electrostatic interaction are shown in Figure 8. The spherical vesicles were disrupted into a nonspecific form in the presence of PAH at a 2.5 M ratio of the unit for the allylamine hydrochloride (AH) to the SpMA. No SpMA-containing vesicles ($PMAA_{216}$-b-$P(MMA_{0.829}$-r-$MAA_{0.171})_{318}$, Dn = 1.63 μm, Dw/Dn = 2.37) retained their spherical shape by the soaking in the solution with the same PAH concentration, although the vesicle size slightly decreased (Dn = 1.16 μm, Dw/Dn = 2.56). Consequently, the disruption of the SpMA-containing vesicles was caused by the insertion of PAH into the hydrophilic surface of the vesicles through the electrostatic attraction between the sulfonium anion and the allylammonium

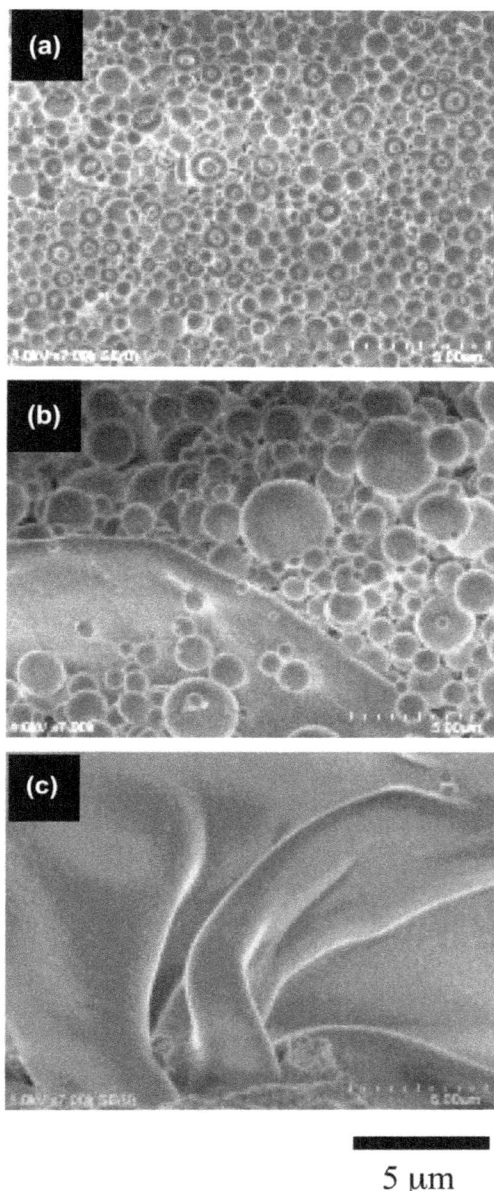

Figure 6. The morphologies of the 5.7-mol% SpMA copolymers with different MMA ratios; (a) BC-31, (b) BC-32, and (c) BC-33.

cation. On the other hand, the vesicles retained their spherical form at a 5.0 AH/SpMA ratio. The average vesicle size slightly decreased (Dn = 1.15 μm) and the size distribution was broadened (Dw/Dn = 2.40) due to the simultaneous formations of much smaller vesicles and partly fused vesicles. Furthermore, still smaller and partly fused vesicles (0.824 μm, Dw/Dn = 1.17) were produced at a 10.0 AH/SpMA ratio. These variations can be accounted for by because PAH at the low concentration disordered the copolymers forming a vesicle due to too short interval between the AH units interacting with the SpMA units in the hydrophilic surface (Scheme 3). On the other hand, PAH at the high concentration held the intervals long enough to retain the arrangement of the copolymers in the vesicle. However, PAH at the still higher concentration expanded the hydrophilic surface area of the vesicles due to the insertion of the higher hydrophilic chains, resulting in a decrease in size of the vesicles. The expansion of the hydrophilic surface area of the critical packing shape of the copolymer into a more sharply truncated conical shape by PAH also caused a decrease in the vesicular size.

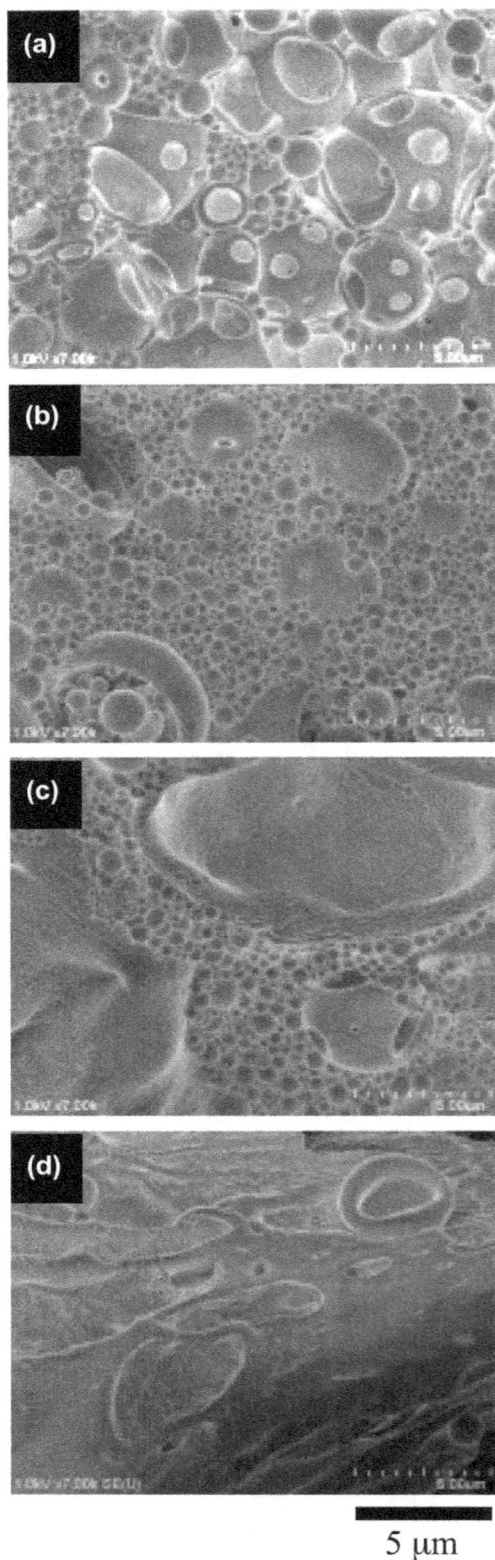

5 μm

Figure 7. The morphologies of the 11.4-mol% SpMA copolymers with different MMA ratios;
(a) BC-41, (b) BC-42, (c) BC-43, and (d) BC-44.

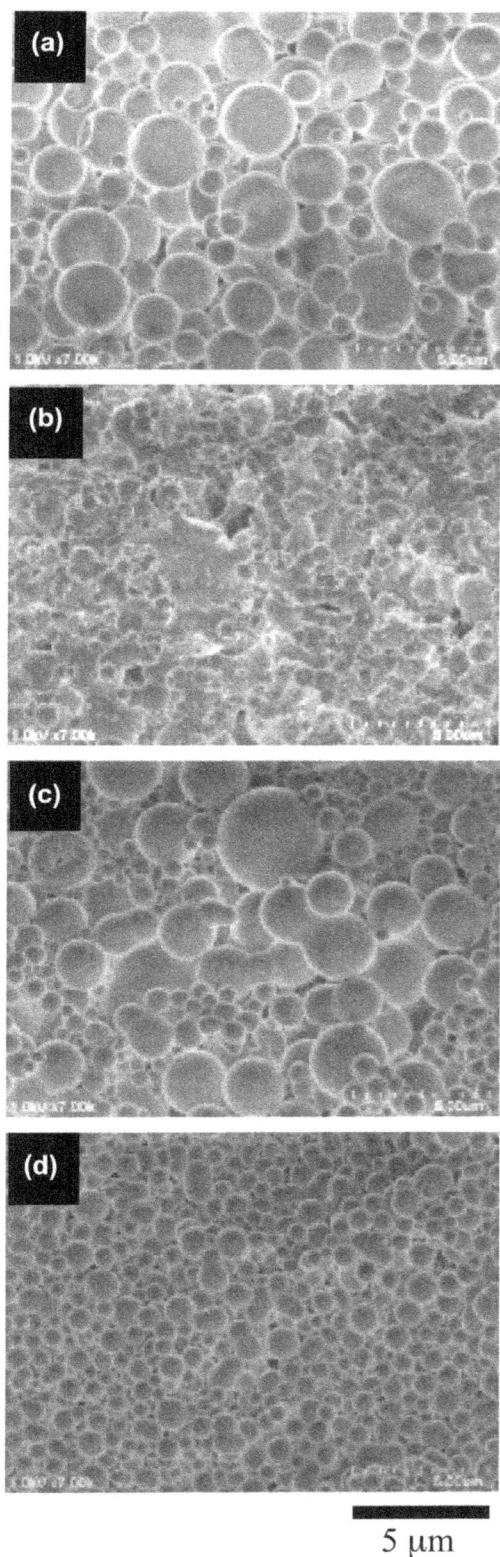

5 μm

Figure 8. The morphological changes in the spherical vesicles of the 1.1-mol% SpMA copolymer (BC-11) by soaked in PAH solutions of different concentrations; AH/SpMA = (a) 0, (b) 2.5, (c) 5.0, and (d) 10.0.

(a)   **(b)**

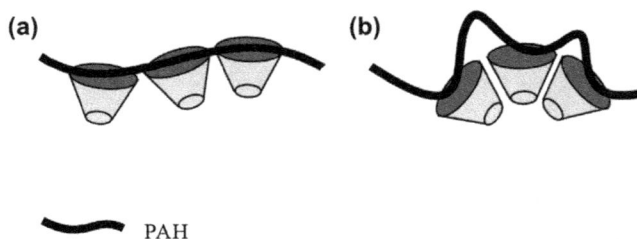

PAH

Scheme 3. The morphological changes by (a) the disruption of the SpMA-containing vesicles at a low PAH concentration and (b) retention of the spherical vesicles at a high concentration.

4. Conclusion

The morphology of the giant spherical vesicles comprised of the amphilhilic diblock copolymer was changed by the incorporation of ionic segments into the hydrophilic block chains. An increase in the ionic segment reduced the size of the vesicles. A decrease in the size was due to the expansion of the optimal hydrophilic surface area of the critical packing shape by the incorporation of the more hydrophilic ionic segments. The ionic segments also delayed the transition from the spherical vesicles into a film-like bilayer by a decrease in the MMA ratio of the hydrophobic block. Furthermore, the insertion of the polyelectrolyte into the hydrophilic surface through the electrostatic interaction caused disruption of the vesicles into a nonspecific form. Whereas a small excess of the polyelectrolyte disrupted the vesicles due to a short segment between the units binding the ionic units in the vesicle surface, its large excess retained the spherical shape, accompanied by the partial fusion of the vesicles.

Funding

This work was financially supported by JSPS Grant-in-Aid for Scientific Research [grant number 25390003].

Author details

Eri Yoshida[1]

E-mail: eyoshida@ens.tut.ac.jp

[1] Department of Environmental and Life Sciences, Toyohashi University of Technology, 1-1 Hibarigaoka, Tempaku-cho, Toyohashi 441-8580, Aichi, Japan.

References

Anderson, R. G. W., Brown, M. S., & Goldstein, J. L. (1977). *Cell* (Vol. 10, p. 351).
http://dx.doi.org/10.1016/0092-8674(77)90022-8

Bigay, J., & Antonny, B. (2012). Curvature, lipid packing, and electrostatics of membrane organelles: defining cellular territories in determining specificity *Developmental Cell, 23*, p. 886–895). doi:10.1016/j.devcel.2012.10.009

Branton, D., Cohen, C. M., & Tyler, J. (1981). *Cell* (Vol. 24, p. 24).
http://dx.doi.org/10.1016/0092-8674(81)90497-9

Cullis, P. R., & De Kruijff, B. (1979). *Biochimica et Biophysica Acta (BBA)–Reviews on biomembranes* (Vol. 559, p. 399).
http://dx.doi.org/10.1016/0304-4157(79)90012-1

Elgsaeter, A., Stokke, B. T., Mikkelsen, A., & Branton, D. (1986). *Science*, (Vol. 234, p. 1217).
http://dx.doi.org/10.1126/science.3775380

Elston, T., Wang, H., & Oster, G. (1998). Energy transduction in ATP synthase. *Nature, 391*, 510–513. doi:10.1038/35185

Ford, M. G. J., Mills, I. G., Peter, B. J., Vallis, Y., Praefcke, G. J. K., Evans, P. R., & McMahon, H. T. (2002). *Nature* (Vol. 419, p. 361). http://dx.doi.org/10.1038/nature01020

Frey, T. G., & Mannella, C. A. (2000). *TRends in biochemical sciences* (Vol. 25, p. 319).
http://dx.doi.org/10.1016/S0968-0004(00)01609-1

Hyvonen, M., Macias, M. J., Nilges, M., Oschkinat, H., Saraste, M., & Wilmanns, M. (1995). Structure of the binding site for inositol phosphates in a pH domain. *EMBO journal, 14*, 4676–4685.

Israelachvili, J. N., Mitchell, D. J., & Ninham, B. W. (1976). *Journal of the Chemical Society, Faraday Transactions II* (Vol. 72, p. 1525).

http://dx.doi.org/10.1039/f29767201525

Itoh, T., Kigawa, T., Kikuchi, A., Yokoyama, S., & Takenawa, T. (2001). *Science* (Vol. 291, p. 1047).
http://dx.doi.org/10.1126/science.291.5506.1047

Johnson, J. E., & Cornell, R. B. (1999). *Molecular membrane biology* (Vol. 16, p. 217).
http://dx.doi.org/10.1080/096876899294544

Kim, J., Mosior, M., Chung, L. A., Wu, H., & McLaughlin, S. (1991). *Biophysical journal* (Vol. 60, p. 135).
http://dx.doi.org/10.1016/S0006-3495(91)82037-9

Kobayashi, S., Uyama, H., Yamamoto, I., & Matsumoto, Y. (1990). *Polymer journal* (Vol. 22, p. 759).
http://dx.doi.org/10.1295/polymj.22.759

Miyazawa, T., Endo, T., Shiihashi, S., & Okawara, M. (1985). *The Journal of Organic Chemistry* (Vol. 50, p. 1332).
http://dx.doi.org/10.1021/jo00208a047

Nishi, T., & Forgac, M. (2002). *Nature reviews molecular cell biology* (Vol. 3, p. 94).
http://dx.doi.org/10.1038/nrm729

Oancea, E., & Meyer, T. (1998). *Cell* (Vol. 95, p. 307).
http://dx.doi.org/10.1016/S0092-8674(00)81763-8

Popot, J. L., & Engelman, D. M. (2000). *Annual review of biochemistry* (Vol. 69, p. 881).
http://dx.doi.org/10.1146/annurev.biochem.69.1.881

Rizo, J., & Sudhof, T. C. (1998). *Journal of biological chemistry* (Vol. 273, p. 15879).
http://dx.doi.org/10.1074/jbc.273.26.15879

Sadava, D. E. (1993). *Cell biology, organelle structure and function* (p. 130). Boston: Jones and Bartlett Publishers.

Yoshida, E. (2009). Photo-living radical polymerization of methyl methacrylate by 2,2,6,6-tetra-methylpiperidine-1-oxyl in the presence of a photo-acid generator. *Colloid and polymer science, 287*, 767–772.
http://dx.doi.org/10.1007/s00396-009-2023-2

Yoshida, E. (2012). Effects of Illuminance and heat rays on photocontrolld/living radical polymerization mediated by 4methoxy2,2,6,6tetramethylpiperidine 1oxyl, *ISRN polymer science*. (Article ID:102186, 6 p.) doi:10.5402/2012/102186

Yoshida, E. (2013). *Colloid and polymer science* (Vol. 291, p. 2733). http://dx.doi.org/10.1007/s00396-013-3056-0

Yoshida, E. (2014a). *Colloid and polymer science* (Vol. 292, p. 1463). http://dx.doi.org/10.1007/s00396-014-3216-x

Yoshida, E. (2014b). *Colloid and polymer science* (Vol. 292, p. 763). http://dx.doi.org/10.1007/s00396-013-3154-z

Yoshida, E. (2014c). *Colloid and polymer science* (Vol. 292, p. 2555). http://dx.doi.org/10.1007/s00396-014-3297-6

Yoshida, E. (2015a). *Colloid and polymer science* (Vol. 293, p. 1835). http://dx.doi.org/10.1007/s00396-015-3577-9

Yoshida, E. (2015b). *Colloid and polymer science* (Vol. 293, p. 649). http://dx.doi.org/10.1007/s00396-014-3482-7

Yoshida, E. (2015c). *Colloid and polymer science* (Vol. 293, p. 2437). http://dx.doi.org/10.1007/s00396-015-3679-4

Yoshida, E. (2015d). *Colloid and polymer science* (Vol. 293, p. 249). http://dx.doi.org/10.1007/s00396-014-3403-9

Yoshida, E. (2015e). *Colloid and polymer science* (Vol. 293, p. 3641). http://dx.doi.org/10.1007/s00396-015-3763-9

Yoshida, E. (2015f). *Supramolecular chemistry* (Vol. 27, p. 274). http://dx.doi.org/10.1080/10610278.2014.959014

Yoshida, E. (2015g). Fabrication of microvillus-like structure by photopolymerization-induced self-assembly of an amphiphilic random block copolymer, *Colloid and polymer science, 293*, 1841–1845. doi:10.1007/s00396-015-3600-1

Permissions

All chapters in this book were first published in CC, by Cogent OA; hereby published with permission under the Creative Commons Attribution License or equivalent. Every chapter published in this book has been scrutinized by our experts. Their significance has been extensively debated. The topics covered herein carry significant findings which will fuel the growth of the discipline. They may even be implemented as practical applications or may be referred to as a beginning point for another development.

The contributors of this book come from diverse backgrounds, making this book a truly international effort. This book will bring forth new frontiers with its revolutionizing research information and detailed analysis of the nascent developments around the world.

We would like to thank all the contributing authors for lending their expertise to make the book truly unique. They have played a crucial role in the development of this book. Without their invaluable contributions this book wouldn't have been possible. They have made vital efforts to compile up to date information on the varied aspects of this subject to make this book a valuable addition to the collection of many professionals and students.

This book was conceptualized with the vision of imparting up-to-date information and advanced data in this field. To ensure the same, a matchless editorial board was set up. Every individual on the board went through rigorous rounds of assessment to prove their worth. After which they invested a large part of their time researching and compiling the most relevant data for our readers.

The editorial board has been involved in producing this book since its inception. They have spent rigorous hours researching and exploring the diverse topics which have resulted in the successful publishing of this book. They have passed on their knowledge of decades through this book. To expedite this challenging task, the publisher supported the team at every step. A small team of assistant editors was also appointed to further simplify the editing procedure and attain best results for the readers.

Apart from the editorial board, the designing team has also invested a significant amount of their time in understanding the subject and creating the most relevant covers. They scrutinized every image to scout for the most suitable representation of the subject and create an appropriate cover for the book.

The publishing team has been an ardent support to the editorial, designing and production team. Their endless efforts to recruit the best for this project, has resulted in the accomplishment of this book. They are a veteran in the field of academics and their pool of knowledge is as vast as their experience in printing. Their expertise and guidance has proved useful at every step. Their uncompromising quality standards have made this book an exceptional effort. Their encouragement from time to time has been an inspiration for everyone.

The publisher and the editorial board hope that this book will prove to be a valuable piece of knowledge for researchers, students, practitioners and scholars across the globe.

List of Contributors

Mark R.StJ. Foreman
Department of Chemistry and Chemical Engineering, Chalmers University of Technology, Gothenburg, Sweden

Mohan Reddy Bodireddy, P.Md. Khaja Mohinuddin, Trivikram Reddy Gundala and N.C. Gangi Reddy
Department of Chemistry, School of Physical Sciences, Yogi Vemana University, Kadapa, Andhra Pradesh 516 003, India

Mahmood Kamali
Faculty of Chemistry, Kharazmi University, 49-Mofetteh Ave.,Tehran, Iran

Nohana C. Ramos, Aurea Echevarria, Arthur Valbon, Guilherme P. Guedes and Cláudio E. Rodrigues-Santos
Departamento de Química, Instituto de Ciências Exatas, Universidade Federal Rural do Rio de Janeiro, 23890-900, Seropédica, RJ, Brazil

Adailton J. Bortoluzzi
Departamento de Química, Universidade Federal de Santa Catarina, 88040-900, Florianópolis, SC, Brazil

E. Chandra Sekhar
Department of Chemistry, Acharya Nagarjuna University, Nagarjuna Nagar, Guntur, Andhra Pradesh, India

K.S.V. Krishna Rao
Polymer Biomaterial Design and Synthesis Laboratory, Department of Chemistry, Yogi Vemana University, Kadapa, Andhra Pradesh, India
Department of Chemical Engineering and Material Science, Wayne State University, Detroit, MI, USA

K. Madhusudana Rao
Nano Information Materials Laboratory, Department of Polymer Science and Engineering, Pusan National University, Busan, South Korea

S. Pradeep Kumar
Department of Microbiology, Yogi Vemana University, Kadapa, Andhra Pradesh, India

Sayed Hossein Siadatifard and Masumeh Abdoli-Senejani
Department of Chemistry, Islamic Azad University-Arak Branch, Arak, Iran

Mohammad Ali Bodaghifard
Faculty of Science, Department of Chemistry, Arak University, Arak, Iran

Djuikom Sado Yanick Gaëlle and Moise Ondoh Agwara
Department of Inorganic Chemistry, University of Yaounde I, Yaounde, Cameroon

Divine Mbom Yufanyi
Department of Chemistry, The University of Bamenda, Bambili, Bamenda, Cameroon

Rajamony Jagan
Sophisticated Analytical Instruments Facility, Indian Institute of Technology, Chennai 600036, India

Refaat M. Mahfouz and Gamal A-W Ahmed
Department of Chemistry, Faculty of Science, Assiut University, Assiut, Egypt

Tahani Al-Rashidi
Department of Chemistry, College of Science, King Saud University, Riyadh, 11451, Kingdom of Saudi Arabia

Sindija Brica, Maris Klavins and Andris Zicmanis
Faculty of Chemistry, University of Latvia, 19, Rainis Boulevard, Riga LV-1586, Latvia

Phillip S. Nejman, Alexandra M.Z. Slawin, Petr Kilian and J. Derek Woollins
EaStChem School of Chemistry, University of St Andrews, St Andrews, Fife KY16 9ST, UK

Feudjio Tsague Chimaine, Amah Colette Benedicta Yuoh, Donatus Bekindaka Eni and Moise Ondoh Agwara
Department of Inorganic Chemistry, University of Yaounde I, Yaounde, Yaounde, Cameroon

Maryam Dehghan, Abolghasem Davoodnia, Mohammad R. Bozorgmehr and Niloofar Tavakoli-Hoseini
Department of Chemistry, Mashhad Branch, Islamic Azad University, Mashhad, Iran

Atefeh Ghasemi, Abolghasem Davoodnia, Mehdi Pordel and Niloofar Tavakoli-Hoseini
Department of Chemistry, Mashhad Branch, Islamic Azad University, Mashhad, Iran

Haixin Yuan
School of Medical Economics & Management, Anhui University of Chinese Medicine, Hefei 230012, China

Kehua Zhang, Jingjing Xia, Xianhai Hu and Shizhen Yuan
Department of Chemistry, Anhui JianZhu University, Hefei 230022, China

Mohamed Mohamady Ghobashy, A. Awad, Mohamed A. Elhady and Ahmed M. Elbarbary
Radiation Research of Polymer Department, National Center for Radiation Research and Technology (NCRRT), Atomic Energy Authority, Nasr City, Cairo, Egypt

Olivia N.J.M. Marasco and Sydney K. Wolny
Department of Chemistry and Biochemistry, University of Lethbridge, Lethbridge, AB, Canada T1K3M4

Jackson P. Knott, Daniel Stuart, Tracey L. Roemmele and René T. Boeré
The Canadian Centre for Research in Advanced Fluorine Technologies, University of Lethbridge, Lethbridge, AB, Canada T1K3M4

Aditi A. Jadhav, Vaishali P. Dhanwe, Prasad G. Joshi and Pawan K. Khanna
Nanochemistry Laboratory, Department of Applied Chemistry, Defence Institute of Advanced Technology (DIAT), Ministry of Defence, Govt. of India, Girinagar, Pune 411 025, India

Ayman H. Ahmed, A.M. Hassan, Hosni A. Gumaa, Bassem H. Mohamed and Ahmed M. Eraky
Faculty of Science, Department of Chemistry, Al-Azhar University, Nasr City, Cairo, Egypt

Eri Yoshida
Department of Environmental and Life Sciences, Toyohashi University of Technology, 1-1 Hibarigaoka, Tempaku-cho, Toyohashi 441-8580, Aichi, Japan

Index

www.ingramcontent.com/pod-product-compliance
Lightning Source LLC
Chambersburg PA
CBHW082013190326
41458CB00010B/3173